Invertebrate Bacteriology

Aurelio Ciancio

Invertebrate Bacteriology

Function, Evolution and Biological Ties

Springer

Aurelio Ciancio
Sustainable Plant Protection Institute
National Research Council
Bari, Italy

ISBN 978-94-024-0882-9 ISBN 978-94-024-0884-3 (eBook)
DOI 10.1007/978-94-024-0884-3

Library of Congress Control Number: 2016948707

Printed on acid-free paper

This Springer imprint is published by Springer Nature
The registered company is Springer Science+Business Media B.V. Dordrecht

Preface

The associations between invertebrates and bacteria represent today a research field particularly active and fertile, yielding a deeper and broader comprehension of many evolutionary mechanisms and life processes. This new horizon has been reached through a large number of specialized research work in many distant fields, extensively explored. They include bacteria associated to insects, nematodes and other invertebrates, present in various habitats from a wide range of environments. Many recent discoveries highlighted the role of bacteria in the evolution of species that have a deep social or economic impact. As a consequence, a review on some fundamental topics may represent an useful tool, helpful for a broad range of researchers and students.

This compendium concerns various invertebrate bacteriology facets, examining processes acting on bacteria that evolved with Metazoa. The term "invertebrate" is indeed a simplification that does not correspond to a common, unique evolutive history or metazoan lineage. In a broad sense, all metazoans lacking a chord or a notochord present in even a larval stage may be included in the category. The definition hence does not fit a given, specific and consistent clade and dates back to the first phylogenetic and morphological categorizations. However, the term is largely used as a synthetic definition, in reference to lineages including metazoan groups that are important for agriculture, for animal and human health or environment ecology. The term is used throughout the volume with this meaning.

Being present in all environments, from deep oceans to plants or animal bodies, the invertebrates represent the most significant and ancient fraction of the eukaryotic biomass present on earth. The evolutive adaptations and links that they established with bacteria are enormous, and occurred over ages and within environmental niches highly diversified. Examining in a single compendium the whole exploration of the invertebrate evolutive radiations, performed by the huge number of bacteria that are (or were) present on earth, is indeed as a very complex, and unaffordable, endeavour. However, aim of this volume is to provide a basic view of most important processes and to produce an excursus about recent advances achieved by most clever studies. One or more aspects of the evolutive paths of prokaryotes and their associates are examined, offering a perspective for current studies on bacterial ecology and evolution. Since the bacteria identified thus far represent only a minimal fraction of the total numbers present on earth, attention has been given to the most fundamental and significant relationships.

Rather than proceeding through the structure of taxonomies and phylogenetic ordering, the volume relies on processes. After two general chapters on bacterial structural organization and evolutive mechanisms, the first Section covers processes found among distant invertebrate Phyla, in trophic niches from diversified environments and ranging from the insect body to deep ocean floors. They include symbiosis, parasitism, vectoring and phoresy that are addressed, at a finer level, focusing on examples from different taxa. In the second Section some molecular mechanisms underpinning the functional and evolutive adaptations of bacteria to invertebrates are examined. They include defense and parasite evasion, main immunity and antibacterial factors, the effects of horizontal gene transfer and the recent advancements in the field of molecular biology and genomics. Finally, in the third Section, the effects of processes related to human activities are reviewed, including climate changes, biological invasions and agriculture, with particular attention to crops and animal transmitted diseases.

I hope the reader will find interest in this work and that the data gathered and analysed will result useful, as an informative basis, for the production and development of advanced research activities and renewed intellectual endeavours.

Bari, Italy Aurelio Ciancio

Acknowledgements

I am indebted with C.M. Adema, O. Giere, V. Girard, O. Gros, J. Hubert, R. Kochevar, D. McKenzie, H. Mehlhorn, M. Siddall and D. Silverman for providing, with other authors and co-authors, part of the volume iconografic materials used in some of the figures. I also acknowledge my colleagues L.C. Rosso, I. Pentimone and M. Colagiero for text reading and suggestions, A. Troccoli for nematode slide loans, and R. Favre and M. Cermola for electron microscopy work. Editor Z. Bernhart is gratefully acknowledged, together with M. van der Stigchel, for providing me all the support and extra-time I required for completing the volume. This book would never have been produced without the help, enthusiastic comments and patient suggestions given by my wife Mariella, that I acknowledge with love.

Contents

Part II Molecular Processes

Part III Applied Approaches

Part I
Structure and General Processes

Chapter 1
The Bacterial Cell

Abstract General concepts and basic informations, with data on structural components and functionning of the bacterial cell are provided, including descriptions of the cell wall, the secretion systems and the cytoskeleton organization. Basic informations about the functionning and organization of the chromosome and plasmids are shown, together with mechanisms of genetic recombination and DNA methylation. Other structural components of the bacterial cell like flagella and pili are also shown. Quorum sensing and functional processes like CRISPR systems involved in adaptation and metabolism are discussed, with basic concepts and data concerning bacterial identification.

Keywords Adaptation • Cell wall • Chromosome • Quorum sensing • Secretion systems

1 Introduction

Bacteria are a very ancient branch of the Tree of life. They successfully evolved to occupy any available niche present on earth, including our body. Bacterial cells can be found in all environments, including the deserts, the deep ocean floors, the Artic and Antartic regions or the high layers of the earth atmosphere. As a consequence, any living species on the planet has to deal with them during its entire life, through a number of interactions that may be beneficial, neutral or can yield severe negative outcomes, threatening in some cases survival and reproduction. Bacteria furthermore carried out (as they still do today), several fundamental functions in the environment. The most important among them are the recycling of elements through the decomposition of any organic matter, the fixation of nitrogen or the release of atmospheric gases including O_2, methane and CO_2.

The diversity of niches inhabited by bacteria is the result of their extreme adaptive capabilities, coupled to their metabolic flexibility and genetic richness and diversity, which arise from their long (in the order of three to four billion years) evolutionary history. Many bacterial lineages were in fact already present on earth well before the appearance of most animal groups, and it is also likely that they will survive to any mass extinction event occurring in the future.

A. Ciancio, *Invertebrate Bacteriology*, DOI 10.1007/978-94-024-0884-3_1

The number of bacterial species estimated thus far is impressive, and only a small fraction of them (roughly 10 %) has been described in detail in the literature. In many environments, the inhabiting bacteria can reach a high level of diversity. In terrestrial habitats the number of taxonomic units may range in the order of 10^4 species per g of soil. An estimate of 10^{14} bacterial cells is present in a human body, tenfolds the number of our own body cells.

Bacteria also covered the widest possible range of food and energetic sources available for life. To resume the power of their capability to explore any given possibility, thanks to their impressive numbers and to the time passed from their appearance, we could consider that if a process (i.e. any reaction, colonization, environmental adaptation and so on) is in theory possible, then it is likely that bacteria already "discovered" it. At this regard, the bacterial life in extreme environments is an example of this capacity. In the last decades, several studies were carried out on species living in extreme environments like the waters around the anoxic hydrothermal vents on the oceanic floor. These observations showed, among other discoveries, a sulphur related bacterial metabolism sustaining entire food chains. They are independent from the solar energy flux that feeds the planet through the photosynthetic C reduction. This was a surprising and unexpected discovery of a major and general significance, that indicated new perspectives about the possible origins of life in light-deprived conditions. These conditions are very different from the commonly observed physical and chemical situations found on the surface of earth. Similar volcanic-thermal environments might also be encountered even outside our planet, thus enlarging the range of situations in which life forms may be expected to arise.

Considering the aims of this volume, it is worth to recall some basic elements peculiar to bacterial cells. In this chapter a general introductory review about their structure and functional organization is given, useful to integrate the processes described in the following chapters.

2 Structure and Functionning

2.1 The Cell Wall

Bacteria are single cell organisms, the majority of species being a few microns wide or long. The classical rod or coccal cell shapes, already considered as typical of bacteria more than a century ago by the pioneers of microbiology, have been progressively integrated by a variation of shapes and morphologies that were identified as far as new members of the Kingdom Bacteria were encountered and described. Some morphologies are indeed very common and not useful for identification. Others instead are relatively typical and reflect bacterial metabolism as well as their fitness and survival strategies, like i.e. the helicoid shape of *Helicobacter*, the spiral shape of Spirochetae, the curved shaped cells (vibrioid type) of the anchored and

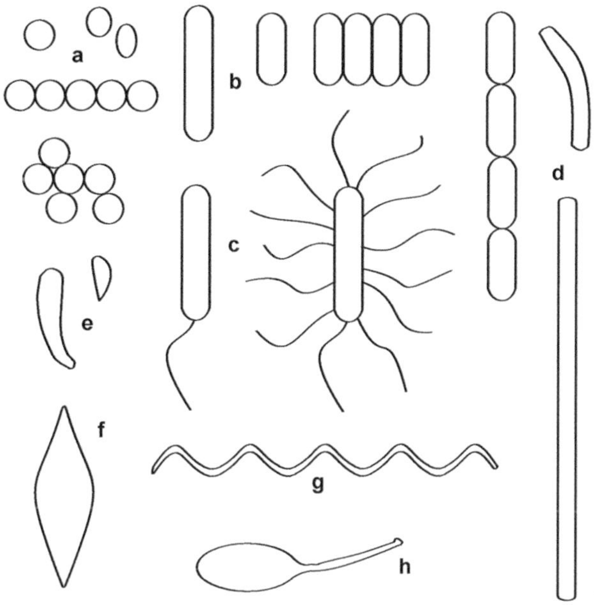

Fig. 1.1 Some of the most common cell morphologies found among bacteria. Cocci (**a**) can be observed as single units, chains or as cell clusters. Rods or bacilli-like cells of different dimensions (**b**) can occur alone, or aggregated in palisades or chains. They can also show single or multiple flagella of different lengths (**c**). Filiform bacteria (**d**) may appear bent or straight (**d**). *Vibrio*-like cells (**e**) can be recognized by their comma-like shape (**e**). Other morphologies include fusiform (**f**), spirochetes (**g**) and *Caulobacter*-like cells (**h**)

pedicellate *Caulobacter* or *Vibrio* spp., or the typical cup-shaped endospores of nematode parasites like *Pasteuria* spp. (Fig. 1.1). Cell morphogenesis is also determined by nutritional requirements and by specific developmental conditions (Cabeen and Jacobs-Wagner 2007). These defined and specific traits may result useful for a first generic identification of the microorganism object of study or at least for an initial, rough assignment of the closest genus or lineage. It is useful to recall that bacterial shape is also the result of physical conditions and of the forces that are exerted on them. This adaptation was shown by Takeuchi et al. (2005) whose experiments revealed that the cells of *Escherichia coli* multiplying in a close, constricting microchamber were able to maintain and transmit their shape for a limited number of generations.

Visual inspection of bacteria and identification in light microscopy is, however, difficult. It may result useful just in a few cases, and only to the experienced microbiologist. Advanced techniques, like Transmission or Scanning Electron Microscopies (TEM, SEM) yield much more details and clearer images. They are obtained at high resolutions from preparations examined in the vacuum, after a number of fixing and staining procedures have been applied. Recently, with the advent of nanoscale resolution microscopies like the Scanning Tunneling (STM) or

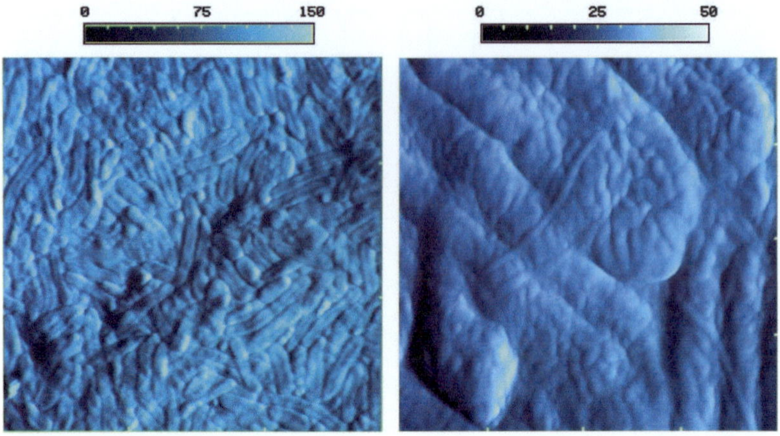

Fig. 1.2 High resolution AFM scans of unfixed, air dried rod cells of *Pseudomonas putida* scanned in force mode, showing shape (*left*), surface wall roughness and flagella (*right*). Scan sizes: 9 μm (*left*) and 2 μm (*right*). *Color bars* show force intensity (nN)

the Atomic Force Microscopy (AFM),[1] higher resolutions can be achieved with simple protocols even on unfixed bacterial cells, living or air dried (Fig. 1.2), or in a condition close to their native state, if they are examined in liquid. Further improvements of TEM analysis like Cryo-electron microscopy also allowed a quasi–atomic scale resolution of biomolecules. These techniques, and their recent advances, allowed the identification at high resolution of structural components present on the bacterial cell surface like the proteins of the secretory system or other transmembrane units (Scheuring et al. 2005; Rico et al. 2011; Bartesaghi et al. 2015).

Several structural arrangements of the cell wall are responsible for the morphology of bacteria and are necessary for their survival. The staining method conceived in 1884 by Hans Christian Gram, still used today after more than one century, is frequently applied to discriminate between two generic bacterial groups, called Gram-positives or Gram-negatives. The staining method reflects the structural differences of the cell wall organization between the two groups (Fig. 1.3). In Gram-negative species the cell wall is surrounded by a thin peptidoglycan layer, covered by a further, outer membrane (OM) containing lipopolysaccharides (LPS). In Gram-positive bacteria the OM is absent but the peptidoglycan layers are thicker than in Gram-negatives, with long anionic polymers of teichoic or teichuronic acids (Schäffer and Messner 2005). Many sources may be found on the internet providing description and technical details about Gram's and other staining procedures, including the use of commercial kits or data provided by manuals and standards of public bodies like Public Health England (2014) or the World Health Organization.

---

[1] *AFM*= a tridimensional scan of a sample surface performed by measuring the vertical deflection induced by the scanned surface on a thin cantilever helding a very sharp tip probe. The vertical movements of the scanning probe, due to repulsive or attractive interactions with the substrate, are detected as intensity changes of a laser beam deflected by the cantilever. Either the force of deflection or the vertical probe displacement are then used to digitally reconstruct the sample surface topography.

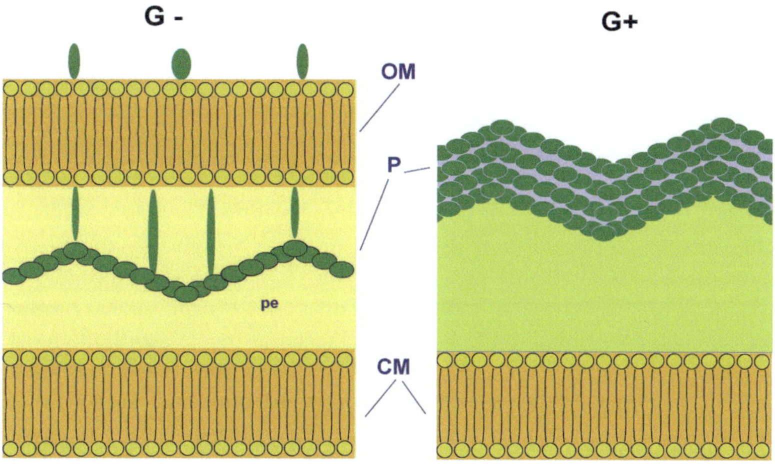

Fig. 1.3 A schematic drawing showing the different structures of the cell wall between Gram – negative (*left*) and Gram-positive bacteria (*right*), due to the presence in the former of an outer membrane (OM) bearing additional porins or enzymes, and to the thickness and organization of the peptidoglycan layers (P). CM cytoplasmic membrane, pe periplasmic space

The bacterial cell envelope is fundamental for survival, as well as metabolism and multiplication. Apart its function to protect the cell content from the external environment, which in many occasions may be hostile and extremely dangerous, it also allows the selective trade off and movement of a wide range of different molecules, mainly proteins, needed for the secretion apparatus, the cell metabolism, or of waste products and other compounds that are excreted.

The OM of Gram-negative bacteria is a strong and very effective, protective barrier (Fig. 1.3). Its LPS layer represents, furthermore, a molecular antigenic signature and is a fundamental target for the first immune system response in either vertebrates and invertebrates. The detection of LPS is in fact the first biochemical signal indicating that an infection is in course, produced by a pathogenic species, since these molecules are not present in the organisms invaded. The OM also contains different types of lipoproteins with additional enzymes and several transmembrane proteins of the β-barrel type. These rolled β-sheets cross the membrane as tubes or cylindrical pores (hence their identification as porins) allowing the passive diffusion of small molecules (sugars, phosphates or amino acids). In *Escherichia coli*, as many as $2.5 \cdot 10^5$ porins of different classes can be found around a cell membrane. Some classes of porins are specialized in the transport of larger molecules, i.e. iron chelates or vitamins (Silhavy et al. 2012).

Peptidoglycan is a polymer of alternating N-acetylglucosamine and N-acetylmuramic acid units, cross-linked by pentapeptide side chains (Vollmer 2008). It is linked to the OM by lipoproteins, conferring rigidity to the bacterial cells. Its organization, all around the cell, largely contributes to their conformation. In Gram-negative bacteria the peptidoglycan layer envelops the periplasmic space, a cell compartment rich in proteins and other molecules, with a density higher than the cytoplasm. It has a defensive function related to the sequestering of noxious enzymes and other molecules. The phospholipid bilayer forming the inner cytoplasmic membrane (CM) is replenished with proteins and enzymes. It is also the cell component on which the majority of the bacterial basic metabolic functions take place (Fig. 1.3).

Bacteria often present further external structures like the S-layers, which play an active role in their life-cycle and underpin their success in a given environment (Silhavy et al. 2012). S-layers are present either in members of the Kingdom Archaea and in Gram-positive or Gram-negative bacteria. They appear as a crystalline array formed by the assemblage of one or more S-layer glycoproteins (SLP) assembled on the outer cell surface and shown in TEM or AFM images as extremely ordered and symmetric arrays of molecules. Different LPS and related proteins may be expressed and assembled in the S-layer in the same species, with the most external protein that, in pathogenic bacteria like *Bacillus antrachis* or *Clostridium difficile*, acts as an antigen. Variations in the expression of one or more different homologue genes for the SLP members represent hence a way for a pathogen to exhibit different domains and to escape the host immune recognition and response.

The several SLP-associated proteins include peptidoglycan hydrolases, putative adhesins, leucine-rich repeats (LRR) and effector-domain proteins. They are exposed to the external bacterium environment and have many functions, suggesting an S-layer role in the bacterium reactivity to the external environment conditions. Secretion of the S-layer subunits is a very challenging task, and during exponential growth a secretion rate of 140 subunits per second is required to keep the integrity of the paracrystalline layer. For these tasks, type I or type II secretion system machineries and accessories are used by different bacterial groups. The S-layer proteins are anchored to the cell surface through LPSs in Gram-negative bacteria and cell wall polysaccharides in Gram-positive species (Fagan and Fairweather 2014).

The bacterial cell morphology is often also dependent on nutritional conditions, and remarkable differences may exist between vegetative cells or resting, dormant spores of the same strains. The spores of i.e. Bacillaceae or *Clostridium* spp. usually show a peculiar morphology, given the presence of a number of protective and resistant external core enveloping layers, often decorated by further components like the parasporal fibers or remnants of the exosporium (see Sect. 2.6.2).

2.2 Secretion Systems (SS)

The interactions of bacteria with other adjacent cells, present either in the host colonized tissues, in different types of symbiotic associations or in consortia of the same species, largely depend on molecules, mainly proteins, that are produced and secreted through the bacterial cell wall. Secretion is a fundamental function of bacteria, and occurs through six different systems present in Gram-negative species (some of them also in common with Gram-positives), with a further, type VII secretion system, found in Gram-positive species (Fig. 1.4). Intracellular bacteria lacking an outer membrane, i.e. some mollicutes deprived of cell wall and some Gram-positive (*Listeria monocytogenes* and *Rhodococcus equi*), show only generic secretion pathways (Tseng et al. 2009).

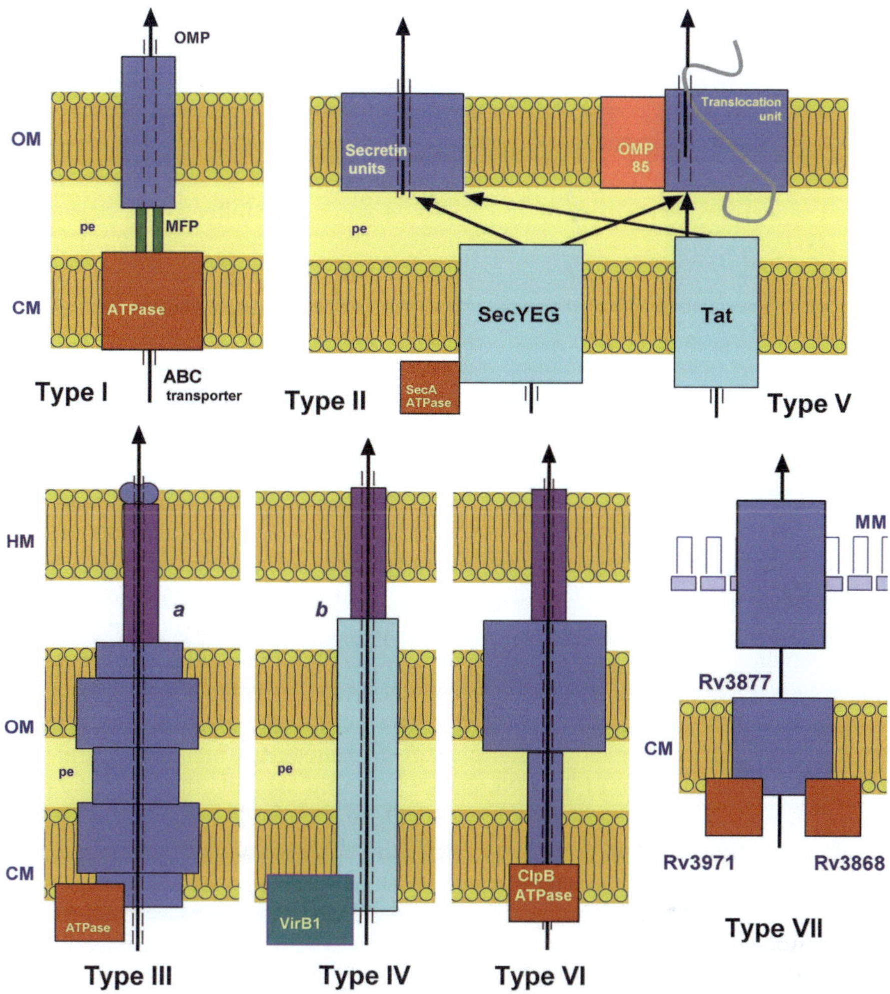

Fig. 1.4 Simplified drawings showing the different components of the bacterial secretion types I to VII. OM: outer membrane, CM cytoplasmic membrane, HM host cell membrane, MM mycobacterial membrane, pe periplasmic space; (**a**) needle, filament or pilus; (**b**) Vir complex. Arrows show direction of effectors secretion. For proteins and functions see text (Adapted and re-drawn from Tseng et al. 2009; Abdallah et al. 2007)

Secretion involves the production and release of several categories of compounds, mainly proteins (which may also be used to assemble external cell components like pili or flagella) as well as enzymes, toxins, antibiotics or other molecules which must be transported to the external cell environment as waste or signalling products. The major components of the type I protein SS are the highly conserved ATP-binding cassette (ABC) transporter, providing energy through two cytoplasmic domains for the hydrolysis of ATP, the Outer Membrane Protein (OMP), a trans-periplasmic channel connected to ABC by the Membrane Fusion Proteins (MFP), which is also found in Gram-positive bacteria.

The OMP secretion and translocation units are present in all Kingdoms. The universal Sec-dependent pathway recognizes a hydrophobic N-terminal sequence on the proteins to be secreted, which are translocated in an unfolded state using ATP and a proton gradient as energy source. The membrane proteins TatA, TatB and TatC recognize a motif rich in basic amino acid residues (S-R-R-x-F-L-K) in the N-terminal end of large co-factors containing the proteins to secrete, that are translocated in a folded state using only the proton gradient (Müller 2005; Papanikou et al. 2007).

Type I SS are involved in the interaction of plant-associated bacteria with their hosts, in the secretion of virulence factors (metalloproteases, adhesins and glycanases) in plant pathogens like *Agrobacterium tumefaciens*, *Pseudomonas syringae* pv. *tomato*, *Ralstonia solanacearum*, *Xanthomonas axonopodis* pv. *citri* and *Xylella fastidiosa*, or in the secretion of specific proteins in the *Rhizobium* and host legume symbiosis (Delepelaire 2004; Holland et al. 2005; Reddy et al. 2007; Tseng et al. 2009).

Type II SS is found only in Gram-negative Proteobacteria, but it includes among its components also the Sec-dependent pathway. It is involved in the secretion of virulence determinants in plant and animal pathogenic species, like the ADP-ribosylating toxins of enterotoxigenic *E. coli* (heat labile toxin), *Vibrio cholerae* (cholera toxin) or *Pseudomonas aeruginosa* (exotoxin A), as well as pectinases and other enzymes of the plant pathogens *Dickeya dadantii* (*Erwinia chrysanthemi*), *Erwinia amylovora* and *X. campestris* pv. *campestris*. The type II SS pore, anchored onto the OM, is composed of 12–15 secretin subunits and is wide enough to allow the passage of large proteins. Further conserved components include integral subunits that anchor the pore to the CM, and pseudopilin subunits that span the periplasm. Energy is provided by an intracellular ATPase regulating the pore functionning, translocating proteins across the inner membrane through the universal Sec or Tat pathways (Cianciotto 2005; Tseng et al. 2009).

The members of the type III SS family, found on Gram-negative bacteria, are formed by several different components assembled to yield an apparatus involved in the secretion of many functional proteins. It is present in species directly interacting with plant or animal cells. The secreted effector products are delivered to the host cytosol through this SS that, acting as a syringe, can penetrate the host cell membrane. In other cases the type III SS (also known as injectisome) can also export proteins to the outer cell environment.

The products transferred through these systems are virulence factors or proteins involved in endosymbiotic relationships, like i.e. those active in the interactions of *Sodalis glossinidius* with its tse-tse fly host (Dale et al. 2001; Galan and Wolf-Watz 2006). Further factors include those present in the association of the mutualist *Photorhabdus luminescens* with entomopathogenic nematodes (Cornelis 2006), or the effectors of rhizobial bacteria interacting with root cells. The injectisome may be formed by up to 25 assembled proteins, and has a common evolutionary origin with the flagellum (Tseng et al. 2009). In the injectisome, a series of basal rings cross the CM and OM, to connect a needle-like structure, a filament or a pilus, each one ending with a terminal translocation pore that penetrates the target cell membrane. These components are assisted by chaperones,[2] further translocation proteins and energy-providing conserved ATPases located at the injectisome base (Cornelis 2006).

Type IV SS is the only system capable to transport either proteins or nucleic acids in cell of plants, animals or other bacteria. It is present in Gram-positive and Gram-negative species of pathogenic or mutualistic bacteria, having in common only the VirB10 (TrbI) protein (Cao and Saier 2003; Christie and Cascales 2005). In *A. tumefaciens*, the model species for this type of SS, the VirB system, exports a DNA-protein complex from the Ti plasmid into the host cell, inducing crown gall tumor tissues in which the opines (C and energy source molecules needed by the bacterium) are produced. Several members of the VirB family are produced and assembled to form the secretory machine, which is fueled by ATP hydrolysis.

Type V SS are based on porins. These ar β-barrel proteins which can originate three sub-classes of SS. The β-barrel units are large proteins formed by twisted and coiled β-sheets, having the first strand connected to the last one by H-bonds. Their general structure and dimensions are determined by the number of strands in the β-sheet and by their degree of stagger, or margin shift (Murzin et al. 1994). In the first subclass T5aSS, the proteins to be secreted show a N-terminal passenger domain (a signal peptide) and reach the periplasmic space through the Sec system, to be directed outwards by the β-barrel present in the OM. The passenger domain may remain attached to the β-barrel protein yielding an external adhesin anchored to the OM, or may be cleaved to yield toxins or enzymes, exported with the intervention of helper proteins (Omp85/YaeT) (Tseng et al. 2009). In the second and third subclasses (T5bSS and T5cSS) the β-barrels are formed by the contribution of pairs of β-barrel and secreted proteins or by the three polypeptides all together (Jacob-Dubuisson et al. 2004).

Type VI SS is an injectisome assemblage similar to the tail of a phage[3] whose first components were initially identified as proteins encoded by *imp* (impaired in N fixation) genes in *Rhizobium leguminosarum* (Bladergroen et al. 2003). The genes encoding the Type VI SS proteins are widespread among proteobacteria, planctomycetes and acidobacteria, which can utilize this system to introduce effector proteins directly into the cytoplasm of other cells. Type VI SS is involved in pathogenesis and virulence, but also in non-antagonistic interactions among bacteria, in which it is used in the transport of proteins lacking N-terminal hydrophobic signal sequences, from a donor cell to an adjacent recipient one (Leiman et al. 2009; Russell et al. 2014).

The system is formed by 13 different proteins and further complements, organized in two main complexes with bridging and cytoplasmic elements like a membrane assembly (including two proteins homologous to members of the type IV SS), a protein similar to the bacteriophage cell-puncturing needle Valine-Glycine Repeat Protein G (or VgrG), and further tube and tail components, like the Hemolysin Coregulated Protein (or Hcp) and the T4 gp25-like protein, respectively.

[2] *Chaperon, chaperonin* = proteins assisting the folding or unfolding of other proteins.

[3] *Bacteriophages, phages* = a large group of specific bacteria-parasitic viruses.

The Hcp1 examers form a tube that spans from the cytoplasmic membrane to the external surface, identical to the bacteriophage T4 tail tube (Leiman et al. 2009). With other components of the Type VI SS, a structure similar to an inverted bacteriophage is assembled with the VgrG providing the cell needle tip, Hcp yielding a tail–tube structure for effector proteins deliver. Other proteins (TssB and TssC) are assembled in a sheath, whose contractions provide the energy for the translocation process (Russell et al. 2014).

The effectors transferred via the Type VI SS range from small single-domain to larger multi-domain proteins that may act as toxins, peptidoglycan targeting enzymes, membrane lipase (phospholipase) effectors or nucleases, targeting nucleic acids of the recipient's cell. In *Pseudomonas aeruginosa* three secreted effector proteins delivered through the Type VI SS have toxicity towards bacteria and are encoded with an adjacent gene that confers immunity to the toxins. Their expression yields, for the secreting cell, a competitive advantage in its antagonistic interbacterial interactions (Russell et al. 2014).

Type VII SS are a clusters of genes and specialized secretion systems found in mycobacteria, a lineage of Gram-positives which are provided with an external lipid layer forming a complex outer membrane (mycomembrane). In *Mycobacterium* spp. the secreted proteins, which lack a signal peptide, are involved in pathogenesis or conjugation. Examples are the virulence and hemolysis effectors of *Mycobacterium marinum*, a species pathogenic in fishes, and the conjugation in *M. smegmatis* or in other pathogens like *Corynebacterium diphtheriae* and *Nocardia* sp., with other clusters also found in more distant Gram-positive groups. Type VII SS is structured with an inner membrane translocation unit formed by the membrane protein Rv3877, and a further mycomembrane channel (Abdallah et al. 2007; Tseng et al. 2009).

2.3 Cytoskeleton

Until recently, bacteria were considered to lack the actin and tubulin proteins forming the cell cytoskeleton of eukaryotes. However, bacterial homologues of eukaryotic proteins have been found, together with new proteins peculiar to the bacterial Kingdom (Jones et al. 2001).

The cytoskeleton is required to maintain either the turgor of the bacterial cell and the integrity of the wall fibers, that face the internal hydrostatic pressure. It also originates an internal network of long-range filaments as polymers of a single class of a self-assembling protein, which are involved in several cell processes and anchor active components, giving the cell its definitive shape (Shih and Rothfield 2006; Cabeen and Jacobs-Wagner 2007).

Three major classes of cytoskeleton proteins are known in bacteria. The bacterial equivalent of the eukaryotic tubulin is FtsZ, which is involved in cell division through the formation of a central ring at the division plane, controlling the time and location of the division structure. In cocci, FtsZ localizes at this area the synthesis of the new cell wall through the recruitment of peptidoglycan and related enzymes. FtsZ recruits several proteins and, in synchrony with the cell separation process, constricts the ring until the division is complete.

FtsZ appears also involved in rods elongation (Cabeen and Jacobs-Wagner 2007). In these bacteria an elongation phase is added to the coccal-type division step by MreB, a further cytoskeleton protein, equivalent of the eukaryotic actin. This protein is involved in cell division and wall elongation, either in bacteria with polar growth after cleavage (like i.e in *Corynebacterium*) and in those with a wall growth along the cylindrical cell sides, the most common process within bacteria.

MreB also controls cell diameter and is responsible for keeping the shape of rod bacteria, through the formation of an internal helicoidal network. In *Bacillus subtilis*, labelled MreB and other elements showed dynamic and rapid movements within the MreB coiled structures, showing an unidirectional movement on a time scale of many seconds (Soufo and Graumann 2004).

The third type of cytoskeletal component is represented by crescentin, an intermediate filament-like protein responsible for the cell curvature, as shown by the cells of *Caulobacter crescentus* (Ausmees et al. 2003; Cabeen and Jacobs-Wagner 2007). An actin homologue, the ParM system, is also involved in the plasmid partitioning (Shih and Rothfield 2006).

2.4 Chromosome and Plasmids

Bacteria have one linear or circular chromosome of variable size. The number of nucleotides (nt) may vary even among closely related species. The actual known range spans from a minimum of 139 kilobase (kb), found in the insect endosymbiont *Candidatus* "Tremblaya princeps", to sizes around 9.1 Mbp, as i.e. observed in *Bradyrhizobium japonicum*. The maximum genome size thus far known for a bacterium was identified in *Sorangium cellulosum* strain So0157-2, which accounts for more than 14.78 Mbp and 11,599 genes (Han et al. 2013; McCutcheon and Moran 2012; Land et al. 2015). In general, the number of genes in bacteria is related to their genome size, at a rate of approx. one gene per kb of DNA, being sometimes significantly reduced in some endosymbionts to a few hundred (Bentley and Parkhill 2004; McCutcheon and Moran 2012; Land et al. 2015).

Several bacteria are also provided with a second repository of genetic information: a linear or circular plasmid representing a further additional DNA unit. Plasmids, whose size vary from 10 to 30 kb, carry useful although non-essential genes. They are transferred at daughter cells at replication, and may also serve for genetic recombinations (Fig. 1.5). This may occur either as plasmids are integrated in the chromosome at replication or exchanged among cells through the pilus during conjugation, a kind of bacterial mating process (Reyes-Lamothe et al. 2012). Plasmids are usually present in multiple copies (small plasmids) or, in case of larger ones, in a low number. In some species, i.e. *Vibrio cholerae* or some α-proteobacteria,

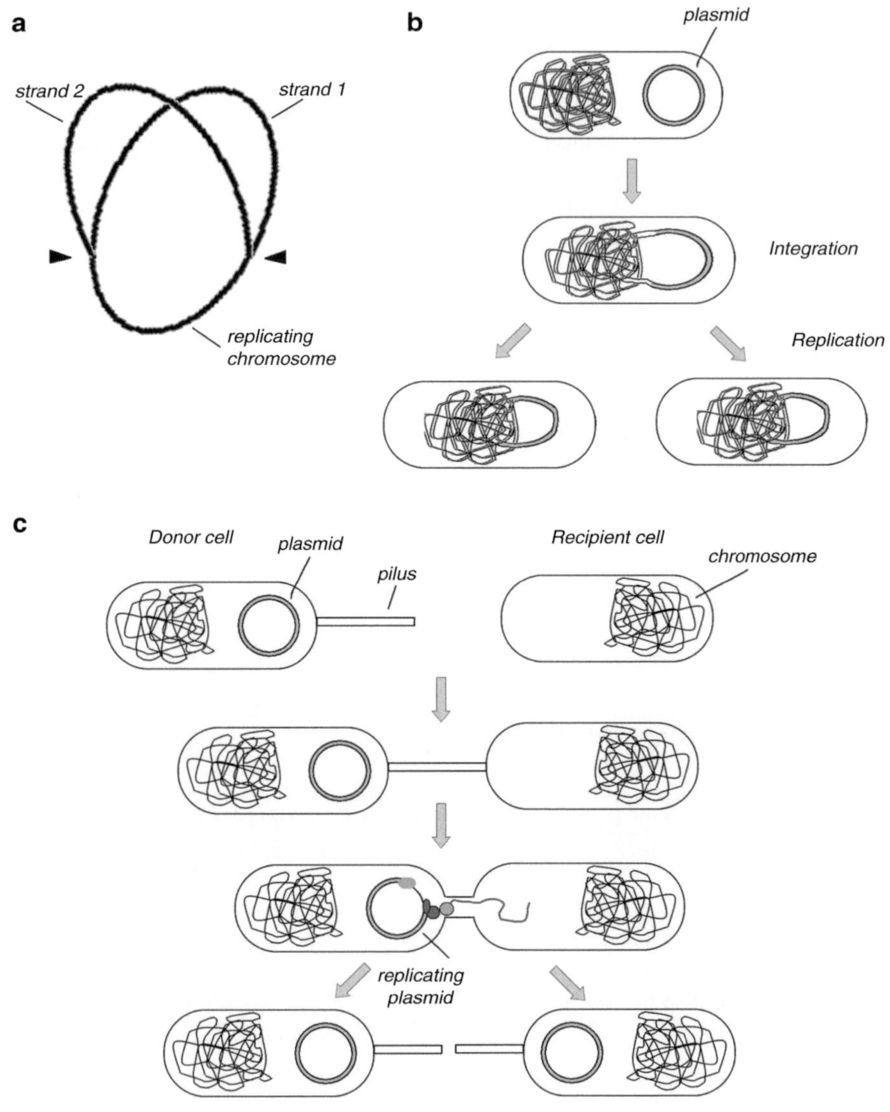

Fig. 1.5 Simplified schemes of chromosome replication fork with origin and termination regions (**a**, *arrowheads*). Plasmid integration (**b**) and pilus-mediated conjugation (**c**)

two chromosomes may be found, with a primary chromosome flanked by a secondary one, a sort of very large plasmid expressing, however, also some essential genes.

As prokaryotes, Bacteria lack a nucleus and their chromosome is tightly packed in the cell environment. Its organization is determined by its negative supercoiling, by the number of nucleoid[4] associated proteins (NAPs), and by further molecules

---

[4] *Nucleoid* = an irregular cell region, deprived of membrane, in which most protein-coated genetic material is compacted.

from the cytoplasm. Either NAPs and other proteins contribute to bridge and organize distant chromosomic regions, and their coding genes are differentially expressed during the bacterial exponential and stationary growth phases (Reyes-Lamothe et al. 2012). Gene transcription occurs outside the nucleoids, which lack the ribosomes on which transcription takes place. Similarly, also the DNA replication is considered to occur outside the nucleoids.

Three families of proteins act in the DNA replication and chromosome segregation. These are spatially and temporally regulated processes, recruiting the replication origin (*ori*) and termination (*ter*) chromosome regions to specific cell locations. Bacterial chromosomes and plasmids have a replication origin which is regulated by the replicon, a unit whose initiation leads to the bidirectional chromosome replication carried out by sister, divergent structures called replisomes (Fig. 1.5).

In *E. coli*, a complex of proteins is involved in replication, starting with the DnaA protein and the ATP complex, that acts as chromosome initiator binding to specific *oriC* sites present in the chromosome. They separate the DNA strands to facilitate synchronous duplication by the replisomes at all *oriC* sites present in the cell. The replisome is assembled by loading divergent DnaB helicases, recruited by the helicase loader DnaC protein. Specific initiator proteins, different from DnaA, are encoded by low-copy number plasmids and secondary chromosomes, with additional mechanisms regulating their copy number. In high-copy number plasmids, further mechanisms (transcription, RNA processing and DNA polymerase I extension) are used. To avoid formation of too many DnaA-ATP filaments or inhibit DnaA transcription, further SeqA proteins bind to the *oriC* at their abundant GATC [5] sites (Reyes-Lamothe et al. 2012).

Chromosome elongation proceeds subsequently through the substitution of DnaC by the DnaG protein and the synthesis of an RNA primer acting as loading site for a ring-shaped β-sliding clamp. This is targeted by up to three polymerases (PolIII), which proceed to increase the DNA replication fork for several kb. Once the following downstream primer is reached, it is removed and the newly formed DNA strands are ligated (Johnson and O'Donnell 2005). When the forks, proceeding from opposite directions, meet at the *ter* sites, the duplication process is terminated by the bound terminator protein Tus (if one fork is delayed the other stops duplication until its arrival). This process is followed by decatenation of cross links by a II topoisomerase, a DNA gyrase, which removes accumulated positive supercoilings, and by a TopoIV that removes interwrappings and catenation links (Neylon et al. 2005; Reyes-Lamothe et al. 2012).

The chromosome dimer resolution and subsequent segregation lead to the complete separation of the genetic material in two distinct cells, achieved by converting dimers to monomers through a FtsK translocase located at the septum (Bigot et al. 2007). Segregation occurs after duplication through a diffusion ratchet mechanism in which three systems, replicon-specific members of the *ParAB-parS* family of partition systems (also used by plasmids), the SMC protein complexes and the FtsK

---

[5] *GATC* = a sequence domain with the four DNA nucleotides guanine, adenine, thymine and cytosine.

translocase, are involved. Cell division then follows, through a machinery called divisome and the completion of the septal ring by the FtsZ cytoskeleton and associated proteins, which coordinate late segregation events with cell division (Reyes-Lamothe et al. 2012).

2.4.1 Genetic Recombination

Sexual genetic crossings were first demonstrated in bacteria in 1946 for the *E. coli* strain K12 by Nobel Prizes J. Lederberg and E. Tatum (Lederberg 1987). The crossing of genetic material between two adjacent bacterial cells (known as conjugation) occurs through the formation of a sort of bridge filament. It requires the presence of a circular F (fertilization) plasmid with a set of approx. 100 genes. The cells carrying the F plasmid, identified as F^+, promote the formation of a thin filament called *pilus* (translation of hair from latin), that gets attached to an adiacent cell to maintain a contact (conjugation) (Fig. 1.5). F^+ cells conjugated with F^- cells (deprived of the F plasmid) transfer in a few minutes through the pilus a linear single-stranded copy of the self replicating F plasmid, which may also include an insertion sequence, a transposable element (transposon) of the cell chromosome. These are segments of DNA carrying one or more genes, that can change their position shifting along the chromosome or may be included in the plasmid, inducing some genes to be activated or, viceversa, inactivated (Graf and Ruby 2000).

In a minor fraction of F^+ cells, the plasmid can also be integrated in the chromosome through a crossover, producing first a long linear DNA and then, through a second crossover, complete the formation of a double strand DNA. In this case the entire donor chromosome may be transferred to a recipient F^- cell, thus producing genetic recombinants. Strains in which the F plasmids are transferred have higher frequencies of conjugation and are hence called High frequency recombination (Hfr) strains. The F plasmid in some Hfr strains may also integrate some chromosomal genes of the recipient cell, moving back to the cytoplasm, from where it may be further transferred, carrying the newly acquired genes with it (Griffiths et al. 1999). Recombination allowed first mappings of bacterial chromosomes, through the application of methods like the interrupted conjugation or the recombinant frequency.

Apart of mutations and genetic drifts, other genetic recombination processes are the transformation, in which bacteria directly acquire large DNA molecules from their surrounding environment, conjugation, in which plasmids or transposons can move long DNA fragments and transduction. In this latter process, a bacteriophage may incorporate, during its assemblage in an infected cell, some host DNA, that is transmitted to a new cell after the host lysis, when a new infective event occurs. The transmitting phage, however, is unable to multiply and to produce a lysis of the new host. The transmitted DNA may hence be incorporated in the recipient cell genome, that acquires new genes or may replace genes already present in its chromosome (Brochier-Armanet and Moreira 2015).

These processes have been active in genes selection and transfer during the bacteria evolution, and control their adaptation to new environment conditions, including the appearance of novel specialized biochemical capabilities (i.e. the degradation of new substrates) or the insurgence of antibiotic-resistant strains.

2.4.2 DNA Methylation

In the cells of bacteria (as for any other organism) most DNA is methylated. This is a fundamental process present in organisms from all Kingdoms, in which a -CH_3 group is covalently linked to the C-5 or N-4 positions of cytosine and the N-6 position of adenine by enzymes known as DNA methyltransferases (MTases), using S-adenosyl methionine as a methyl donor. Main function of methylation is the protection of DNA, which otherwise will be degraded by the restriction endonucleases active in the cell, as occurs for foreign transposons or viral DNA, once introduced. A second important function of methylation is the correction of mismatches found on newly replicated DNA, which is checked by the cell mismatch repair system before methylation occurs and, in case of errors, eventually repaired (Cooper et al. 1993; Low et al. 2001).

Being reversible and inactivating the transcription of methylated genes, this process is considered as an epigenetic mechanism of gene regulation. Different MTases are known, most important being those active in cell regulatory events and involved in bacterial virulence, like the DNA cytosine MTase (Dcm, methylating cytosine at the C-5 position in CC[A/T]GG sequences), the DNA adenine methylase (Dam, methylating adenine at N-6 in GATC), and the cell cycle-regulated methylase (CcrM, which methylates at N-6 the adenine in GAnTC) (Low et al. 2001).

Dam adenine methylation affects interactions between regulatory proteins and DNA, inhibiting the expression of some genes (i.e. *trpR* and Tn*10* transposase) or enhancing their transcription. This activity affects genes active in fundamental cell growth processes like DNA replication, segregation and mismatch repair, as well as DNA transposition. Dam overexpression in many mutant pathogens also reduced their virulence (Low et al. 2001). Competition between Dam methylation and DNA binding proteins inhibiting methylation yields hemimethylated regions (i.e. the *oriC* required in DNA replication), originating DNA methylation patterns (DMPs) (Braaten et al. 1994).

Dam epigenetic regulation may involve repressor or activator genes. In *E. coli*, GATC methylation sites control operons like the *pap* and *agn*, inhibiting their regulatory proteins Lrp and OxyR, thus affecting expression of the secreted proteins of the Pap pili and the OM, respectively. In *pap* and related operons, methylation targets two GATC sites 102 bp apart, regulating Lrp binding. In *agn*, methylation of three close GATC sites inhibits OxyR binding. The DMPs and related proteins have an inheritable control of gene expression, whose effects can be also observed in the cell progeny. DMPs may be transmitted to daughter cells if GATC methylation is blocked, originating distinct epigenetic states propagated by a positive feedback loop allowing the appearence of distinct epigenetic lineages (Low et al. 2001; Casadesús and Low 2006).

CcrM is a β DNA methylase originally described in *Caulobacter* and also present in α-Proteobacteria, essential in the maintenance of hemimethylated DNA, acting as a global regulator of gene expression. Either Dam and CcrM are strictly regulated. The 130 Dam molecules present in fast growing cells, methylating all GACT sites during replication, are activated by five promoters controlled by the cell growth rate. CcrM transcription is under control of the cell cycle transcription regulator, methylating the chromosome at the onset of replication (Rasmussen et al. 1995; Low et al. 2001).

2.5 *Flagella and Pili*

The bacterial flagellum is a complex structure which propels the cell by rotating at high speed. It represents a true biological rotary engine, fueled by electric charges passing through the plasma membrane at its basal parts, built through the assemblage of several proteins. The flagellar function is controlled by complex regulatory pathways, allowing the bacterium to respond to positive chemiotactic signals, to be propelled towards a nutrient rich environment.

The bacterial flagellum is coded by more than 50 genes that produce the structural protein components or control their assemblage. Phylogenetic analysis of the genes clusters from 11 bacterial phyla showed that the homologs derive from an ancient group of 24 related genes whose similarity suggests their origin from a few, or even single, ancient precursor (Liu and Ochman 2007). The structural components of the *E. coli* K-12 MG1655 flagellum and their assembling pathways are available at the KEGG online database at http://www.genome.jp/kegg-bin/show_pathway?eco02040.

The pili (plural of *pilus*) or fimbriae (latin for 'fibers') are filaments extending from the bacterial cell surface. They are formed through the assembage of pilins, 15–25 kDa proteins that also act at their terminus as adhesins, mediating adhesion to host cells through the recognition of specific adhesion domains. These may be oligosaccharide residues of glycoprotein or glycolipids. By this way the bacterial cell can adhere to its substrate without being affected by the negative repulsive charge of its target cell. Pili are also involved in other functions, like the DNA transfer, biofilm formation and cells aggregation, twitching motility and pathogenesis. They are important virulence factors and targets for the development of vaccines. Pili evolved in several Gram-negative (with robust, retractile Type I and Type IV pili, also involved in DNA uptake), and Gram-positive lineages (thinner, and mainly involved in cell adhesion), being instrumental in the evolution of their pathogenic, environmental and nutritional needs (Proft and Baker 2009).

2.6 *Quorum Sensing*

The ability of bacteria to recognize other cells in their proximity and react to their density through a number of biochemical signals after a given threshold is achieved is called '*Quorum sensing*' (QS; *quorum* from latin: 'of the others'). It aims at adapting and regulating gene pathways and cascades that affect and steer the cell metabolism and behaviour on a global scale in the community. QS may affect the success of colonial bacteria like endosymbionts or pathogens, controlling the formation of their colonies and biofilms.[6]

The main class of signal-carrying molecules are membrane-diffusible products (autoinducers), whose accumulation controls the expression/repression of a number of target genes, in a cell density-dependent way. QS was originally found in the bioluminescent *Vibrio fisheri*, an endosymbiont of the squid *Euprymna scolopes*. It controls, among other processes, the level of bacterial light emission through small N-acylated homoserine lactone (AHL) molecules, a fundamental signal either for the survival of the symbiont, when reaching high densities, as for the host (Eberhard et al. 1981; Boettcher and Ruby 1995; McFall-Ngai and Ruby 2000). The system is based on LuxI-type synthases producing AHL and on LuxR-type binding receptor proteins (Diggle et al. 2007; Galloway et al. 2012).

Being involved in the control of key genes like those coding for virulence factors, QS has been extensively studied in human pathogens, and different types of autoinducers have been identified. In the Gram-positive opportunistic pathogen *Staphyloccocus aureus*, the autoinducer oligopeptide (termed AIP) is induced by an accessory gene regulator (*agr*) locus, which regulates 70 genes. These include several virulence factors of two classes, involved in host attachment and immune system evasion, or in host invasion and toxin production, respectively. Since *agr* is present in four variants due to genetic polymorphisms, the corresponding AIPs can bind to all *agr* receptors, but can activate only their own specific receptor. This suggested a possible way for cross inhibition (termed '*Quorum quenching*') of the virulence determinants pertinent to one or more *agr* loci, as shown in an insect infection model (Fleming et al. 2006). The expression of *agr* is involved in biofilm formation and in cell clusters detachment, since its repression showed a *S. aureus* enhanced adhesion capability. This function has direct consequences, being likely that biofilm regulation and *agr* activation are required in an initial low pathogenicity strategy aiming at avoiding the host immune response, followed by an increase of virulence, during the subsequent host invasion phase (Dobretsov et al. 2009; Antunes et al. 2010).

Other processes controlled by QS signals are cell swarming, stress resistance and production of toxins or other secondary metabolites. Apart of oligopeptide autoinducers and AHL, other molecular products active as QS signals are some organic acids, furanosyl borate diesters or furane derivatives. In marine environments, Gram-negative bacteria rely on AHL signals and luxR gene cluster proteins that activate transcription of QS genes responsible for bioluminescence, biofilm formation or virulence. In environments where biofouling (the rapid colonization of a new surface by bacteria and invertebrates) depends on biofilm formation and maintenance, QS plays a key role in the colonization of submersed surfaces, since its interruption affects the sequential colonization by feeding invertebrate larvae. The production of AHLs is widespread among invertebrates, as it was also observed in

[6] *Biofilm* = a sessile community of bacteria adhering to each other and to a surface.

bacteria associated to sponges like *Mycale laxissima* and *Ircinia strobilina* (Taylor et al. 2004; Dobretsov et al. 2009).

Similarly, also in plant pathogenic or symbiotic bacterial associations, AHL detection showed the occurrence of QS-based systems. Products and processes controlled included extracellular polysaccharides, enzymes, antibiotics, siderophores, pigments, secreted proteins, or the transfer of plasmids, biofilm formation, cell motility and fitness. Interference appeared as a possible low-environmental impact strategy to impair and control bacteria-induced diseases or regulate beneficial phenotypes of symbionts. Several *in vitro* tested AHL-based modulators have been applied, to control the causal agent of soft rot *Pectobacterium* (*Erwinia*) *carotovora*, the corn leaf pathogen *Pantoea stewartii*, the plant tumor inducer *Agrobacterium tumefaciens* or other plant pathogens like *Pseudomonas syringae*. Although retaining the inhibitory activities on the QS systems tested, AHL degradation and other inhibitory assays showed that in natural conditions many aspects of these interactions should be identified, and that the timing and dose of application are crucial factors for effective virulence control (von Bodman et al. 2003; Galloway et al. 2012).

2.7 CRISPR-CAS[7] System

The response to pathogens based on an adaptive defense immunity, which was considered until a few years ago as unique for eukaryotes, has been reported and its functionning recently demonstrated also in Bacteria and Archaea (Ishino et al. 1987; Jansen et al. 2002; Barrangou et al. 2007). This system called CRISPR-CAS integrates the innate defense response of bacteria, recognizing conserved features of their viral pathogens.

The "memory" of a CRISPR and CRISPR associated system (CAS) is stored as small DNA sequences originated from the viral pathogens, called "spacers". These are interleaved within the CRISPR DNA repeats and flanked by a set of CAS protein coding genes. The system works through a first adaptation step, in which the Cas proteins recognize and acquire a spacer fragment from the target invading pathogen DNA. This fragment is then integrated into the CRISPR array to form its "memory". In a second step, identified as expression and maturation stage, the CRISPR array is transcribed to originate a series of CRISPR RNAs that are flanked by a part of the repeat sequence. These are then combined with one or more CAS proteins to form a complex. In the final interference stage, foreign DNA is recognized into the cell by base-pairing with the RNA region of the complex, which then leads to the degradation of the foreign DNA by nuclease proteins.

This system has been adopted for editing bacterial genomes through a number of engineered nucleases (Sander and Joung 2014). For CRISPR-CAS review see Amitai and Sorek (2016).

[7] *CRISPR*=clustered, regularly interspaced, short palindromic repeat.

3 Metabolism

The metabolism of bacteria is extremely complex and variable, reflecting the natural history of life on earth and the overlapping of different evolutive paths eventually discovered and followed on a scale of ages. Although the understanding of the metabolic interactions is actually far from being complete, studies *in vivo* on a number of model species represent the best approach for deciphering the diversity of biochemical networks and products, and their functional contributions. A global systemic view of bacterial metabolism thus identified a central core of pathways related to C catabolism, biosynthesis of nucleotides, amino acids, vitamins and cofactors. A second level of metabolic integration concerns pathways unique to specific lineages, like photosynthesis, iron oxidation, N fixation or methylotrophy. However, more data from studies based on phenotypic changes and mutations are needed to decipher the complexity of bacterial metabolism, or to discover new pathways or components (Downs 2006).

In general, metabolism is classified using a simple primary scheme based on terms identifying the energy (*photo-* or *chemo-*), electron donors (*-organo-* or *-litho-*) and C (*-heterotroph* or *-autotroph*) sources. The combinations of terms for these three general processes yield a metabolic classification pertinent to a particular organism, including bacteria. Plants are for example photolithoautotrophs, since they use light as energy source (photo) with inorganic donors (litho) and fix the CO_2 (autotroph). The Cyanobacteria are photoautotrophs, using light and water as organic donor by fixing the CO_2.

Starting from these bases, the diversity and richness of the biochemical processes achieved by bacteria and of the eventual ecological functions they deploy are very large and remain difficult to categorize. A modular approach however, may help in categorizing the regulatory circuits and the building blocks of the bacterial adaptive metabolism (Wang and Levin 2009; Chubukov et al. 2014). Being present in any environment, bacteria perform such a wide range of services that the entire biosphere would be unsustainable and undoubtly different, if they were absent. Many species are also adapted to particular trophic niches, as for example those involved in the anaerobic degradation of oil and hydrocarbons, playing by this way a beneficial bioremediation role in disasters like oil spill accidents, eventually clearing the devastating pollution effects produced (Heider et al. 1999). In this perspective an even more beneficial activity is deployed by these bacteria when they mineralize widespread hydrocarbon pollutants present in low amounts but on wide areas, in seawaters and coastal environments. Bacteria-based bioremediation is actually an economic activity targeting the market of contaminated areas deprived of an indigenous degrading microflora.

Bacteria evolved several persistence strategies highly effective to sustain their survival in the environment for long periods of time, once the external conditions for their life show drastic changes. This occurs when i.e. they have to face a reduction in nutrients, or one or more long term changes in external physical conditions (i.e. increased temperature, dryness or salt concentration), or when suitable hosts are absent, due to a specific adaptation to the symbiotic life style. The evolution of specialized metabolism in extremophiles allowing new niche colonization and/or the development of resting and durable spores represent two examples of successful strategies exploited in various bacterial lineages.

Readers wishing to deepen the study on more specific aspects of bacterial metabolism, cell structure and molecular functions can address useful online databases like KEGG, the Kyoto Encyclopedia of Genes and Genomes (http://www.genome.jp/kegg/), Microme (http://www.microme.eu/), MetaCyc (http://metacyc.org/) or other available tools that contributed to the development of actual research approaches of system biology (Heinemann and Sauer 2010; Tang 2011). Further general microbiology treatises, including *Bergey's Manual of Systematic Bacteriology* (Boone et al. 2001), Dworkin (2006) and Madigan et al. (2014), may also be considered, together with the reviews herein cited.

4 Identification

For more than one century the bacterial taxonomy mainly relied on *in vitro* cultures and/or careful slide examinations in light microscopy, as well as on the morphological study or histochemical analyses and staining, for comparison with known species phenotypes. Later on, biochemistry investigations and TEM or SEM analyses provided further ultrastructural and biochemical data, that coupled with immunocytochemistry and antibody labelling techniques, yielded clearer views of bacterial phylogenetic and taxonomic organizations. Further progress has been achieved through the advent of fluorescent labelling methods, including fast *in situ* hybridization (FISH) techniques, labelling cells, organs and tissues with specific antibody or DNA-based fluorescent probes. Actually, culturable species are identified and described through a number of molecular tools and approaches including the use of specific immobilized antibodies (ELISA), the identification of main biochemical profiles or through the sequencing of one or more genes or even of their entire genome.

The genes most routinely sequenced for identification purposes are the ribosomal genes 5S, 16S and 23S (Fig. 1.6). They are universally present in all species and necessary for the translation machinery of the ribosomes, responsible for the synthesis of proteins. Furthermore, they are consistent as concerns function and size. Their precursor 30S RNA is cleaved after transcription. Subsequent cleavage

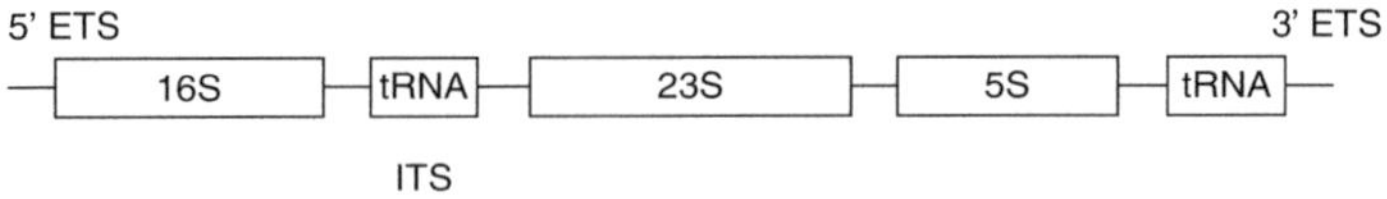

Fig. 1.6 A simplified scheme of bacterial ribosomal genes, with transfer RNAs, external transcribed spacers (ETS) at the 5′ and 3′ positions and internal transcribed spacer (ITS). Inverted repeats (*internal lines*) originate stem structures recognized and cleaved by *RNAseIII*

and trimming result in the three RNA molecules and in two tRNAs (Lafontaine and Tollervey 2001).

The small subunit 16S rRNA ribosomal gene is the most common sequence required for the description of new species, and is also applied for detection or identification at the species or subspecies levels (Pace et al. 1986; Yarza et al. 2014).

The ribosomal genes are present in multiple copies and represent a relatively stable part of the genetic code, essential for cell functionning and acting as scaffolds for the correct spatial arrangement of the ribosomal proteins (Woese 1987). The 16S gene includes several conserved regions, and nine areas of genetic divergence termed "hypervariable regions". In most bacterial lineages the approx. 1550 bp long 16S rRNA ribosomal gene is widely sequenced, being informative enough to determine the position of an isolate in relation to other species or strains. These data are also suitable as molecular tools to measure the evolutionary distances established among taxa. They provide the guidelines to classify the bacterial evolutionary groupings and divergence, as adopted in reference texts like the *Bergey's Manual of Systematic Bacteriology* (Clarridge 2004; Chakravorty et al. 2007).

Since the first applications of the polymerase chain reaction (PCR) to the study of the bacterial rRNA genes, a number of primers either universal or specific for given lineages have been progressively made available. The primers' nucleotide positions are indicated by their numbering in reference to the 16S of *E. coli* nucleotides. A large amount of rRNA ribosomal gene sequence data are available in the literature and public databases. For primers specific to bacterial lineages see Lane et al. (1985) and Wilson et al. (1990).

PCR is largely applied for bacterial identification, by sequencing and comparing amplified 16S ribosomal gene data with the ca. 10^6 sequences of the same fragment from known or unknown species, already deposited and made available on the NCBI GenBank (www.ncbi.nlm.nih.gov/genbank/), the Ribosomal Database Project (RDP) (http://rdp.cme.msu.edu/) or the EMBO databases. Queries are available through on line bioinformatic tools like BLAST (http://blast.ncbi.nlm.nih.gov/Blast.cgi). Although the entire 16S gene sequence is required for a bacterium description, the first 500 bp are usually sufficient for identification, being a region with higher frequencies of nucleotidic diversity. Ribosomal genes are also used as preferential targets in massive gene sequencing projects (Klingworth et al. 2013).

The identification of a bacterium in culture is, however, not always an easy and simple task, as even the species concept in bacteria is debated (Konstantinidis et al. 2006). The boundary for species determination is conventionally agreed at around 97 % of nucleotidic identity. Below this threshold the species examined may be considered different. PCR amplification from environmental samples or plant or animal tissues may also yield undescribed species.

The efforts aiming at the identification of a bacterium, in particular when applying PCR to few unculturable cells as often encountered in invertebrate tissues, may often yield undefined results. This may occur if close comparison data are lacking (as for new, undescribed and/or uncultured species). The correct placement of a bacterium in a phylogenetic context largely depends on the quality, length and accuracy of the gene sequencing and informations produced. This work also benefits from data proceeding from the identification works carried out in other laboratories, and made available in databases. In many circumstances, however, due to the insurgence of mutations, deletions, insertions or other mechanisms, assessing the taxonomic position or discriminating among different strains of the same species may result uncertain. Introns[8] have been found also in the 16S gene of some bacterial lineages like in large sulfur bacteria of the genus *Thiomargarita*. Their presence may render difficult sequencing this region if working with close undescribed or sibling species (Salman et al. 2012). It is hence wise to compare also the data proceeding from other unique or more specific genes, if their sequence composition is known. This is important for example when identifying isolates characterized by different levels of virulence, not coded by their ribosomal genes.

When close species cannot be discriminated through comparison of their 16S sequences, other ribosomal gene regions may hence be used. 16S data are generally accepted for identification up to the genus level, but often they do not result resolutive, when determining a species identity (Zeigler 2003; Sundquist et al. 2007). Other targets for sequencing include the 16S–23S rRNA internal transcribed spacer or the 65 kDa heat shock protein encoding gene (Goh et al. 1996; Clarridge 2004). For any gene selected, however, enough deposited sequences must be made available in databases for effective and reliable comparisons.

A gene with potential for bacteria identification at the species level is *cpn60*, coding for a 60 kDa universal chaperonin. It is used as a target for detection and quantitative determination, for bacteria barcoding or for the assessment of metagenome profiles in different types of samples (Dumonceaux et al. 2006; Links et al. 2012). Barcoding relies on a number of gene sequences showing a inter-species distance higher than their intra-specific variation (the so-called "barcode gap"). It thus allows a reliable identification at the species level (Links et al. 2012). The *cpn60* gene has been used to overcome some limitations of the 16S-based identification approach.

Recently, powerful methods have been developed based on massive or whole genome sequencing approaches, that have also been made available as commercial services, worldwide. These are Roche 454™ Pyrosequencing, Illumina™ Next Generation Sequencing (NGS), and other mass sequencing methods, often identified through their corresponding device or technology (PacBio™, Ion Torrent™, Nanopore sequencing). These approaches are particularly useful since they provide broad and abundant sequence data on cultured species. They also are helpful when the bacterial cells cannot be multiplied *in vitro*, as in the so-called "fastidious" bacteria, due to some specific limitations or to the lack of knowledge about their nutritional requirements. These bacteria are indeed very common in any environment and represent a significant fraction of the species actually present on earth, which are yet undiscovered and/or undescribed (estimated at around 90 % of all species present on the planet). Also in this case, many manuals and a large amount of data

---

[8] *Introns* = non-coding nucleotidic regions inserted mostly within eukaryotic genes. During gene expression they are not trascribed nor translated into aminoacid sequences.

proceeding from publications or consultable databases (NCBI, EMBO) provide substantial support for comparison of the sequences produced and for genes/species identification.

The -omics approaches applied thus far to the study of bacteria are rapidly increasing our knowledge about their biology through the still growing number of genome-wide studies and comparative sequencing. Actually, at least 14,000 bacterial genomes have been sequenced partially or in total, an impressive result that opens the way to a new era in microbiology (Land et al. 2015). Progress has also been achieved in RNA sequencing and gene expression studies (transcriptomics), in the application of metagenomic approaches to the study of bacterial ecology and, in particular, of the species composition and diversity in well defined body segments (i.e. gut metagenomics) or environments (i.e. soil, ocean). Although not pertinent to the aims of this volume, many of these studies and related discoveries will be analized and reported throughout the following chapters for their direct and indirect contributions to the bacteriology of invertebrates.

References

Abdallah, A. M., et al. (2007). Type VII secretion–mycobacteria show the way. *Nature Reviews Microbiology, 5*, 883–891.

Amitai, G., & Sorek, R. (2016). CRISPR–Cas adaptation: Insights into the mechanism of action. *Nature Reviews Microbiology, 14*, 67–76.

Antunes, L. C. M., Ferreira, R. B. R., Buckner, M. M. C., & Finlay, B. B. (2010). Quorum sensing in bacterial virulence. *Microbiology, 156*, 2271–2282.

Ausmees, N., Kuhn, J. R., & Jacobs-Wagner, C. (2003). The bacterial cytoskeleton: An intermediate filament-like function in cell shape. *Cell, 115*, 705–713.

Barrangou, R., et al. (2007). CRISPR provides acquired resistance against viruses in prokaryotes. *Science, 315*, 1709–1712.

Bartesaghi, A., et al. (2015). 2.2 Å resolution cryo-EM structure of β-galactosidase in complex with a cell-permeant inhibitor. *Science, 348*, 1147–1151.

Bentley, S. D., & Parkhill, J. (2004). Comparative genomic structure of prokaryotes. *Annual Review of Genetics, 38*, 771–791.

Bigot, S., Sivanathan, V., Possoz, C., Barre, F. X., & Cornet, F. (2007). FtsK, a literate chromosome segregation machine. *Molecular Microbiology, 64*, 1434–1441.

Bladergroen, M. R., Badelt, K., & Spaink, H. P. (2003). Infection-blocking genes of a symbiotic *Rhizobium leguminosarum* strain that are involved in temperature-dependent protein secretion. *Molecular Plant-Microbe Interactions, 16*, 53–64.

Boettcher, K. J., & Ruby, E. G. (1995). Detection and quantification of *Vibrio fischeri* autoinducer from symbiotic squid light organs. *Journal of Bacteriology, 177*, 1053–1058.

Boone, D. R., Castenholz, R. W., & Garrity, G. M. (Eds.). (2001). *The Archaea and the deeply branching and phototrophic bacteria* (Bergey's manual of systematic bacteriology, Vol. I). New York: Springer, 721 pp.

Braaten, B. A., Nou, X., Kaltenbach, L. S., & Low, D. A. (1994). Methylation patterns in pap regulatory DNA control pyelonephritis-associated pili phase variation in *E. coli*. *Cell, 76*, 577–588.

Brochier-Armanet, C., & Moreira, D. (2015). Horizontal gene transfer in microbial ecosystems. In J. C. Bertrand, P. Caumette, P. Lebaron, R. Matheron, P. Normand, & T. Sime-Ngando (Eds.),

Environmental microbiology: Fundamentals and applications: Microbial ecology (pp. 445–481). Dordrecht: Springer.
Cabeen, M. T., & Jacobs-Wagner, C. (2007). Skin and bones: The bacterial cytoskeleton, cell wall, and cell morphogenesis. *The Journal of Cell Biology, 179*, 381–387.
Cao, T. B., & Saier, M. H. (2003). The general protein secretory pathway: Phylogenetic analyses leading to evolutionary conclusions. *Biochimica et Biophysica Acta, 1609*, 115–125.
Casadesús, J., & Low, D. (2006). Epigenetic gene regulation in the bacterial world. *Microbiology and Molecular Biology Reviews, 70*, 830–856.
Chakravorty, S., Helb, D., Burday, M., Connell, N., & Alland, D. (2007). A detailed analysis of 16S ribosomal RNA gene segments for the diagnosis of pathogenic bacteria. *Journal of Microbiological Methods, 69*, 330–339.
Christie, P. J., & Cascales, E. (2005). Structural and dynamic properties of bacterial type IV secretion systems (review). *Molecular Membrane Biology, 22*, 51–61.
Chubukov, V., Gerosa, L., Kochanowski, K., & Sauer, U. (2014). Coordination of microbial metabolism. *Nature Reviews Microbiology, 12*, 327–340.
Cianciotto, N. P. (2005). Type II secretion: A protein secretion system for all seasons. *Trends in Microbiology, 13*, 581–588.
Clarridge, J. E. (2004). Impact of 16S rRNA gene sequence analysis for identification of bacteria on clinical microbiology and infectious diseases. *Clinical Microbiology Reviews, 17*, 840–862.
Cooper, D. L., Lahue, R. S., & Modrich, P. (1993). Methyl-directed mismatch repair is bidirectional. *Journal of Biological Chemistry, 268*, 11823–11829.
Cornelis, G. R. (2006). The type III secretion injectisome. *Nature Review Microbiology, 4*, 811–825.
Dale, C., Young, S. A., Haydon, D. T., & Welburn, S. C. (2001). The insect endosymbiont *Sodalis glossinidius* utilizes a type III secretion system for cell invasion. *Proceedings of the National Academy of Science USA, 98*, 1883–1888.
Delepelaire, P. (2004). Type I secretion in gram-negative bacteria. *Biochimica et Biophysica Acta, 1694*, 149–161.
Diggle, S. P., Gardner, A., West, S. A., & Griffin, A. S. (2007). Evolutionary theory of bacterial quorum sensing: When is a signal not a signal? *Philosophical Transactions of the Royal Society B, 362*, 1241–1249.
Dobretsov, S., Teplitski, M., & Paul, V. (2009). Mini-review: Quorum sensing in the marine environment and its relationship to biofouling. *Biofouling, 25*, 413–427.
Downs, D. M. (2006). Understanding microbial metabolism. *Annual Review of Microbiology, 60*, 533–559.
Dumonceaux, T. J., et al. (2006). Enumeration of specific bacterial populations in complex intestinal communities using quantitative PCR based on the chaperonin-60 target. *Journal of Microbiological Methods, 64*, 46–62.
Dworkin, M. (Ed.). (2006). *The prokaryotes* (3rd ed., Vol. 7). New York: Springer.
Eberhard, A., et al. (1981). Structural identification of autoinducer of *Photobacterium fischeri* luciferase. *Biochemistry, 20*, 2444–2449.
Fagan, R. P., & Fairweather, N. F. (2014). Biogenesis and functions of bacterial S-layers. *Nature Reviews Microbiology, 12*, 211–222.
Fleming, V., et al. (2006). Agr interference between clinical *Staphylococcus aureus* strains in an insect model of virulence. *Journal of Bacteriology, 188*, 7686–7688.
Galan, J. E., & Wolf-Watz, H. (2006). Protein delivery into eukaryotic cells by type III secretion machines. *Nature, 444*, 567–573.
Galloway, W. R., Hodgkinson, J. T., Bowden, S., Welch, M., & Spring, D. R. (2012). Applications of small molecule activators and inhibitors of quorum sensing in Gram-negative bacteria. *Trends in Microbiology, 20*, 449–458.
Goh, S. H., et al. (1996). HSP60 gene sequences as universal targets for microbial species identification: Studies with coagulase-negative staphylococci. *Journal of Clinical Microbiology, 34*, 818–823.

Graf, J., & Ruby, E. G. (2000). Novel effects of a transposon insertion in the *Vibrio fischeri glnD* gene: Defects in iron uptake and symbiotic persistence in addition to nitrogen utilization. *Molecular Microbiology, 37*, 168–179.

Griffiths, A. J. F., et al. (1999). *Modern genetic analysis*. New York: W. H Freeman.

Han, K., et al. (2013). Extraordinary expansion of a *Sorangium cellulosum* genome from an alkaline milieu. *Scientific Reports, 3*, 2101. doi:10.1038/srep02101.

Heider, J., Spormann, A. M., Beller, H. R., & Widdel, F. (1999). Anaerobic bacterial metabolism of hydrocarbons. *FEMS Microbiology Reviews, 22*, 459–473.

Heinemann, M., & Sauer, U. (2010). Systems biology of microbial metabolism. *Current Opinion in Microbiology, 13*, 337–343.

Holland, I. B., Schmitt, L., & Young, J. (2005). Type 1 protein secretion in bacteria, the ABC-transporter dependent pathway (review). *Molecular Membrane Biology, 22*, 29–39.

Ishino, Y., Shinagawa, H., Makino, K., Amemura, M., & Nakata, A. (1987). Nucleotide sequence of the *iap* gene, responsible for alkaline phosphatase isozyme conversion in *Escherichia coli*, and identification of the gene product. *Journal of Bacteriology, 169*, 5429–5433.

Jacob-Dubuisson, F., Fernandez, R., & Coutte, L. (2004). Protein secretion through autotransporter and two-partner pathways. *Biochimica et Biophysica Acta, 1694*, 235–257.

Jansen, R., Embden, J. D., Gaastra, W., & Schouls, L. M. (2002). Identification of genes that are associated with DNA repeats in prokaryotes. *Molecular Microbiology, 43*, 1565–1575.

Johnson, A., & O'Donnell, M. (2005). Cellular DNA replicases: Components and dynamics at the replication fork. *Annual Reviews in Biochemistry, 74*, 283–315.

Jones, L., Carballido-Lopez, R., & Errington, J. (2001). Control of cell shape in bacteria: Helical actin-like filaments in *Bacillus subtilis. Cell, 104*, 913–922.

Klingworth, A., et al. (2013). Evaluation of general 16S ribosomal RNA gene PCR primers for classical and next-generation sequencing-based diversity studies. *Nucleic Acids Research, 41*, e1.

Konstantinidis, K. T., Ramette, A., & Tiedje, J. M. (2006). The bacterial species definition in the genomic era. *Philosophical Transactions of the Royal Society B, 361*, 1929–1940.

Lafontaine, D. L. J., & Tollervey, D. (2001). The function and synthesis of ribosomes. *Nature Reviews Molecular Cell Biology, 2*, 514–520.

Land, M., et al. (2015). Insights from 20 years of bacterial genome sequencing. *Functional & Integrative Genomics, 15*, 141–161.

Lane, D. J., Pace, B., Olsen, G. J., Stahl, D. A., Sogin, M. L., & Pace, N. R. (1985). Rapid determination of 16S ribosomal RNA sequences for phylogenetic analyses. *Proceedings of the National Academy of Science USA, 82*, 6955–6959.

Lederberg, J. (1987). Genetic recombination in bacteria: A discovery account. *Annual Reviews of Genetics, 21*, 23–46.

Leiman, P. G., et al. (2009). Type VI secretion apparatus and phage tail-associated protein complexes share a common evolutionary origin. *Proceedings of the National Academy of Science USA, 106*, 4154–4159.

Links, M. G., Dumonceaux, T. J., Hemmingsen, S. M., & Hill, J. E. (2012). The chaperonin-60 universal target is a barcode for bacteria that enables de novo assembly of metagenomic sequence data. *PLoS ONE, 7*, e49755.

Liu, R., & Ochman, H. (2007). Stepwise formation of the bacterial flagellar system. *Proceedings of the National Academy of Science USA, 104*, 7116–7121.

Low, D. A., Weyand, N. J., & Mahan, M. J. (2001). Roles of DNA adenine methylation in regulating bacterial gene expression and virulence. *Infection and Immunity, 69*, 7197–7204.

Madigan, M. T., Martinko, J. M., Bender, K. S., Buckley, D. H., Stahl, D. A., & Brock, T. (2014). *Brock biology of microorganisms* (14th ed.). San Francisco: Pearson Benjamin-Cummings.

McCutcheon, J. P., & Moran, N. A. (2012). Extreme genome reduction in symbiotic bacteria. *Nature Reviews Microbiology, 10*, 13–26.

McFall-Ngai, M. J., & Ruby, E. G. (2000). Developmental biology in marine invertebrate symbioses. *Current Opinion in Microbiology, 3*, 603–607.

Müller, M. (2005). Twin-arginine-specific protein export in *Escherichia coli. Research in Microbiology, 156*, 131–136.

Murzin, A. G., Lesk, A. M., & Chothia, C. (1994). Principles determining the structure of β-sheet barrels in proteins. I. A theoretical analysis. *Journal of Molecular Biology, 236*, 1369–1381.

Neylon, C., Kralicek, A. V., Hill, T. M., & Dixon, N. E. (2005). Replication termination in *Escherichia coli*: Structure and antihelicase activity of the Tus-Ter complex. *Microbiology and Molecular Biology Reviews, 69*, 501–526.

Pace, N. R., Olsen, G. J., & Woese, C. R. (1986). Ribosomal RNA phylogeny and the primary lines of evolutionary descent. *Cell, 45*, 325–326.

Papanikou, E., Karamanou, S., & Economou, A. (2007). Bacterial protein secretion through the translocase nanomachine. *Nature Review Microbiology, 5*, 839–851.

Proft, T., & Baker, E. N. (2009). Pili in gram-negative and gram-positive bacteria – Structure, assembly and their role in disease. *Cellular and Molecular Life Sciences, 66*, 613–635.

Public Health England. (2014). *Staining procedures. UK standards for microbiology investigations*. TP 39 Issue 1.2. http://www.hpa.org.uk/SMI/pdf.

Rasmussen, L. J., Lobner-Olesen, A., & Marinus, M. G. (1995). Growth rate-dependent transcription initiation from the dam P2 promoter. *Gene, 157*, 213–215.

Reddy, J. D., Reddy, S. L., Hopkins, D. L., & Gabriel, D. W. (2007). TolC is required for pathogenicity of *Xylella fastidiosa* in *Vitis vinifera* grapevines. *Molecular Plant-Microbe Interactions, 20*, 403–410.

Reyes-Lamothe, R., Nicolas, E., & Sherratt, D. J. (2012). Chromosome replication and segregation in bacteria. *Annual Review of Genetics, 46*, 121–143.

Rico, F., Su, C., & Scheuring, S. (2011). Mechanical mapping of single membrane proteins at submolecular resolution. *Nano Letters, 11*, 3983–3986.

Russell, A. B., Peterson, S. B., & Mougous, J. D. (2014). Type VI secretion system effectors: Poisons with a purpose. *Nature Reviews Microbiology, 12*, 137–148.

Salman, V., Amann, R., Shub, D. A., & Schulz-Vogt, H. (2012). Multiple self-splicing introns in the 16S rRNA genes of giant sulfur bacteria. *Proceedings of the National Academy of Science, 109*, 4203–4208.

Sander, J. D., & Joung, J. K. (2014). CRISPR-Cas systems for editing, regulating and targeting genomes. *Nature Biotechnology, 32*, 347–355.

Schäffer, C., & Messner, P. (2005). The structure of secondary cell wall polymers: How gram-positive bacteria stick their cell walls together. *Microbiology, 151*, 643–651.

Scheuring, S., Lévy, D., & Rigaud, J. L. (2005). Watching the components of photosynthetic bacterial membranes and their in situ organisation by atomic force microscopy. *Biochimica et Biophysica Acta, 1712*, 109–127.

Shih, Y. L., & Rothfield, L. (2006). The bacterial cytoskeleton. *Microbiology and Molecular Biology Reviews, 70*, 729–754.

Silhavy, T. J., Kahne, D., & Walker, S. (2012). The bacterial cell envelope. In L. Shapiro & R. Losick (Eds.), *Additional perspectives on cell biology of bacteria* (pp. 1–16). USA: Cold Spring Harbor Laboratory Press.

Soufo, H. J. D., & Graumann, P. L. (2004). Dynamic movement of actin-like proteins within bacterial cells. *EMBO Reports, 5*, 789–794.

Sundquist, A., et al. (2007). Bacterial flora-typing with targeted, chip-based pyrosequencing. *BMC Microbiology, 7*, 108.

Takeuchi, S., DiLuzio, W. R., Weibel, D. B., & Whitesides, G. M. (2005). Controlling the shape of filamentous cells of *Escherichia coli. Nano Letters, 5*, 1819–1823.

Tang, J. (2011). Microbial metabolomics. *Current Genomics, 12*, 391–403.

Taylor, M. W., et al. (2004). Evidence for acyl homoserine lactone signal production in bacteria associated with marine sponges. *Applied and Environmental Microbiology, 70*, 4387–4389.

Tseng, T. T., Tyler, B. M., & Setubal, J. C. (2009). Protein secretion systems in bacterial-host associations, and their description in the Gene Ontology. *BMC Microbiology, 9*(Suppl. 1), S2.

Vollmer, W. (2008). Structural variation in the glycan strands of bacterial peptidoglycan. *FEMS Microbiology Reviews, 32*, 287–306.

von Bodman, S. B., Bauer, W. D., & Coplin, D. L. (2003). Quorum sensing in plant-pathogenic bacteria. *Annual Reviews of Phytopathology, 41*, 455–482.

Wang, J. D., & Levin, P. A. (2009). Metabolism, cell growth and the bacterial cell cycle. *Nature Reviews Microbiology, 7*, 822–827.

Wilson, K. H., Blitchington, R. B., & Greene, R. C. (1990). Amplification of bacterial 16S ribosomal DNA with polymerase chain reaction. *Journal of Clinical Microbiology, 28*, 1942–1946.

Woese, C. R. (1987). Bacterial evolution. *Microbiological Reviews, 51*, 221–271.

Yarza, P., et al. (2014). Uniting the classification of cultured and uncultured bacteria abd archaea using 16S rRNA gene sequences. *Nature Reviews Microbiology, 12*, 635.

Zeigler, D. R. (2003). Gene sequences useful for predicting relatedness of whole genomes in bacteria. *International Journal of Systematic Evolutive Microbiology, 53*, 1893–1900.

Chapter 2
Bacteria and Eukaryotes Evolution

Abstract Basic aspects of the origins of bacteria and invertebrates are discussed. Evolutionary processes are reviewed concerning their positioning in the Tree of Life, including the available fossil records and phylogenetic reconstructions. Processes underpinning bacterial diversity within actual lineages are shown, together with main aspects distinguishing Archaea and Bacteria. The distinctive traits of Eukarya and Metazoa evolution are briefly discussed, together with their phylogenetic relationships.

Keywords Archaea • Bacteria • Evolution • Metazoa • Phylogenesis • Tree of life

1 Introduction

The biosphere has a very ancient biological and geochemical history and its evolutionary process is still at work today. The appearance of life is estimated to have started with the emergence of first primordial cells, dating back to around 3.8–4.5 Gyr ago (Trevors 2003; Martin and Russell 2002; Battistuzzi et al. 2004). Understanding the origin of life is a fundamental cultural issue whose philosophical and ethical implications remain out of the aims of this volume. However, investigative efforts developed on these basic questions produced an intense and productive debate, originating a number of reflexions and experimental or analytical studies. The latter aimed at understanding the physical and chemical conditions present on Earth at the time of appearance of the first life forms (Trevors 2003). Their goal was to infer the earliest evolutionary processes at the origin of life, mainly relying on the application of modern phylogenetic and molecular approaches (Margulis 1996; Martin and Russell 2002; Battistuzzi et al. 2004).

Emergence of life in primordial Earth possibly occurred on either surface or subsurface protected environments. To allow the self-assembly of first primordial cell components it is considered necessary that external environmental conditions had to remain stable for a very long time in order to sustain enzymatic catalysis of the first, simple basic elements. Also, these reactions needed to be sustained by a continuous and stable energy flow. Starting from these considerations, some hypotheses consider the origin of life as an event directly linked to subsurface environments.

A. Ciancio, *Invertebrate Bacteriology*, DOI 10.1007/978-94-024-0884-3_2

These spaces had to be rich in basic, fundamental elements and deep enough to ensure protection of primordial cells from the harmful cosmic radiations of that time (Trevors 2002). As a consequence, first life events may have occurred in a subsurface environment characterized by a hydrophobic layer and a water interface, including hydrocarbons or oil of non-biological origin. Other hypotheses consider a network of self-sustaining molecular replicators, or even the possibility that meteorites or other spatial objects could have acted as primordial selective forces or as spreaders of first life building blocks or forms, during the early ages of heavy Earth bombardment from outer space (Line 2002).

The origin of the primordial cells probably resided in a combination of factors allowing the conservation of sufficient amounts of starting building blocks and products. One hypothesis, consistent with the physical conditions of early Earth, considers appearance of primordial cells within tri-dimensional shelters originated by iron sulfide scaffolds, formed through precipitation from hydrotermal vents. A series of chemoautotrophic reactions characterized by a progressive complexity may have occurred within the internal compartments of these geological scaffolds, providing a continuous energy flow and leading to the assemblage and subsequent exit of primordial living forms (Martin and Russell 2002). A more recent experimental study showed, however, that the building blocks required for the first self assembly of life (nucleic acids, proteins and lipids) may have originated from simpler chemical compounds like H_2S and HCN, fueled by UV radiation (Patel et al. 2015).

Planetary processes, including orbital precessions, continental drifts and tides, as well as collisions with other planetoids or asteroids significantly contributed to construct the early scenarios in which primordial life-forms originated on Earth. In general, the whole process of life appearance, evolution and subsequent extinctions, is not inferred as a continuous, linear sequence of changes, rather it can be conceived as a complex process alternating phases of intense evolutionary and extinctions times to periods of evolutionary stasis (Cavalier-Smith 2006).

The generally accepted idea that Archaea and Bacteria derive from a unique, basic root of the tree of life was recently questionned by whole genome studies and by the identification of multiple mechanisms of genetic exchange, that gave rise to alternative evolutionary hypotheses (Bapteste et al. 2009).

Whether the origin of life on Earth corresponds to the origin of bacteria is indeed an active matter of reflexion. Apart of "how" life originated, also the question about "how many times" did this event occur represents in fact an active object of discussion. The "dilemma of origins" considers indeed the unicity or plurality of first ancestral forms, e.g. if a single common ancestor existed, originating all life forms later present on earth, or whether replicated appearance events produced the first building blocks of life, that subsequently ignited the early evolutionary scenario (Martin and Russell 2002, Delaye and Moya 2008).

2 Evolutionary Processes

The most ancient fossils reminiscent of a bacterial colony organization are the ancient (Pre-Cambrian) stromatolites. These are organo-sedimentary structures of microbial origin, formed by accretion of benthic calcareous deposits often in the form of laminated, columnar bodies (Fig. 2.1). Stromatolites originate from mineral and microcristalline carbonate deposits embedding sediments and microbial cells. Stromatolites were identified in 1908 by Kalkowsky and the oldest example, a deposit from Western Australia, is dated around 3.55 Gyr ago (Riding 1999).

The second source of bacterial fossils, more recent but already of interest for this volume due to their possible associations with invertebrates, is given by the amber inclusions, dating back up to the relatively "recent" Paleozoic era (540–250 Myr ago). These fossil resins were produced by coniferous trees as their non-volatile residues polymerized at the issues of plants growth processes or injuries. During this reaction, many organisms became entrapped or engulfed by the viscous, polymerizing resins, where they remained preserved forever. Amber records often show a complex of preserved organisms representative of the paleoenvironments characterizing primordial forests, including well conserved insects and other invertebrates (Schmidt et al. 2010).

The inclusions are precious sources of informations about the evolution of many organisms as they also include fungi, protozoa and other microorganisms. First records of bacteria embedded in amber were reported in the late nineteenth and early twentieth centuries, mainly from material collected in Pennsylvania and Baltic areas (Schmidt and Schäfer 2005). Records often show filamentous bacterial cells and Cyanobacteria-like microorganisms (Fig. 2.2) (Girard et al. 2009; Girard and Adl 2011).

Knowledge on bacterial evolution from amber inclusions is often biased by the limited information obtainable by visual inspections and morphometric data.

Fig. 2.1 Modern, intertidal stromatolites from Hamelin Pool (Western Australia) (Adapted from Reid et al. 2003)

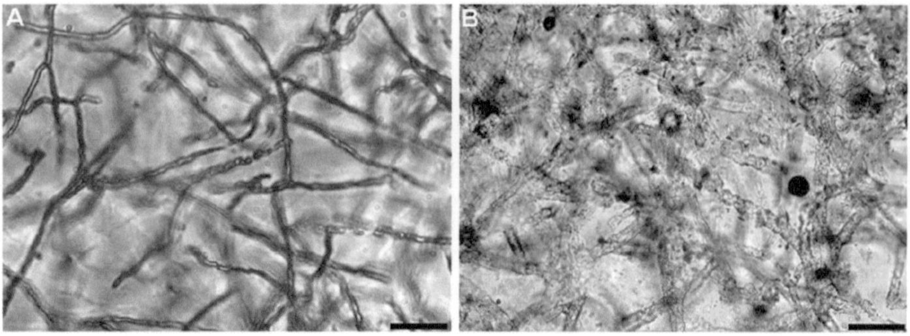

Fig. 2.2 Fossil cells of actinomycete-like bacteria (**a**) and of *Palaeocolteronema cenomanensis* (Cyanobacteria, (**b**) embedded in mid Cretaceous ambers from France (scale bars = 10 (**a**) and 50 (**b**) μm) (Adapted from Girard and Adl 2011)

DNA-RNA investigations are difficult but not impossible, related to the degradation conditions of the original nucleic acids and the contemporary risks of external contaminations (Schmidt and Schäfer 2005). There are reports in the scientific literature concerning the isolation of bacterial cells from amber and even their subsequent cultivation (Poinar and Poinar 1994; Cano and Borucki 1995; Greenblatt et al. 1999, 2004). Santiago-Rodriguez et al. (2014) reported the isolation of a number of preserved bacteria from a 25 to 40 Myr old Dominican amber, whereas Lambert et al. (1998) described a novel species, *Staphylococcus succinus*, from Dominican amber.

Bacteria isolated from a Dominican sample and a 120 Myr old Israeli amber included Gram positive species of the genera *Bacillus* and *Staphylococcus*, together with other actinobacteria and anaerobic cocci, classified through their 16S rRNA ribosomal gene sequence and FAMS (Fatty Acid Methyl Ester) analysis (Greenblatt et al. 2004). The 16S rRNA phylogenetic data also showed that the amber isolates were closer to the branch nodes than their modern counterparts, thus supporting the hypothesis about their ancestral status. In many studies of this kind, however, the possibility of sample contamination(s) occurring at different times after completion of the amber polymerization process cannot be excluded, and either the stringent application of surface sterilization procedures and the replication of results represent an obligate approach for validation.

Apart of adoption of most stringent microbiological procedures, a further required precaution is the careful sample inspection for the occurrence of fissures or cracks allowing burial of more recent bacterial cells in the sample, a process that may take place many times and at different ages (Greenblatt et al. 2004). In any case, even if a sample contamination occurred at later times varying on a scale of 10^3–10^5 years rather than some Myr, these studies provided impressive knowledge about the survival strength of some bacterial lineages (Greenblatt et al. 2004). They are also useful for the identification of new metabolites from ancient species (Cano and Borucki 1995). They also are helpful for the estimation of time divergence parameters and the reconstruction of ancestral bacterial phylogenies (Santiago-Rodriguez et al. 2014).

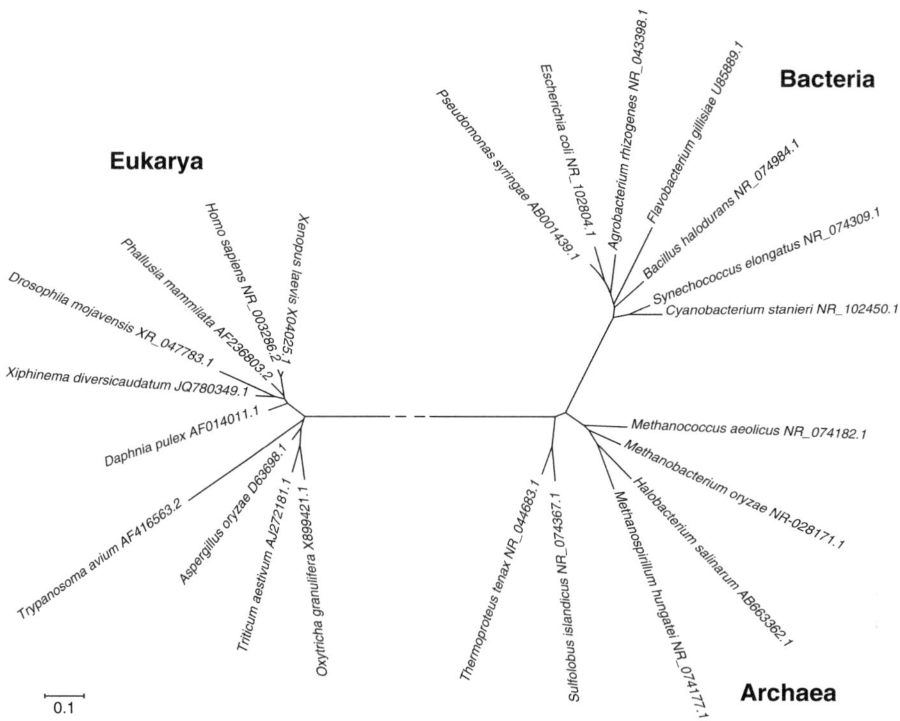

Fig. 2.3 Neighbor-Joining evolutionary tree of 23 species representative of the three Kingdoms of life, inferred using 16S and 18S ribosomal RNA gene sequences and computed using the Jukes-Cantor method based on the number of base substitutions per site (*Scale bar = 0.1 substitutions*)

Due to their relatively "recent" age, amber records do not provide enough informations as to clearly establish the early evolutionary history of bacteria, as well as their subsequent interactions with primordial pluricellular organisms. At this regard, basic questions concerning "when and how" (or at the issue of which mechanisms) the Bacteria and Eukarya Kingdoms diverged are still open and in many circumstances the answers on this subject remain in the field of inferential research. In particular, the separation of Eukarya from Bacteria and Archaea (Fig. 2.3) is still controversial. Considering the evolution of the cytoskeleton and the cell nucleus, various possible hypotheses are actually debated by the scientific community, including (*i*) the separation of the three evolutive paths from a common, more ancient progenitor, (*ii*) the origin of Eukarya through the acquisition, by an archeon, of a Bacteria member or (*iii*) the descendence of Archaea and Eukarya from a common bacterial ancestor (Erickson 2007; Godde 2012).

A key process recognized in the Eukarya evolution is the capacity of endocytosis or phagocytosis, behaviours allowing a cell to introgress or engulf a solid particle or another cell as well. This view, initially proposed by Stanier (1970), has been largely supported in recent years (Cavalier-Smith 2002a; Doolittle 2000; Hartman and Fedorov 2002; Kurland et al. 2006). It is possible that this capacity paved the way to the assemblage of cellular components leading to more complex organizations. This was a fundamental step in the evolution of pluricellular organisms, considering the importance, at the onset of the Eukarya appearance, of bacterial cells that became, at the issues of their successful introgression, structural eukaryotic cell components known today as mitochondria, chloroplasts and hydrogenosomes.

The biochemical machinery and the arsenal of functional proteins that emerged in the early Earth environments reflected the composition and physical properties of the atmosphere, as well as its subsequent biogeochemical changes. It is possible that actual living organisms reflect the conditions met by the first common ancestor(s) at the onset of life appearance. The physics of early Earth originated indeed a number of selective pressures that determined the following evolution of life. In their review on the origin of cells, Martin and Russell (2002) list a number of biochemical components that the ancestral progenitor cell(s) had to possess to survive and multiply. These include tRNA, and related proteins, DNA, RNA and their polymerases as well as translation factors and ATPases. Furthermore, the pathways considered as obligate evolutive steps had to include the biosynthesis of nucleotides and aminoacids as well as the carbon metabolism for related biosynthetic precursors and cofactors (Martin and Russell 2002).

Differences between Eukarya and Archaea are consistent with dissimilar origins. Glycolytic metabolic pathways, from glucose to pyruvate, are ubiquitous among eukaryotes and are mirrored by homologues among several bacterial genomes, but cannot be found in Archaea, in which alternative pathways exist. Similarly, they also differ for their lipids and glycerol pathways: eubacterial fatty acids are linked to *sn*-glycerol-3-phosphate by an ester bond and have no equivalent among archean lipids, whose isoprenoids have a very different biochemical origin and are connected to *sn*-glycerol-1-phosphate through ether bonds. Differences also concern glycerol isomers synthetic enzymes, as well as the cell wall composition of the two lineages (Martin and Russell 2002).

There is a general consensus about the insurgence, in primordial Bacteria, of a light-dependent biochemistry (pigments, photosynthesis) at a later stage, whose evolution anticipated the appearance of photosynthesis and largely contributed to its subsequent success, through the evolution of algae and plants. This process largely relied on horizontal gene transfer (HGT), a key evolutionary mechanism that, adding to vertical gene transmission and natural selection, significantly contributed to the diversification of many, if not all, living species. Basically, a species may acquire, through HGT, one or more operational genes from an unrelated foreign organism. These transferred genes are then transmitted to the progeny when they are functionally integrated into the recipient cells (Aravind et al. 1998; Doolittle 2000). Although not directly postulated at the time of its discovery, HGT was originally described in strains of *Corynebacterium diphtheriae* that became virulent through phage-mediated transduction (Freeman 1951).

This genetic exchange and enrichment mechanism was later confirmed by many genome-level studies (Martin 1999; Brochier et al. 2000; Wiedenbeck and Cohan 2011) and is actually accepted as a strong and very influential evolutive force. An example of the HGT contribution to the evolution of bacterial lineages is given by the occurrence of photosynthesis-related proteins among five distinct bacterial phyla, as shown by whole-genome analyses (Raymond et al. 2002). A further confirmation of the importance of HGT is given by the presence of Archaea-related genes in some bacterial genomes (Aravind et al. 1998). For further informations about HGT in bacteria-invertebrate associations see Chap. 8 of this volume.

Although past HGT events can interfere with the reconstruction of the phylogenetic signals underlying bacterial evolution, the identification of highly conserved gene sets and proteins may overcome the dominance of these events. Phylogenetic reconstruction of the evolutive divergence among bacterial lineages based on a set of 32 concatenated proteins from representative species (54 Eubacteria, 15 Archaebacteria and three eukaryotes) allowed the estimation of the time of divergence within each domain (Battistuzzi et al. 2004).

These results were obtained by the application of a molecular clock, an analysis based on phylogenies constructed using diversification rates obtained through known or inferred times of mutational events. Using several calibration points and rates varying on the different tree branches, the analyses carried out by Battistuzzi et al. (2004) showed an estimated separation time between Archaea and Bacteria dating back to 4 Gyr ago. They also showed the intervals of separation of main eubacterial groups, that occurred around 3.2–2.5 Gyr ago, with a later appearance of Cyanobacteria at the onset of the period characterized by increasing concentrations of atmospheric oxygen, estimated to have occurred around 2.3 Gyr ago. In another phylogenetic reconstruction study, Sheridan et al. (2003) found a Bacteria-Archaea last common ancestor occurring around 4.29 Gyr ago, and a separation of the two Kingdoms dating back to 3.46 Gyr ago.

3 Bacterial Diversity

The bacterial evolutive mechanisms previously described largely depended and interacted with the changes and instabilities that characterized the early planetary biogeochemical history. The resulting evolutive radiations gave raise to a huge number of bacterial lineages that explored and colonized, in different ages, any possible niche on the planet, accounting for modern biodiversity found on Earth. From the perspective of this volume it is worth to note that bacteria have a much longer and more ancient evolutionary history than pluricellular organisms including the invertebrates, and that their first evolutive steps were much more complex and constructive than those characterizing their recent evolution (Line 2002). Furthermore, considering that the invertebrates represent a complex of phyla and heterogeneous phylogenetic groups that appeared on Earth at different, although more recent, evolutionary times, it is reasonable to consider that the first bacteria-invertebrate associations and interactions could in theory date back to the early times of metazoan appearance, around 800–900 Myr ago (see Sect. 4 in this chapter).

The levels of biodiversity within actual bacterial lineages is impressive: there is no environmental niche or living form that does not include or interact with bacteria, and although some groups are still cryptic or underepresented due to the difficulties in their study or culturing, it is possible to infer some estimates about their numbers. Recent advances in massive genomic sequencing from environmental samples provided in fact an evaluation about the density and biodiversity levels present in many environments, based on informative gene or genome-wide data. However, first estimates from Whitman et al. (1998) already showed that the prokaryotes present in ocean, soil or oceanic and terrestrial subsurfaces range in the order of $4–6\times10^{30}$ cells, with a production rate of 1.7×10^{30} cells per year. Numbers for the first 200 m of oceanic waters consider a total of 3.6×10^{28} cells, 2.9×10^{27} of which are autotrophs, whereas deeper oceanic layers are inhabited by 6.5×10^{28} cells. Similarly, huge densities can be encountered in terrestrial habitats, where bacteria numbers range from 3×10^{9} to $1–1.5\times10^{10}$ cells per g of soil (Torsvik et al. 1990, 1996; Polyanskaya et al. 2013). In any case, only a minimal fraction of the species found in any census can be cultivated. Considering these numbers and the limited amounts of species cultured, it is evident that the global biodiversity of modern bacterial genomes is enormous and that the species actually known represent just the tip of an iceberg.

Bacterial high multiplication and mutation rates warrant in many cases a fast reaction time to the changes occurring in their surrounding environments. Other lineages explored or adapted to life in specific environmental niches, like thermophilic, acidophilic or halophilic groups. This extreme flexibility, coupled to the bacterial genetic plasticity and HGT, underpins their impressive capacity for colonization and adaptation to new trophic niches, as well as their metabolic efficiency and environmental persistence.

The number of bacterial taxa identified thus far is in the order of $1 - 4\times10^{6}$, a level probably underestimated since this number can reach up to a billion. Only a fraction, in the order of 3×10^{4} species, is available in culture (Curtis et al. 2002; Schloss and Handelsman 2004; Dykhuizen 2005). Identifying the best species definition boundaries for bacteria is often a critical step, but it is widely recognized that a minimal 3 % difference in sequence identity, usually but possibly not limited to 16S rRNA genes, is needed for this purpose (Stackebrandt and Goebel 1994).

Actually, 60 phyla are recognized from the Bacteria or Archaea kingdoms, but many are only represented by a reduced set of sequences (McDonald et al. 2012; Walker 2014). They are characterized by very distinct evolutive and metabolic pathways (Tables 2.1 and 2.2). This number is expected to increase due to new candidate phyla arising from uncultured lineages found through the application of new technologies, i.e. single cell genome sequencing of unculturable species, or due to the discovery of new taxa and/or lineages identified solely on the basis of sequences assembled from environmental or metagenomic samplings (Hugenholtz et al. 1998; Sheridan et al. 2003; Hugenholtz and Kyrpides 2009; Brochier-Armanet et al. 2011; Wrighton et al. 2012; Rinke et al. 2013).

Many bacterial genome sequencing projects are actually in course, and their data are expected to shed light on several unresolved questions concerning evolution and diversity (see http://www.genomesonline.org/ for review).

Table 2.1 Archaea lineages as defined through genome-wide sequencing or environmental studies

Phyla	Main traits
Crenarchaeota	RNA polymerase subunit Rpb8; DNA Topoisomerase IA; CdvABC eukaryotic-like cell division system; aerobic or anaerobic heterotrophs; chemolithoautotrophs; methanogenesis; hydrogen and sulfur metabolism; phototrophic halophiles
Euryarchaeota	Two methanogenic, two thermophilic, one acidophilic and one halophilic lineages; DNA Topoisomerase IA; bacteria-like FtsZ cell division system; methanogens; halophilic, acidophilic, thermophilic and hyperthermophilic species
Korarchaeota	Many biosynthetic pathways are absent; RNA polymerase subunit Rpb8; DNA Topoisomerase IA; bacteria-like FtsZ cell division system; uncultured; hot springs, aquatic habitats; likely chemolithotrophic
Thaumarchaeota	Mesophiles or thermophiles, some members with aerobic NH_4^+ oxidation; DNA Topoisomerase IB; bacteria-like FtsZ and CdvABC eukaryotic-like cell division systems; mesophilic

For reviews see Brochier-Armanet et al. (2011), Dawson et al. (2006), and Pester et al. (2011)

Table 2.2 Main phyla of Bacteria as revealed by genome and environmental studies

Phyla	Main traits
Acidobacteria	Frequent in soil or many water habitats; majority of species uncultured; Gram-negative
Actinobacteria	Frequent in soil and water; pathogens (*Mycobacterium*) or symbionts of plants or animals; N fixation in roots (*Frankia*); production of antibiotics (*Streptomyces*); some taxa thermophilic and γ-ray resistant; Gram-positive, some species spore-forming
Aquificae	Hyperthermophiles, acidophiles or obligate anaerobes; found in extreme environments: hot springs, sulfur pools and thermal deep ocean vents; autotrophs; Gram-negative
Armatimonadetes	Former candidate division OP10; from environmental samples, geothermal soil or acquatic habitats; aerobic, pigmented, thermophilic; chemoheterotrophic; Gram-negative
Bacteroidetes	From soil and water habitats; associated to animals; some periodontal pathogens (*Bacteroides*); anaerobic; Gram-negative
Caldiserica	Former candidate phylum OP5; thermophilic; anaerobic and chemoheterotrophic; filamentous; Gram-negative
Chlamydiae	Obligate intracellular human and animal parasites; no ATP synthesis; sequences found also in water and soil habitats; considered Gram-negative; form an infective spore analogous (elementary body)
Chlorobi	Green sulphur bacteria; anaerobic obligate photo-autotrophs that oxidize sulphide in chlorosomes; includes also facultative anaerobic organoheterotrophs and an aerobic photoheterotroph that cannot oxidize sulfide; Gram-negative
Chloroflexi	Green non-sulfur filamentous bacteria; obligate or facultative, anoxygenic anaerobic phototrophs; frequent in soil and wastewater; also included a chemolithoautotrophic nitrite oxidizer; Gram-negative
Chrysiogenetes	Anaerobic, mesophilic, use arsenate as electron acceptor; Gram-negative; one species only

(continued)

Table 2.2 (continued)

Phyla	Main traits
Cyanobacteria	Oxygenic photosynthesis with chlorophyll *a* (blue pigment); N fixation; frequent in water or land; Gram-negative; some with spore-like resistant structures
Deferribacteres	Anaerobic, chemoorganotrophs; use Fe(III), Mn(IV) or nitrate as electron acceptor; mesophilic or thermophilic; Gram-negative
Deinococcus – Thermus	Resistant to extreme γ-irradiation, dessiccation or high temperatures; Gram-positive
Dictyoglomi	Anaerobic, chemoorganotrophs; thermophilic; Gram-positive
Elusimicrobia	Former candidate phylum TG1; obligate endosymbiont in termite guts; sequences in soil, aquifers; Gram-negative
Fibrobacteres	Cellulose degradation in herbivore intestines; also found in soil and aquifers; obligate anaerobic; Gram-negative.
Firmicutes	Include Bacilli and Clostridia; aerobic or anaerobic; frequent in soil, water and animals; with thermotolerant, dessiccation-resistant endospores; mostly Gram-positive
Fusobacteria	Obligate anaerobic; found in mammals intestine and oral cavity; dental pathogens; Gram-negative
Gemmatimonadetes	Former candidate division BD (syn. KS-B); aerobic; found in sewage sludge, sequences from soil; Gram-negative
Lentisphaerae	Marine, freshwater heterotrophs; anaerobic digesters found in faeces; secrete exopolysaccharides; Gram-negative
Nitrospirae	Aerobic chemolithotrophic, nitrite oxidizers; some forms magnetotactic; *Thermodesulfovibrio* obligate acidophilic, anaerobic; in water, soil habitats, sludge; Gram-negative
Planctomycetes	Frequent in water, deep ocean and soil habitats; lack peptidoglycan; budding forms; cells with compartments and a nucleoid; aerobic chemoheterotrops; anaerobic chemoautotrophs ammonium oxidizers; Gram-negative
Proteobacteria	Largest, highly diversified phylum; 5 lineages: α, β γ, δ and ε; chemoorganotrophs, chemolithotrophs or phototrophs; aerobic and anaerobic, facultative aerobes or microaerophiles; many of medical, industrial or agricultural value; include purple nonsulfur bacteria and rhizobia; mostly mesophilic, some thermophilic or psychrophilic; many motile by diverse mechanisms; freeliving, symbionts, pathogens; Gram-negative
Spirochaetes	Include also animal pathogens (*Borrelia*, *Leptospira*, *Treponema*); motile through inner axial filaments; spiral shaped; resting cystic or granular forms; Gram-negative
Synergistetes	Anaerobic; rod-shaped bacteria; from humans, animals, terrestrial and oceanic habitats; Gram-negative
Tenericutes	Former Mollicutes (*Mycoplasma*, *Phytoplasma*); lack a cell wall; with a cytoskeleton; Gram-negative
Thermodesulfobacteria	Thermophilic; found in thermal habitats, hot springs, digestors; anaerobic; sulfate reducing chemoheterotrophs; Gram-negative
Thermotogae	Thermophilic; anaerobic; with external envelope sheat; Gram-negative
Verrucomicrobia	Chemoheterotrophs, some thermoacidophilic methylotrophs; in soil or aquatic habitats; in hot springs some acidophilic methane oxidizers; cell with compartments and nucleoid; some endosymbionts in nematodes; Gram-negative

For reviews see: Cho et al. (2004), Garrity and Holt (2001a, b, c, d, e, f), Fuerst and Sagulenko (2011), Griffiths et al. (2005), Hedlund (2001), Jumas-Bilak et al. (2009), Kersters et al. (2006), Lee et al. (2009, 2013), Liu et al. (2012), Mori et al. (2009), Nishida et al. (2011), Quaiser et al. (2003), Reysenbach (2001), Sorokin et al. (2012), Vandekerckhove et al. (2000)

Considering cell morphology and structural organization, the first ancestor of the kingdom Bacteria is thought to be a Gram-positive, rod-like cell. This view is based on a number of observations and considerations proposed by various research groups. They concern the organization of the cellular membrane and wall, as well as a number of physical selective constraints acting on the early cell division process.

The radiations of Gram-negative species and cocci are thought to derive from this initial lineage, also in consideration of their parasitic habits on animals and plants, which appeared later, during the course of evolution (Koch 2003). However, the view of a simple ancestral bacterial cell surrounded by a peptidoglycan layer challenges previous considerations about early emergence of Gram-negatives at the basis of the Bacteria radiation (Cavalier-Smith 2002b).

The evolution of the kingdom Bacteria shows lineages with many distinctive traits, often characterized by phylogenies difficult to reconstruct. They include taxa with aerobic or anaerobic, photosynthetic, cyanotrophic or heterotrophic metabolism, as well as extremophiles with acidophilic or halotrophic adaptations. Some of them are very important for their relationships with plants or animals as symbionts or pathogens (Table 2.2).

4 Eukaryota and Metazoa Evolution

The distinctive traits of Eukarya are resumed in the etimology of the kingdom name (Ancient Greek: *ευ κάρυον*, clear core or nut), reminiscent of the presence of a cell nucleus delimited by a membrane. This trait distinguishes them from Archaea and Bacteria. Eukaryotic cells also include various extranuclear organelles, which result from a series of multiple acquisitions that occurred during their evolutionary history, through endosymbiosys or endocytosis.

The eukaryotic cell is indeed a complex machinery resulting from an ancient assemblage of different pieces, each contributing to a number of functional and successful adaptive changes, the best known being the aerobic mitochondrial metabolism. The mitochondrion is recognized today as the descendent of an acquired endosymbiotic ancestor related to modern species of α-Proteobacteria, putatively considered as an ancient progenitor of the actual order Rickettsiales (Lang et al. 1999; Martin et al. 2001; Koonin 2010; Godde 2012). Other organelles found in eukaryotic cells are the chloroplasts and the hydrogenosomes. However, a further, fundamental contribution to the subsequent diversification of eukaryotic organisms was given by the acquisition of several genes proceeding either from Archaea and Bacteria, through multiple HGT events (Godde 2012).

The origin of the nucleus has been debated for many years, and its endosymbiotic origin, already proposed by Mereschkowsky in 1905 (but later on dismissed) was re-evaluated by Margulis in 1970. Many authors proposed various theories to explain the emergence of eukaryotic cells, examining the nature of endocytosis or, in alternative, the occurrence of cell fusion events. Two Euryarchaeota were proposed as putative candidates for the first nucleated progenitor and recipient cell: a member of the *Thermoplasma* lineage and a methanogen. The former was considered due to the presence of histones and cytoskeleton, as well as for the lack of a cell wall (Searcy 1992). The latter was proposed based on the hypothesis of a hydrogen-dependent, anaerobic methanogenic host, harbouring an α-proteobacterium mitochondrion and a hydrogenosome precursor (Martin and Müller 1998). This view was challenged by a further hypothesis considering a Crenarchaeota host cell, based on structural similarities between archeal and eukaryotic ribosomes (Godde 2012).

Most recent hypothesis on the archaean origin of the nucleus consider a member of the phylum Thaumarchaeota as a guest for a bacterium, based on the presence in this lineage of proteins similar to the eukaryotic DNA topoisomerase IB (Brochier-Armanet et al. 2008). Other similarities between Archaea and eukaryotes include a large RNA polymerase A subunit, two types of single stranded DNA binding proteins and the presence of histones (Forterre 2011; Godde 2012).

The debate about the origin of the eukaryotic cell is still open, and deciphering the contributions of Archaea and Bacteria to the emergence of Eukarya is still an active and fertile field of study, also considering the fallouts that this knowledge may produce in medical or other research fields (Koonin 2010; Alvarez-Ponce and McInerney 2011). It is worth to underline the significance of some similarities shown by eukaryotes to the archaeal informational and, more important, operational cell systems (cell divison, protein secretion, energy , vesicles etc.), as opposed to previous views of a bacterial derivation for the latter systems. This scenario is, however, balanced by the closer similarities found between eukaryotic and bacterial membrane lipids (Ettema and Bernander 2009; Pereto et al. 2004; Gribaldo et al. 2011).

Recent new hypotheses consider the eukaryotic cell as resulting from the assemblage of the thermophilic Archaea *Ignicoccus* and *Nanoarchaeum* with a mitochondrion α-proteobacterium precursor, by means of transfer or fusion events occurred within a hydrothermal vent environment (Godde 2012). The eukaryotic evolutive radiation was initially dated by protein divergence or molecular clock analyses to a period ranging between 2.0 and 0.9 Gyr ago (Douzery et al. 2004; Hedges et al. 2004). More accurate estimates with calibrated microfossil records, however, placed the first eukaryotic radiation around 1.1 Gyr ago, at the Mesoproterozoic–Neoproterozoic boundary (Berney and Pawlowski 2006). The phylogeny was compatible with trees obtained with conserved proteins by Baldauf et al. (2000) and placed the Metazoa radiation at around 810 Myr ago, between the preceding divergence of green and red algae (late Mesozoic, around 930 Myr ago) and before that of land plants (early Paleozoic, around 510 Myr ago) (Berney and Pawlowski 2006).

The Metazoa include all pluricellular animals, forming a wide polyphyletic lineage within the Eukarya which thus includes all phyla related to the invertebrates (Fig. 2.3). The term "invertebrate" represents indeed a generic simplification that does not correspond to a common, unique evolutive history of metazoan lineages. In a broad sense, all metazoans deprived of a vertebral system or a notochord present in even a larval stage (the latter are instead classified in the phyla "-chordata" within the Deuterostomia, including ascidians, cephalochordates and in Vertebrata, see Fig. 2.4) may be included in the invertebrate category. The definition hence does not

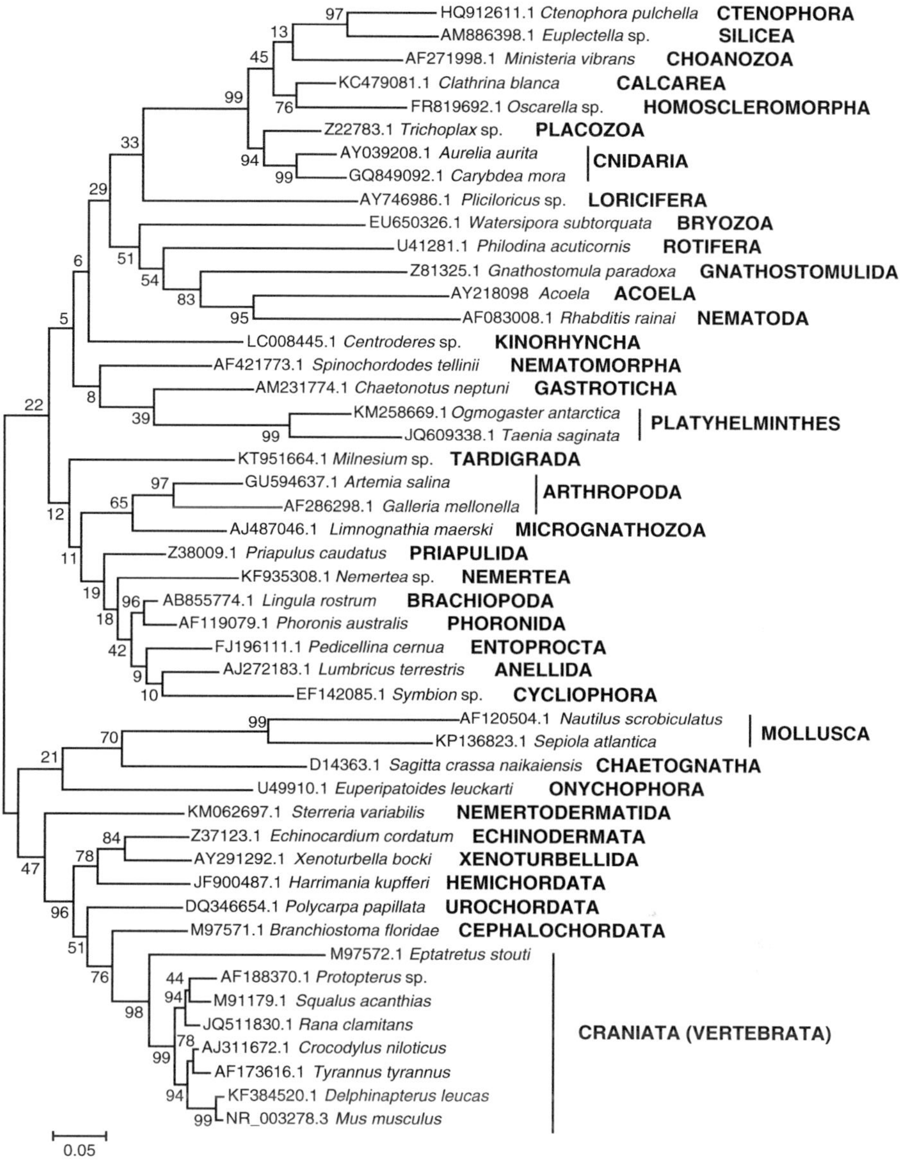

Fig. 2.4 A phylogenetic tree for the evolutive radiations of Metazoa (including single cell Choanozoa), based on 48 aligned sequences of the 18S ribosomal RNA gene (see GenBank codes), for selected representative species. Molecular phylogenetic analysis was performed by inferring evolutionary history with the Maximum Likelihood method based on the Jukes-Cantor model. The tree with highest log likelihood (−47131.62) is shown with the percentage of trees in which the taxa clustered next to branches. Initial trees for heuristic search were obtained by the Neighbor-Join and BioNJ algorithms using the Maximum Composite Likelihood (MCL) approach, selecting the topology with superior log likelihood value. Branch lengths measured in the number of substitutions per site (see scale). Analyses performed with ClustalW, BioEdit and MEGA6 (Thompson et al. 1994; Hall 2011; Tamura et al. 2013). For details of phyla (shown in capital letters) see Edgecombe et al. (2011)

correspond to a given, specific and homogeneous clade and dates back to the early phylogenetic studies and morphological categorizations, lacking a taxonomic precision. However, the term still remains largely in use as a synthetic definition, mainly in reference to a number of metazoan lineages, some of which are important for agriculture as well as for human and animal health or environment ecology. The term will hence be used throughout this volume with reference to this meaning.

References

Alvarez-Ponce, D., & McInerney, J. O. (2011). The human genome retains relics of its prokaryotic ancestry: Human genes of archaebacterial and eubacterial origin exhibit remarkable differences. *Genome Biology and Evolution, 3*, 782–790.

Aravind, L., Tatusov, R. L., Wolf, Y. I., Walker, D. R., & Koonin, E. V. (1998). Evidence for massive gene exchange between archaeal and bacterial hyperthermophiles. *Trends in Genetics, 14*, 442.

Baldauf, S. L., Roger, A. J., Wenk-Siefert, I., & Doolittle, W. F. (2000). A kingdom-level phylogeny of eukaryotes based on combined protein data. *Science, 290*, 972–976.

Bapteste, E., et al. (2009). Prokaryotic evolution and the tree of life are two different things. *Biology Direct, 4*, 34.

Battistuzzi, F. U., Feijao, A., & Blair Hedges, S. (2004). A genomic timescale of prokaryote evolution: Insights into the origin of methanogenesis, phototrophy, and the colonization of land. *BMC Evolutionary Biology, 4*, 44.

Berney, C., & Pawlowski, J. (2006). A molecular time-scale for eukaryote evolution recalibrated with the continuous microfossil record. *Proceedings of the Royal Society B, 273*, 1867–1872.

Brochier, C., Philippe, H., & Moreira, D. (2000). The evolutionary history of ribosomal protein RpS14: Horizontal gene transfer at the heart of the ribosome. *Trends in Genetics, 16*, 529–533.

Brochier-Armanet, C., Gribaldo, S., & Forterre, P. (2008). A DNA Topoisomerase IB in Thaumarchaeota testifies for the presence of this enzyme in the last common ancestor of archaea and eucarya. *Biology Direct, 3*, 54.

Brochier-Armanet, C., Forterre, P., & Gribaldo, S. (2011). Phylogeny and evolution of the Archaea: One hundred genomes later. *Current Opinion in Microbiology, 14*, 274–281.

Cano, R. J., & Borucki, M. K. (1995). Revival and identification of bacterial spores in 25- to 40-million-year-old Dominican amber. *Science, 268*, 1060–1064.

Cavalier-Smith, T. (2002a). The phagotrophic origin of eukaryotes and phylogenetic classification of Protozoa. *International Journal of Systematic and Evolutionary Microbiology, 52*, 297–354.

Cavalier-Smith, T. (2002b). The neomuran origin of Archaebacteria, the negibacterial root of the universal tree and the bacterial megaclassification. *International Journal of Systematic Evolutionary Microbiology, 52*, 7–76.

Cavalier-Smith, T. (2006). Cell evolution and earth history: Stasis and revolution. *Philosophical Transactions of the Royal Society of London B, 361*, 969–1006.

Cho, J. C., Vergin, K. L., Morris, R. M., & Giovannoni, S. J. (2004). *Lentisphaera araneosa* gen. nov., sp. nov, a transparent exopolymer producing marine bacterium, and the description of a novel bacterial phylum, Lentisphaerae. *Environmental Microbiology, 6*, 611–621.

Curtis, T. P., Sloan, W. T., & Scannelli, J. W. (2002). Estimating procaryotic diversity and its limits. *Proceedings of the National Academy of Sciences USA, 99*, 10494–10499.

Dawson, S. C., Delong, E. F., & Pace, N. R. (2006). Phylogenetic and ecological perspective on uncultured Crenarchaeota and Korarchaeota. In M. Dworkin, S. Falkow, E. Rosenberg, K. H.

Schleifer, & E. Stackebrandt (Eds.), *The prokaryotes: Vol. 3: Archaea. Bacteria: Firmicutes, Actinomycetes* (pp. 281–289). New York: Springer.

Delaye, L., & Moya, A. (2008). Evolution of reduced prokaryotic genomes and the minimal cell concept: Variations on a theme. *BioEssays, 32*, 281–287.

Doolittle, W. F. (2000). Uprooting the tree of life. *Scientific American, 282*, 90–95.

Douzery, E. J. P., Snell, E. A., Bapteste, E., Delsuc, F., & Philippe, H. (2004). The timing of eukaryotic evolution: Does a relaxed molecular clock reconcile proteins and fossils? *Proceedings of the National Academy of Sciences USA, 101*(15), 386–391.

Dykhuizen, D. (2005). Species numbers in bacteria. *Proceedings of the California Academy of Science, 56*(Suppl. 1), 62–71.

Edgecombe, G. D., et al. (2011). Higher-level metazoan relationships: Recent progress and remaining questions. *Organisms Diversity & Evolution, 11*, 151–172.

Erickson, H. P. (2007). Evolution of the cytoskeleton. *BioEssays, 29*, 668–677.

Ettema, T. J., & Bernander, R. (2009). Cell division and the ESCRT complex: A surprise from the archaea. *Communicative & Integrative Biology, 2*, 86e88.

Forterre, P. (2011). A new fusion hypothesis for the origin of eukarya: Better than previous ones, but probably also wrong. *Research in Microbiology, 162*, 77–91.

Freeman, V. J. (1951). Studies on the virulence of bacteriophage-infected strains of *Corynebacterium diphtheriae. Journal of Bacteriology, 61*, 675–688.

Fuerst, J. A., & Sagulenko, E. (2011). Beyond the bacterium: Planctomycetes challenge our concepts of microbial structure and function. *Nature Reviews Microbiology, 9*, 403–413.

Garrity, G. M., & Holt, J. G. (2001a). Phylum BIII. Thermodesulfobacteria *phy. nov*. In D. R. Boone & R. W. Castenholz (Eds.), *Bergey's manual of systematic bacteriology. Volume one. The Archaea and the deeply branching and phototrophic bacteria* (pp. 389–393). New York: Springer.

Garrity, G. M., & Holt, J. G. (2001b). Phylum BV. Chrysiogenetes *phy. nov*. In D. R. Boone & R. W. Castenholz (Eds.), *Bergey's manual of systematic bacteriology. Volume one. The Archaea and the deeply branching and phototrophic bacteria* (pp. 421–425). New York: Springer.

Garrity, G. M., & Holt, J. G. (2001c). Phylum BVI. Chloroflexi *phy. nov*. In D. R. Boone & R. W. Castenholz (Eds.), *Bergey's manual of systematic bacteriology. Volume one. The Archaea and the deeply branching and phototrophic bacteria* (pp. 427–446). New York: Springer.

Garrity, G. M., & Holt, J. G. (2001d). Phylum BVII. Nitrospirae *phy. nov*. In D. R. Boone & R. W. Castenholz (Eds.), *Bergey's manual of systematic bacteriology. Volume one. The Archaea and the deeply branching and phototrophic bacteria* (pp. 451–464). New York: Springer.

Garrity, G. M., & Holt, J. G. (2001e). Phylum BIX. Deferribacteres *phy. nov*. In D. R. Boone & R. W. Castenholz (Eds.), *Bergey's manual of systematic bacteriology. Volume one. The Archaea and the deeply branching and phototrophic bacteria* (pp. 465–471). New York: Springer.

Garrity, G. M., & Holt, J. G. (2001f). Phylum BXI. Chlorobi *phy. nov*. In D. R. Boone & R. W. Castenholz (Eds.), *Bergey's manual of systematic bacteriology. Volume one. The Archaea and the deeply branching and phototrophic bacteria* (pp. 601–623). New York: Springer.

Girard, V., & Adl, S. M. (2011). Amber microfossils: On the validity of species concept. *Comptes Rendus Palevolution, 10*, 189–200.

Girard, V., Breton, G., Brient, L., & Néraudeau, D. (2009). Sheathed prokaryotic filaments, major components of Mid-Cretaceous French amber microcoenoses. *Journal of Paleolimnology, 42*, 437–447.

Godde, J. (2012). Breaking through a phylogenetic impasse: A pair of associated archaea might have played host in the endosymbiotic origin of eukaryotes. *Cell & Bioscience, 2*, 29.

Greenblatt, C. L., Davis, A., Clement, B. G., Kitts, C. L., Cox, T., & Cano, R. J. (1999). Diversity of microorganisms isolated from amber. *Microbial Ecology, 38*, 58–68.

Greenblatt, C. L., Baum, J., Klein, B. Y., Nachshon, S., Koltunov, V., & Cano, R. J. (2004). *Micrococcus luteus* – survival in amber. *Microbial Ecology, 48*, 120–127.

Gribaldo, S., Forterre, P., & Brochier-Armanet, C. (2011). Editorial: Archaea. *Research in Microbiology, 162*, 1–4.

Griffiths, E., Petrich, A. K., & Gupta, R. S. (2005). Conserved indels in essential proteins that are distinctive characteristics of Chlamydiales and provide novel means for their identification. *Microbiology, 151*, 2647–2657.
Hall, T. (2011). BioEdit: An important software for molecular biology. *GERF Bulletin of Biosciences, 2*, 60–61.
Hartman, H., & Fedorov, A. (2002). The origin of the eukaryotic cell: A genomic investigation. *Proceedings of the National Academy of Sciences, USA, 99*, 1420–1425.
Hedges, S. B., Blair, J. E., Venturi, M. L., & Shoe, J. L. (2004). A molecular timescale of eukaryote evolution and the rise of complex multicellular life. *BMC Evoutionary. Biology, 4*, 2.
Hedlund, B. P. (2001). Phylum XIII. Verrucomicrobia *phy. nov*. In D. R. Boone & R. W. Castenholz (Eds.), *Bergey's manual of systematic bacteriology. Volume one. The Archaea and the deeply branching and phototrophic bacteria* (pp. 785–793). New York: Springer.
Hugenholtz, P., & Kyrpides, N. C. (2009). A changing of the guard. *Environmental Microbiology, 11*, 551–553.
Hugenholtz, P., Goebel, B. M., & Pace, N. R. (1998). Impact of culture-independent studies on the emerging phylogenetic view of bacterial diversity. *Journal of Bacteriology, 180*, 4765–4774.
Jumas-Bilak, E., Roudière, L., & Marchandin, H. (2009). Description of '*Synergistetes*' phyl. nov. and emended description of the phylum '*Deferribacteres*' and of the family *Syntrophomonadaceae*, phylum '*Firmicutes*'. *International Journal of Systematic and Evolutionary Microbiology, 59*, 1028–1035.
Kersters, K., De Vos, P., Gillis, M., Swings, J., Vandamme, P., & Stackebrandt, E. (2006). Introduction to the proteobacteria. In M. Dworkin, S. Falkow, E. Rosenberg, K. H. Schleifer, & E. Stackebrandt (Eds.), *The prokaryote., Volume 5: Proteobacteria: Alpha and beta subclasses* (pp. 3–37). New York: Springer-Verlag.
Koch, A. L. (2003). Were Gram-positive rods the first bacteria? *Trends in Microbiology, 11*, 166–170.
Koonin, E. V. (2010). The origin and early evolution of eukaryotes in the light of phylogenomics. *Genome Biology, 11*, 209.
Kurland, C. G., Collins, L. J., & Penny, D. (2006). Genomics and the irreducible nature of eukaryote cells. *Science, 312*, 1011–1014.
Lambert, L. H., et al. (1998). *Staphylococcus succinus* sp. nov., isolated from Dominican amber. *International Journal of Systematic Bacteriology, 48*, 511–518.
Lang, B. F., Gray, M. W., & Burger, G. (1999). Mitochondrial genome evolution and the origin of eukaryotes. *Annual Review of Genetics, 33*, 351–397.
Lee, K. C., et al. (2009). Phylum *Verrucomicrobia* representatives share a compartmentalized cell plan with members of bacterial phylum *Planctomycetes*. *BMC Microbiology, 9*, 5.
Lee, K. C. Y., Herbold, C. W., Dunfield, P. F., Morgan, X. C., McDonald, I. R., & Stott, M. B. (2013). Phylogenetic delineation of the novel phylum *Armatimonadetes* (Former candidate division OP10) and definition of two novel candidate divisions. *Applied Environmental Microbiology, 79*, 2484–2487.
Line, M. A. (2002). The enigma of the origin of life and its timing. *Microbiology, 148*, 21–27.
Liu, Z., et al. (2012). '*Candidatus* Thermochlorobacter aerophilum:' An aerobic chlorophotoheterotrophic member of the phylum *Chlorobi* defined by metagenomics and metatranscriptomics. *The ISME Journal, 6*, 1869–1882.
Margulis, L. (1996). Archaeal-eubacterial mergers in the origin of Eukarya: Phylogenetic classification of life. *Proceedings of the National Academy of Sciences, USA, 93*, 1071–1076.
Martin, W. (1999). Mosaic bacterial chromosomes: A challenge en route to a tree of genomes. *Bioessays, 21*, 99–104.
Martin, W., & Müller, M. (1998). The hydrogen hypothesis for the first eukaryote. *Nature, 392*, 37–41.
Martin, W., & Russell, M. J. (2002). On the origins of cells: A hypothesis for the evolutionary transitions from abiotic geochemistry to chemoautotrophic prokaryotes, and from prokaryotes to nucleated cells. *Philosophical Transactions of the Royal Society of London* B *358*, 59–85.

Martin, W., Hoffmeister, M., Rotte, C., & Henze, K. (2001). An overview of endosymbiotic models for the origins of eukaryotes, their ATP-producing organelles (mitochondria and hydrogenosomes), and their heterotrophic lifestyle. *Biological Chemistry, 382*, 1521–1539.

McDonald, D., et al. (2012). An improved Greengenes taxonomy with explicit ranks for ecological and evolutionary analyses of bacteria and archaea. *The ISME Journal, 6*, 610–618.

Mori, K., Yamaguchi, K., Sakiyama, Y., Urabe, T., & Suzuki, K. (2009). *Caldisericum exile* gen. nov., sp. nov., an anaerobic, thermophilic, filamentous bacterium of a novel bacterial phylum, *Caldiserica* phyl. nov., originally called the candidate phylum OP5, and description of *Caldisericaceae* fam. nov., *Caldisericales* ord. nov. and *Caldisericia* classis nov. *International Journal of Systematic and Evolutionary Microbiology, 59*, 2894–2898.

Nishida, H., Beppu, T., & Ueda, K. (2011). Whole-genome comparison clarifies close phylogenetic relationships between the phyla Dictyoglomi and Thermotogae. *Genomics, 98*, 370–375.

Patel, B. H., Percivalle, C., Ritson, D. J., Duffy, C. D., & Sutherland, J. D. (2015). Common origins of RNA, protein and lipid precursors in a cyanosulfidic protometabolism. *Nature Chemistry, 7*, 301. doi:10.1038/nchem.2202.

Pereto, J., Lopez-Garcia, P., & Moreira, D. (2004). Ancestral lipid biosynthesis and early membrane evolution. *Trends in Biochemical Sciences, 29*, 469e477.

Pester, M., Schleper, C., & Wagner, M. (2011). The Thaumarchaeota: An emerging view of their phylogeny and ecophysiology. *Current Opinion in Microbiology, 14*, 300–306.

Poinar, G. O., & Poinar, R. (1994). *The quest for life in amber*. New York: Addison-Wesley.

Polyanskaya, L. M., Gorodnichev, R. B., & Zvyagintsev, D. G. (2013). Sizes of bacterial cells in soils determined by cascade filtration technique. *Biology Bulletin, 40*, 130–137.

Quaiser, A., et al. (2003). Acidobacteria form a coherent but highly diverse group within the bacterial domain: Evidence from environmental genomics. *Molecular Microbiology, 50*, 563–575.

Raymond, J., Zhaxybayeva, O., Gogarten, J. P., Gerdes, S. Y., & Blankenship, R. E. (2002). Whole-genome analysis of photosynthetic prokaryotes. *Science, 298*, 1616–1620.

Reid, R. P., James, N. P., Macintyre, I. G., Dupraz, C. P., & Burne, R. V. (2003). Shark Bay stromatolites: Microfabrics and reinterpretation of origins. *Facies, 49*, 299–324.

Reysenbach, A. L. (2001). Phylum BII. Thermotogae *phy. nov*. In D. R. Boone & R. W. Castenholz (Eds.), *Bergey's manual of systematic bacteriology. Volume one. The Archaea and the deeply branching and phototrophic bacteria* (pp. 369–387). New York: Springer.

Riding, R. (1999). The term stromatolite: Towards an essential definition. *Lethaia, 32*, 321–329.

Rinke, C., et al. (2013). Insights into the phylogeny and coding potential of microbial dark matter. *Nature, 499*, 431–437.

Santiago-Rodriguez, T. M., et al. (2014). *luxS* in bacteria isolated from 25- to 40-million-year-old amber. *FEMS Microbiology Letters, 350*, 117–124.

Schloss, P. D., & Handelsman, J. (2004). Status of the microbial census. *Microbiology and Molecular Biology Reviews, 68*, 686–691.

Schmidt, A. R., & Schäfer, U. (2005). Leptotrichites resinatus new genus and species: A fossil sheathed bacterium in alpine Cretaceous amber. *Journal of Paleolimnology, 79*, 175–184.

Schmidt, A. R., et al. (2010). Cretaceous African life captured in amber. *Proceedings of the National Academy of Sciences, USA, 107*, 7329–7334.

Searcy, D. G. (1992). Origins of mitochondria and chloroplasts from sulfur-based symbioses. In H. Hartman & K. Matsuno (Eds.), *Origins and evolution of prokaryotic and eukaryotic cells* (pp. 47–87). Singapore: World Scientific Publishing.

Sheridan, P. P., Freeman, K. H., & Brenchley, J. E. (2003). Estimated minimal divergence times of the major bacterial and archaeal phyla. *Geomicrobiology Journal, 20*, 1–14.

Sorokin, D. Y., et al. (2012). Nitrification expanded: Discovery, physiology and genomics of a nitrite-oxidizing bacterium from the phylum *Chloroflexi*. *The ISME Journal, 6*, 2245–2256.

Stackebrandt, E., & Goebel, B. M. (1994). Taxonomic note: A place for DNA-DNA reassociation and 16s rRNA sequence analysis in the present species definition in bacteriology. *International Journal of Systematic Bacteriology, 44*, 846–849.

Stanier, R. Y. (1970). Some aspects of the biology of cells and their possible evolutionary significance. *Symposium of the Society for General Microbiology, 20*, 1–38.

Tamura, K., Stecher, G., Peterson, D., Filipski, A., & Kumar, S. (2013). MEGA6: Molecular evolutionary genetics analysis version 6.0. *Molecular Biology and Evolution, 30*, 2725–2729.

Thompson, J. D., Higgins, D. G., & Gibson, T. J. (1994). CLUSTAL W: Improving the sensitivity of progressive multiple sequence alignment through sequence weighting, position-specific gap penalties and weight matrix choice. *Nucleic Acids Research, 22*, 4673–4680.

Torsvik, V., Goksøyr, J., & Dane, F. L. (1990). High diversity of DNA in soil bacteria. *Applied and Environmental Microbiology, 56*, 776–781.

Torsvik, V., Sorheim, R., & Goksoyr, J. (1996). Total bacterial diversity in soil and sediment communities a review. *Journal of Industrial Microbiology, 17*, 170–178.

Trevors, J. T. (2002). The subsurface origin of microbial life on the Earth. *Research in Microbiology, 153*, 487–491.

Trevors, J. T. (2003). Early assembly of cellular life. *Progress in Biophysics & Molecular Biology, 81*, 201–217.

Vandekerckhove, T. T., Willems, A., Gillis, M., & Coomans, A. (2000). Occurrence of novel verrucomicrobial species, endosymbiotic and associated with parthenogenesis in *Xiphinema americanum*-group species (Nematoda, Longidoridae). *International Journal of Systematic and Evolutionary Microbiology, 50*, 2197–2205.

Walker, A. (2014). Adding genomic 'foliage' to the tree of life. *Nature Microbiology, 12*, 78.

Whitman, W. B., Coleman, D. C., & Wiebe, W. J. (1998). Prokaryotes: The unseen majority. *Proceedings of the National Academy of Sciences, USA, 95*, 6578–6583.

Wiedenbeck, J., & Cohan, F. M. (2011). Origins of bacterial diversity through horizontal genetic transfer and adaptation to newecological niches. *FEMS Microbiology Reviews, 35*, 957–976.

Wrighton, K. C., et al. (2012). Fermentation, hydrogen, and sulfur metabolism in multiple uncultivated bacterial phyla. *Science, 337*, 1661–1665.

Chapter 3
Symbiotic Relationships

Abstract Different types of symbiotic associations link bacteria and invertebrates. Paths related to evolutionary processes leading to symbiosis are described on the basis of genome data and phylogenetic analyses. Endosymbiotic bacteria and their location in different hosts are reviewed, together with the role of bacteriocytes and other functional adaptations of ecto- and endosymbionts. Main traits related to symbiosis include functional reproductive manipulation, the effects of multiple concomitant symbionts, mechanisms of acquisition and specificity. Different benefits provided by symbiosis improve fitness and/or yield selective advantages in different host types, including nutritional, defense and detoxification mechanisms.

Keywords Acquisition • Bacteriocyte • Benefit • Defense • Ectosymbiont • Endosymbiont • Fitness • Host interactions • Pathogens • Reproductive manipulation • *Wolbachia*

1 Introduction

Symbiosis is a widespread and ubiquitous process active on evolutionary time scales that largely affected the diversification and success of many bacterial and eukaryotic lineages. Symbionts from several distinct bacterial taxa inhabit cells or tissues of invertebrates, or live in association with them. This process occurred in widely diversified host phyla, ranging from sponges to corals, molluscs, anellids, nematodes or insects, as well as in many disparate environments, varying from deep ocean vents to coral reefs or soil and vegetation habitats. In a general view, symbiosis is a key process in evolution, acting as a force that contributed to the emergence and diversification of Eukarya and pluricellular organisms (Douglas 2014a). As stated in Chap. 2, the most ancient and studied endosymbionts are the mitochondria and chloroplasts, whose irreversible introgression into eukaryotic precursors allowed the appearence of organisms provided with aerobic energetic metabolism and photosynthesis. Given their role and position in the early evolutionary history of Eukarya, however, they will not be considered here. For reviews on their origins and introgression see Chap. 2.

A. Ciancio, *Invertebrate Bacteriology*, DOI 10.1007/978-94-024-0884-3_3

Symbiotic relationships have been defined as the associations of two or more organisms (symbionts) that closely live together. Originally the term "*symbiosis*" (coined from Greek: $\sigma\acute{\upsilon}\nu$ = together, $\beta\acute{\iota}\omega\varsigma$ = life, living) was proposed by the plant pathologist De Bary in 1879, for a wide category that, apart of *mutualism* (in which both organisms receive a mutual benefit), also included other widespread relationships like *parasitism* (in which one organism is affected by the life of another associated species, living at its expenses in intimate contact with it). A further relationship considers the lack of either apparent beneficial or detrimental effects among symbionts, usually referred to as *commensalism* (Douglas 1994). In fact, the original definition of symbiosis was mainly based on the close, physical association of cells belonging to two or more species, whatever its outcome. In more recent studies and literature, however, its meaning became restricted to simple, mutualistic associations. Since in many cases the associations evolved with host or environment-dependent patterns, like i.e. the endosymbionts that switched from a parasitic life towards mutualism (Bordenstein et al. 2009), for clarity of this review the boundaries of the term "symbiosis" should be defined. It is herein intended sensu De Bary, retaining its original, wider and more general association perspective (Douglas 1994; Wilkinson 2001). However, due to a large number of practical issues concerning the possible exploitation of bacteria for biological control of many pests or other invertebrates management, parasitism will be treated in Chap. 4, whereas the relationships between invertebrate vectors and bacteria will be discussed in Chap. 5 and phoresy in Chap. 6.

Symbiotic associations can be classified on the basis of the guest bacterium location that can be internal (endosymbiosis), or in some way related to an external environment (ectosymbiosis). Endosymbionts can be found as intra- or extracellular guests in many host lineages, within specialised cells or one or more tissues (i.e. gonads, midgut or fat bodies of insects), respectively. Ectosymbionts include instead a partial stage or a whole life-cycle in an open body environment (i.e. the surface of insect eggs or other hosts organs like the oral aperture, various gut cavities or i.e. the gills of mussels or crustaceans). Other specifications concern the partial or total dependence of symbionts on the host metabolism, giving rise to obligate or facultative relationships (Haynes et al. 2003; Burke et al. 2009). Finally, also the transmission mechanisms of symbiotic bacteria may provide the basis for a classification: endosymbiotic bacteria may be transmitted through an inheritable introgression into the hosts maternal (sometimes also paternal) germ lines and through the embryos (vertical transmission) or through a vectoring or other environment-based transfer mechanism active from host to host (horizontal transmission of ectosymbionts) (Gehrer and Vorburger 2012). A combination of both processes was also observed (Haselkorn et al. 2009; Bright and Bulgheresi 2010; Ebert 2013; Henry et al. 2013; Duron et al. 2014). The latter mechanism, known as mixed-mode transmission (MMT) has been recognized for many symbiotic bacteria including *Hamiltonella defensa*, *Regiella insecticola*, *Paenibacillus larvae*, *Wolbachia* (α-Proteobacteria, Rickettsiales), *Asaia*, *Rickettsia* or *Borrelia* spp. (Gonella et al. 2012; Ebert 2013).

Multiple infections by distinct bacterial lineages are common within invertebrates. They are often characterized by differences in the mechanisms of host interaction and guest bac-

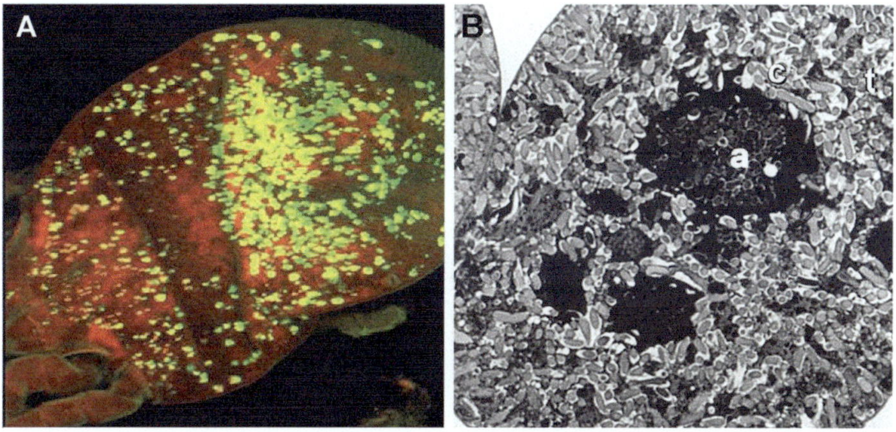

Fig. 3.1 Bacterial symbionts from the parenchyma of the mite *Dermatophagoides farinae* (**a**), shown by fluorescent *in situ* hybridization. Light micrograph (**b**) of the lantern bug *Pyrops candelaria* bacteriome, showing three different endosymbiotic bacteroids (*a*, *c*, *t*) (Adapted (**a**) from Hubert et al. (2012) and (**b**) from Wang and Cheung (1998))

teria manipulation or by the occurrence of a facultative metabolism (Fig. 3.1). These secondary associations often arose at the issue of symbiotic events that occurred on the same hosts, but at different evolutionary times (Thao et al. 2000; Thao and Baumann 2004a, b; Baumann et al. 2006; Perlman et al. 2006; Oliver et al. 2010; White et al. 2011; Six 2013). Coinfection may produce different outcomes, including interference exerted by a species on another one. Nested endosymbiosis, with endosymbionts harboring further bacterial cells (see Sect. 2.2.2. in this chapter) was also reported, associated to intense HGT-mediated genetic flows (Husnik et al. 2013), gene losses and evolution of reduced genomes (McCutcheon and Moran 2012).

In many bacterial lineages the endosymbiotic associations are widespread. However they also resulted difficult to study, often due to the lack of suitable culturing methods. Before the advent of modern molecular biology tools, most common methodologies applied for symbionts detection and study depended on light microscopy, including fluorescence *in-situ* hybridization techniques (Fig. 3.1a) and TEM (Wang and Cheung 1997). The number of species discovered as involved in symbiotic associations increased thanks to modern sequencing technologies, accounting to a significant fraction of total sequences (as shown by the data made available on the NCBI database, http://www.ncbi.nlm.nih.gov/) (Fig. 3.2). This representation appears still affected by the difficulties inherent the collection and study of many endosymbionts, whose biodiversity and biology remain largely unknown. This assumption appears true indeed for any bacterial lineage. It is widely accepted today that the total number of known species identified, after almost two centuries of microbiology research, accounts only for a small fraction (covering an approx. 10 %) of the total array of species estimated to occur on the planet (see Chap. 2).

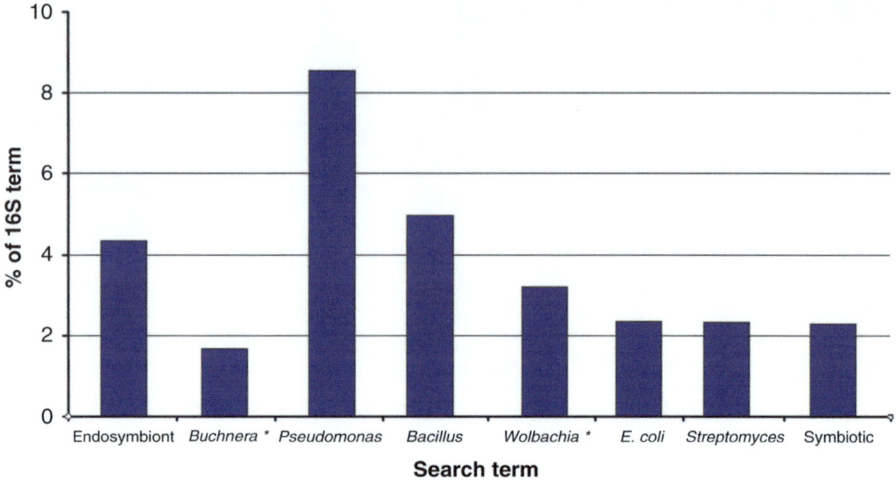

Fig. 3.2 Number of sequences of most studied invertebrate endosymbiotic genera (*marked with an asterisk*) with other bacteria and search terms, expressed as percentage of the total number of ribosomal 16S sequences (Source: NCBI GenBank (July 2014))

The research on symbiosis is a fertile field of study as investigative efforts on bacteria and invertebrate associations may provide insights on peculiar, adaptive biochemical pathways and processes, which evolved at the issue of the specific functionalities that characterize these relationships. Furthermore, many bacteria hold potential for the discovery and application of new molecules, given their ability to produce compounds with high biotechnological value (Haygood et al. 1999; Müller et al. 2004; Chaves et al. 2009). Among the biotechnological fallouts of high potential value expected by these studies it is worth to mention also the use of endosymbionts as possible tools to reduce the viral loads in a number of insect vectors of either human or animal diseases (Bian et al. 2010).

The lack of suitable cultivation methods and the difficulties inherent the isolation and study of endosymbionts, apart of problems related to their identification and description, hinder exploitation unless specific gene sequences are directly identified from the host and cloned. Advances in genomic, metagenomic and transcriptomic approaches through NGS methods offer possible alternative routes to the study of symbionts. NGS approaches allowed the study of entire genomes or functional genes expressed in endosymbiotic associations within widely diversified hosts. They are opening at the same time new perspectives for the subsequent cloning of genes of interest, for research or exploitation purposes. Prokaryote genomes listed by NCBI (http://www.ncbi.nlm.nih.gov/genome/browse/) accounted thus far (Nov. 2014) for more than $2.9 \cdot 10^4$ bacterial lines sequenced at various depths, from 5052 species. Among other advancements, NGS data and experimental approaches are providing new and unexpected perspectives on the evolution of many symbiont genomes, extending our knowledge on their occurrence, environmental spread as well as host adaptation and specificity (Robidart et al. 2008; Shigenobu and Wilson 2011; McCutcheon and Moran 2012; Saha et al. 2012; Webster 2014).

In this chapter some bacteria and invertebrate symbiotic partnerships will be examined, describing functional processes and adaptive traits resulting from their specific relationships. They highlight the nature of the evolutive forces that characterized this fundamental biological process. Among the wide array of host clades, insects provide many examples of successful endosymbiotic associations, found almost in all families. However, they also represent one of the most studied and sampled group of invertebrates, suggesting that a similar deep scan within other phyla may also reveal an equivalent diversity and frequency of symbiotic associations, evolutive links and adaptive traits. Potential applications deriving from the sometimes complex biochemistry of symbionts and their hosts are also discussed.

2 Symbiosis and Evolution

2.1 The Age of Symbiotic Associations

In order to reconstruct evolutionary processes leading to successful endosymbiotic relationships, researchers mostly relied on the analysis of modern genomes through available bioinformatic and statistical tools. In many cases the sequencing and comparative study of genomic data yielded significant insights on the age of first symbionts evolution, which is usually measured on a time scale of several Myrs. It is also hypothesized that the first associations may possibly date back to the early appearence of the corresponding host invertebrate ancestors. For example, by comparing the nucleotide divergence measured between two sequenced genomes of the aphid endosymbiont *Buchnera aphidicola* (γ-Proteobacteria) proceeding from two different hosts (*Schizaphis graminum* and *Acyrthosiphon pisum*), Tamas et al. (2002) showed that the relationship was first estabilished at least 150 Myrs ago and that during the last 50–70 Myrs the two bacterial genomes remained mostly unchanged, as shown by their high degree of conservation. The genomes stability was revealed by the minimal genetic rearrangements of the inferred gene architectures, which remained stable in spite of the nucleotidic divergence occurring in time (Tamas et al. 2004). For discussion about the putative occurrence of deleterious genetic drifts or changes affecting genome stability of endosymbionts see Baumann (2005).

Buchner (1965) recognized two main types of endosymbionts in insects, named the P-endosymbionts, ascribed to a conserved morphological type and characterized by a strict obligate relationship with the host, and the S-endosymbionts, which were morphologically diverse and facultative guests (Baumann 2005). Many P-endosymbionts, like *Tremblaya* spp. from mealybugs, were considered to derive from a single, ancestral infection event that led to the integration and vertical transmission of the bacterium and to its subsequent coevolution with the host. Many durable evolutionary relationships have been observed in insects lineages, i.e. the carpenter ants (*Camponotus* spp.) and their γ-Proteobacteria endosymbionts (Sauer et al. 2000), as well as in tsetse flies, aphids or cockroaches (Bandi et al. 1995; Chen et al. 1999; Baumann 2005; Urban and Cryan 2012). These coevolutive paths are shown by the similarity and correspondence of phylogenetic trees based on ribosomal or other genes of both symbiotic bacteria and hosts (Gruwell et al. 2007; Kölsch and Pedersen 2010; Urban and Cryan 2012). Similar matchings of phylogenetic trees are not usually found when examining S-endosymbionts, whose phylogenies do not appear congruent with those of their hosts (Baumann 2005).

For many P-endosymbiotic associations, molecular clock analyses yielded evolutionary times close to the age of the host appearance, indicating that symbiosis can be often considered as an early event in the evolutionary history of a species or lineage. The association of termites and cockroaches with their *Flavobacterium–Bacteroides* symbionts was dated back to 135–250 Myrs ago (Bandi et al. 1995), whereas the association of the european wasp *Philanthus triangulum* with a *Streptomyces* symbiont located in the female antennal glands (from which it can be also secreted), was inferred to date back 26–67 Myrs ago, around the time of differentiation of the genus *Philanthus* (Kaltenpoth et al. 2006).

Congruent phylogenies are frequent among endosymbionts and their hosts due to vertical transmission and cellular confinement, showing that cospeciation can be one important driver of bacterial species diversity (Sachs et al. 2011). The association time and phylogenies are useful to clarify the evolution of both organisms, and often fit the onset age of the corresponding hosts evolutive radiations. For insects, the oldest lineage, the Collembola, appeared around 400 Myrs ago (Engel and Grimaldi 2004). Further examples include the cockroaches of the genus *Cryptocercus* and *Blattabacterium* associations (Maekawa et al. 2005; Clark and Kambhampati 2003), the armored scales (Hemiptera: Diaspididae) and their Bacteroidetes endosymbiont, *Candidatus*[1] "Uzinura diaspidicola" (Gruwell et al. 2007) or the reed beetles (Coleoptera: Donaciinae) and its endosymbiotic bacterium from Enterobacteraceae, which is required for construction of the insect underwater pupation cocoon (Kölsch and Pedersen 2010).

Based on deepest divergence within clades and a 2–4 % rate of sequence changes per 100 Myrs, the age of association for phloem sucker insects and their P-endosymbionts was estimated to be 100–250 Myrs ago for *Carsonella rudii* (in psyllids), 100–200 Myrs for *Portiera aleyrodidarum* (in whiteflies) and *Ca.* "Tremblaya princeps" (in mealybugs), and 115–230 Myrs for *B. aphidicola* (in aphids). Similarly, the β-proteobacterium *Vidania fulgoroideae*, endosymbiotic in Auchenorrhyncha (Rhynchota, including planthoppers, leafhoppers, spittlebugs and cicadas), was ofter found to co-occur in association with many other (up to six) ensosymbionts present in the same hosts, the most frequent being *Sulcia muelleri* (phylum Bacteroidetes). The latter represents a monophyletic clade descending from an early infection event originated around 230 Myrs ago, spreading through

[1] *Candidatus* (shortened: *Ca.*) = the term refers to a described but unculturable bacterium, for which a deposited culture is not available. Since taxonomic rules still require a deposited culture to validate a species, the bacterium is indicated in this way as a "candidate" to species. The possibility to change these rules at the issue of whole genome sequences available has been proposed and is debated.

vertical transmission within the subsequent Auchenorrhyncha evolutive radiation (Moran et al. 2005). The phylogenies of *V. fulgoroideae* and related hosts were instead congruent with a more recent coevolutive relationship, whose time of association was estimated around 130–200 Myrs ago, fitting the age estimated for first planthoppers fossils (Urban and Cryan 2012).

For *Baumannia cicadellinicola*, endosymbiotic in xylem sap feeders (sharpshooters, Cicadellidae), the early association event was estimated to have occurred around 90–180 Myrs ago (Baumann 2005).

Processes similar to those observed in the evolution of insect symbiosis can be also found in other phyla. In the association involving the gutless flatworm *Paracatenula* (Platyhelminthes) and its tiotrophic endosymbiont *Ca.* "Riegeria" (α-Proteobacteria), a strict coevolutive relationship was found when comparing congruent cladograms based on the symbiont 16S rRNA and the host concatenated 18S and 28S rRNAs, with an estimated association time dating back to 500–620 Myrs ago. This appears to be one of the oldest known mutualistic relationships, occurred at the onset of the early Cambrian evolution of bilaterian diversity. The endosymbiotic association was estimated as already established at the onset of platyhelminths evolution (Gruber-Vodicka et al. 2011; Leisch et al. 2011).

Ancient and specific associations of vertically transmitted endosymbionts belonging to the *Chloroflexi* and *Acidobacteria* have been reported from the sponge *Svenzea zeai* (Lee et al. 2009). Sponges are among the oldest Metazoa and their association with bacteria is estimated to have a Precambrian origin (Wilkinson 1984). Wide range phylogenetic studies on sponge endosymbiotic Cyanobacteria (including *Synechococcus* and *Prochlorococcus* spp.) showed either the occurrence of polyphyletic symbionts that proceeded through multiple and independent associative events, and the evolution of sponge-associated bacterial lineages with a much longer and host-specific coevolutionary history (Steindler et al. 2005).

Convergent evolution of two monophyletic and closely related chemoautotrophic bacterial groups in the marine gutless oligochaete *Inanidrilus leukodermatus* (Anellida) and the nematode *Laxus* sp. (Stilbonematinae) appeared to have evolved in a relatively more recent time, estimated around 20–40 Myrs ago (Dubilier et al. 1995, 2008).

2.2 Endosymbiosis

2.2.1 Location and Bacteriocytes

Endosymbiotic bacteria settle within their hosts body in a regulated way. Their sites are considered to follow specific "rules" which reflect their evolutive history and host relationships, as well as the metabolic links often of a nutritional nature, underpinning successful associations. In many cases, involving – but not limited to – insects, endosymbiotic bacteria can be found in the cytoplasm of specialized cells called *bacteriocytes* (also known as 'mycetocytes'), provided with several host-derived vesicles enclosing the bacterial cells and often clustered in a specialized tissue called *bacteriome* (sometimes also cited as 'mycetome') (Douglas and Dixon 1987; Douglas 1998, 2014b; Baumann 2005). Bacteriocytes evolved repeatedly, at the issue of many diverse mutualistic associations, in members of disparate phyla living in different environments and characterized by variable levels of complexity. Functionally and morphologically similar bacteriocytes can be found in fact in distant phyla varying from insects (i.e. aphids, ants and almost all lineages) to sponges, anellids, nematodes, cnidaria, molluscs or platyhelminths (Schuett et al. 2007) (Fig. 3.3).

Bacteriocytes represent the convergent evolutive solution allowing a successful and stable host-symbiont adaptation process, based on the integration of fundamental functions yielding one or more mutual benefits. These processes are subject to positive selection, enhancing the host metabolic/nutritional or defense fitness. The

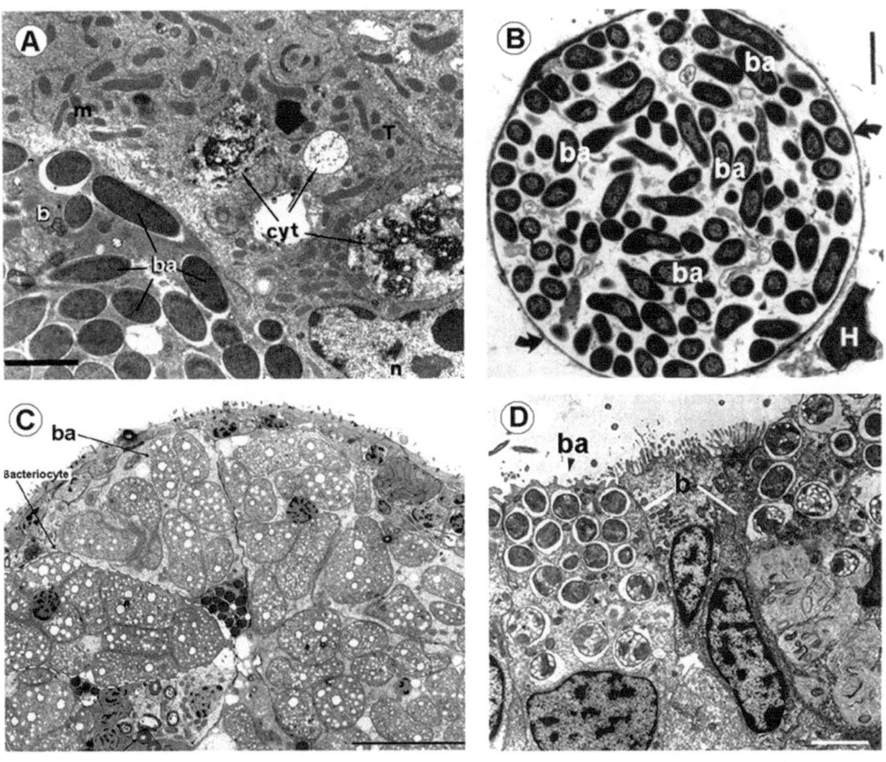

Fig. 3.3 Transmission electron micrographs of cross sections showing bacteriocytes (*b*) with endosymbiotic bacteria (*ba*) from (**a**) the cockroach Blattella germanica (*m* mitochondria); (**b**) the echinoderm Antedon bifida (*arrows*: bacteriocyte membrane, *H* host cell nucleus); (**c**) the plathyhelmint Paracatenula cf. polyhymnia (*dotted line*: symbiotic bacteria; *dashed line*: bacteriocytes); (**d**) the gill filament of the mussel *Bathymodiolus* sp. *Scale bars*: **a**, **b**, **d** = 2 μm, **c** = 10 μm (Adapted from Yin et al. 2001 (**a**); Kelly and McKenzie 1995 (**b**); Leisch et al. 2011 (**c**); Kochevar et al. 1992 (**d**))

endosymbionts location is hence determined by and related to the functional process involved in the association. An example of functional association is given by gill bacteriocytes of the bivalve *Codakia orbicularis*, that are crowded with endocellular bacteria which metabolize sulphur compounds intercepted from the surrounding marine environment to which the gills are exposed (Frenkiel and Mouëza 1995). A similar process appeared and evolved also in other phyla, as shown by i.e. the bacteriocyte cells found in the gill tissue of the clam *Calyptogena magnifica*. Transmission electron microscopy (TEM) examinations showed that these cells were replenished with reproducing bacteria characterized, also in this case, by a sulphur-based metabolism. The bacteriocytes were subject to a cyclic resorption by the surrounding host tissues, thus playing a fundamental nutritional role (Fiala-Médioni and Métivier 1986).

In some insect endosymbionts, i.e. *Blochmannia*, the bacteria can be found in the cytoplasm of specific host cells, whereas other bacteria, like *Wigglesworthia*, have an extracellular location within a specific host tissue (Attardo et al. 2008; Stoll et al. 2010). In other associations, i.e. the olive fly *Bactrocera oleae* and its endosymbiont, *Ca*. "Erwinia dacicola", the guest bacterium changes its site after metamorphosis, switching from an intracellular lifestyle in midgut cells to an extracellular location in the foreguts (Estes et al. 2009).

Wolbachia pipientis (Rickettsiales) is one of the most widespread and studied endosymbiotic bacterium. Members of this lineage were first observed in the reproductive system of *Culex pipiens* from which it was originally described (Hertig and Wolbach 1924). It was later found in organs or tissues of other insects (Lo et al. 2007; Saridaki and Bourtzis 2010), among which it is almost ubiquitous, with an estimate of more than 70 % of species interested to this association (Hilgenboecker et al. 2008; Saridaki and Bourtzis 2010).

Although considered as a single species, *Wolbachia* represents a group of lineages, given the wide diversification of forms infecting very distant invertebrate hosts. Members of the *Wolbachia* group often occur in combination with other endosymbionts. In specimens of *Drosophila melanogaster*, hosting a *Wolbachia* sp. and a further *Spiroplasma* sp., the bacteria were mostly observed in specific locations from distinct organs of the host, *Wolbachia* being located in the Malpighian tubules whereas *Spiroplasma* found within the ovaries (Goto et al. 2006; Haselkorn et al. 2009).

The vertically transmitted endosymbionts of aphids, present in almost any group, have been extensively studied, describing in depth many details of fundamental genetic and biochemical processes. Aphids bacteriocytes containing cells of *B. aphidicola* are located in the host hemocel, the body interior. They were estimated to reach up to 10 % of the aphid body mass, accounting for up to 10^5 cells · mg^{-1} of host wet weight (Baumann et al. 2006). In the vetch aphid *Megoura viciae*, around 80 bacteriocytes were counted at birth, which later increased in size, rather than number, during the whole insect lifespan (Douglas 1998). Although bacteriocytes represent a successful evolutive construct, other endosymbiont locations are, however, possible. These may also include a direct contact with the host cells, like i.e. the polymorphic member of the *Cardinium* lineage (*Ca*. "Paenicardinium endonii") discovered in adults and juveniles of the plant-parasitic nematode *Heterodera glycines*. The cells of this endosymbiont are located within the host pseudocoelom and intestine or hypodermal chords, as well as amid ovary walls, oocytes and spermatozoa (Noel and Atibalentja 2006).

In many obligate endosymbionts the formation of a bacteriocyte or similar structures, and the resulting vertical bacterium transmission, are part of sophisticated mechanisms often synchronous with the host development and lifecycle, as i.e. for vertical transmission of *B. aphidicola* in *Acyrthosiphon pisum* (Koga et al. 2012).

In case of horizontal transmission, a strict, environment-dependent relationship may link the symbiont to its host. The bacterium location in the host body may be in fact determined at the issue of specific ecological roles played by the symbiont, whose presence provides a functional, selective advantage. This is the case, for example, of the bioluminescent gram-negative *Vibrio fischeri*, an endosymbiont hosted in the light organ of the Hawaiian bobtail squid *Euprymna scolopes* (Visick and Mcfall-Ngai 2000). The cephalopod is a shallow water species that relies on bioluminescence to avoid predation during night feeding. The bacterium introgression into the squid induces the formation of a sophisticated light organ, located in the center of the body cavity. The bacterial light is emitted and modulated under host control at the same intensity and wavelenght of moon or starlight. The light organ has a complex anatomy (see Sect. 4.2) and its bioluminescence aims at baffling and confounding marine predators, avoiding the projection of a shadow by mimicking the background, incident light (Mcfall-Ngai 1999; Visick and Mcfall-Ngai 2000).

Differences in patterns of host location may also depend on the populations or lineages of the same endosymbiont, often considered as the result of adaptive processes. Being the most studied endosymbiont, *Wolbachia* offers many examples of this flexibility, reflected by its evolution within distinct host phyla (Table 3.1). Its different locations reflect specific metabolic or reproductive adaptations, including switching from mutualism to parasitism (Bordenstein et al. 2009).

A further example of adaptive mechanism is given by the ants of the genus *Camponotus*, which harbor the intracellular mutualistic bacterium *Blochmannia floridanus*. The symbiont is hosted in bacteriocytes whose locations and distribution change in relation to the host life-stage. In *Camponotus*, vertical transmission proceeds through the emergence of *Blochmannia* cells from bacteriocytes, which later invade the oocytes. During the whole host lifecycle the bacterium location is subject to a complex re-arrangement, adapting its cells to inhabit distinct body sites, from the larval through the pupal, up to the final adult stage. During metamorphosis, the bacteriocytes are located in the host outer layer of the midgut epithelium. These layers are then fragmented in clusters interspersed with bacteria-free midgut cells in the last instar larva, to form again a continuous layer of bacteriocytes in the gut lumen during the late pupal stage. The bacteria increase in numbers and spread also to enterocytes and in what appears to be a sort of bacteriome. Their density and the number of bacteriocytes then decrease in later adult stages, when the cells are also digested by the host (Stoll et al. 2010).

Tropism, the symbiont ability to colonize different host organs and tissues and to translocate in the host body throughout the different phases of its lifecycle, was also

Table 3.1 Host body location of some *Wolbachia* endosymbionts

Host	Organ tissue or cells	References
Acromyrmex echinatior (Arthropoda: Insecta: Hymenoptera)	Haemolymph, fat body, hindgut, midgut, faeces, ovary	Frost et al. (2014)
Asobara tabida (Insecta: Hymenoptera)	Oocyte posterior cytoplasm; thorax	Dedeine et al. (2001)
Bemisia tabaci (Insecta: Rhynchota)	In bacteriocytes, with primary symbionts; in abdomen, outside bacteriocytes	Skaljac et al. (2010) and Gottlieb et al. (2008)
Callosobruchus chinensis (Insecta: Coleoptera)	In salivary glands, fat body, gut and brain	Ijichi et al. (2002)
Cimex lectularius (Insecta: Rhynchota)	In gonad-associated bacteriocyes	Hosokawa et al. 2010
Drosophila melanogaster (Insecta: Diptera)	Thoracic muscles, retina, ovary, embryo	Min and Benzer (1997) and Veneti et al. (2004)
Drosophila melanogaster (Insecta: Diptera)	Testis, spermatogonia and spermatocytes; oocytes	Clark et al. 2002 and Ferree et al. 2005
Drosophila simulans (Insecta: Diptera)	Head, gut, Malpighian tubules	Osborne et al. (2012)
Linyphiidae (Arthropoda: Arachnida): *Bathyphantes gracilis*, *Erigone atra*, *E. dentipalis*, *E. longipalpis*	Legs, abdomen	Goodacre et al. (2006) and Goodacre and Martin (2013)
Brugia malayi, *B. pahangi*, *Dirofilaria immitis* (Nematoda: Onchocercidae)	Lateral hypodermal chords of adults, secretory-excretory canals, reproductive system (ovaries, oocytes, morulae, microfilariae)	Kramer et al. (2003) and Landmann et al. (2010)
Mansonella (*Cutifilaria*) *perforata* (Nematoda: Onchocercidae)	Gonad epithelium, intestine wall cells	Ferri et al. (2011)
Radopholus similis (Nematoda: Tylenchida)	Ovary	Haegemann et al. (2009)

observed in ticks of the genus *Rhipicephalus*, hosting a *Coxiella* sp. and a *Rickettsia*-like symbionts (Rickettsiales) vertically transmitted via the female ovary (Lalzar et al. 2014). Using specific probes, cells of *Coxiella* sp. were found in clusters in the female reproductive system, in the interstitial cells of unfed females and, after feeding, between primary oocytes and inside maturing and fully developed oocytes, as resulting by a polar clustering process occurring during oogenesis. *Coxiella*, however, also occupied the Malpighian tubules of adults, packed in small vesicles mainly located in the distal parts. They enlarged once the hosts received a blood meal, always keeping the lumen of the tubules free of bacteria (Lalzar et al. 2014). Malpighian tubules, which are connected to the alimentary tract and intestine of insects and ticks, are involved in detoxification processes and represent a further insect or tick body part that is often interested by symbiotic associations (Kölsch and Pedersen 2010; Osborne et al. 2012).

Endosymbiosis also evolved in organisms inhabiting extreme environments. Deep sea hydrothermal vents along the Galapagos Rift and the East Pacific Rise (EPR) in the Eastern Pacific Ocean yielded the first known deep sea endosymbiotic association of a thioautotrophic bacterium with the Vestimentiferan *Riftia pachyptila* (Anellida). These mouthless and gutless tubeworms are anchored to the base of vents and sides. They are sheltered in a chitin, schlerotized tube which they produce, from which a branchial gill-like plume protrudes. The individuals aggregate in dense colonies and can reach up to 3 m in lenght (Stewart and Cavanaugh 2006). Their nutrition entirely depends on the presence of a chemosynthetic γ-proteobacterium living endocellularly inside a host-derived vacuole and included in bacteriocytes. These form a specific, vascularized bacteriome (trophosome) located in the body trunk wall. The endosymbiont, described as *Ca.* "Endoriftia persephone" (Robidart et al. 2008), provides the host with a primary nutritional source through the oxidation of reduced sulfur compounds and carbon fixation via the Calvin cycle, receiving back from the tubeworm the nutrients intercepted in the vent surrounding waters by its plume. The pleomorphic[2] endosymbionts, whose density can reach up to 10^9 cells per g of fresh trophosome – corresponding to 15–35 % of the host body volume (Stewart and Cavanaugh 2006) – are environmentally transmitted, being ingested by the *R. pachyptila* juveniles, the only stages provided with mouth and gut (Harmer et al. 2008).

Studies on *R. pachyptila* and other related anellids showed that their endosymbiotic bacteria are genetically homogeneous and belong to a single phylotype. This includes the endosymbionts of a further deep sea tubeworm, *Tevnia jerichonana*, living in deep sea environments with higher sulphur concentrations, and those of the closely related species *Oasisia alvinae* and *Ridgeia piscesae* (Gardebrecht et al. 2012). Differences in their living environments suggest that the tubeworms are able to maintain stable internal conditions, estabilishing and sustaining a successful association (Gardebrecht et al. 2012).

---

[2] *Pleomorphism* = occurrence of different cell morphologies in the same bacterial species, due to different nutritional or environmental factors (*syn.* = polymorphic).

Endosymbionts develop and multiply within bacteriocytes, in what appears to be a stable and confined niche environment in intimate contact with the host body, responding to evolutive pressures selecting for a functional separation from the guests' cells. This process may be related to the host needs to keep symbiont cells distinct from other possible invasive pathogens and/or to protect them from its own defensive reactions and immune response. This hypothesis is supported by i.e. the high expression levels observed for an invertebrate-type lysozyme coding gene, found in the trascriptome of *A. pisum* bacteriocytes. Such levels of antibiotic production were considered instrumental to the exclusion of other bacterial intruders and/or to the degradation of impaired *B. aphidicola* cells (Nakabachi et al. 2005).

From a general perspective, bacteriocytes are functional to processes from which either the host and the bacterium receive a benefit, including vertical transmission, the enhancement of guest bacteria metabolic and nutritional efficiency and/or the maintenance or protection of the bacterial genomic organization. These goals appear easier to achieve possibly within a confined body space, limiting access to external recombination events or reducing the frequencies of genetic flows or gene enrichment mechanisms (i.e. HGT, intrusion of bacteriophages or spread of transposition elements). In some cases the bacterial host is subject to a metabolic specialization, i.e. the production of essential amino acids, and to the reduction of its genome, fully integrating by this way its existence with that of the host (Moran 2002; Feldhaar et al. 2007; Hosokawa et al. 2010). In others, i.e. *Wolbachia*, the bacteriocytes are functional to higher rates of guest reproduction, inducing strong steering effects active on the host sex ratio even altering its speciation, through mechanisms like i.e. cytoplasmic incompatibility, induced pathenogenesis, higher male mortality rates or feminization (Telschow et al. 2007).

2.2.2 Multiple Endosymbionts

The functional 'compartment' solution found for many symbionts at the issue of evolution was indeed such a successful strategy that happened to be replicated many times, even on the same host lineage. Insects show many examples of multiple endosymbiotic invasions, and several groups also harbor distinct facultative and/or obligate secondary endosymbionts. One example is given by the morphologically different endosymbiontic γ-Proteobacteria found in the same *Adelges* spp. hosts. These pests, which feed on conifers in Europe, America and Asia, harbor two endosymbionts, *Ca.* "Steffania adelgidicola" and *Ca.* "Ecksteinia adelgidicola", which are vertically transmitted, each confined in a specific bacteriocyte type (Toenshoff et al. 2012).

Among insects, various secondary bacteria can be found in separated body parts or even nested one inside the other, as in the case of the mealybug *Planococcus citri*, which harbors the β-proteobacterium endosymbiont *Ca.* "Tremblaya princeps", whose cells in turn host a further γ-proteobacterium endosymbiont, *Ca.* "Moranella endobia" (von Dohlen et al. 2001; Husnik et al. 2013). *Ca.* "Tremblaya princeps" was shown to possess the smallest bacterial genome (139 kb) known thus far, in which only a few essential genes were conserved. This complex tripartite relationship resulted by the functional integration and conservation of genes proceeding from organisms from up to six different lineages, which were introgressed during the host evolutionary history through a series of successful HGT events (McCutcheon and von Dohlen 2011; McCutcheon and Moran 2012).

Up to three endosymbionts (*Wolbachia*, *Arsenophonus* and *Rhizobiales*) were reported in populations of the ant *Anoplolepis gracilipes* from areas around the Pacific and Indian Oceans. Prevalence levels and number of multiple associations were shown to depend on the host population sampling sites, with *Wolbachia* and *Arsenophonus* displaying higher prevalence levels. In spite of the distance among sampling sites, which ranged in the order of several thousand km, all symbionts showed a low level of genetic variability (Sebastien et al. 2012).

Multiple co-infecting endosymbionts were reported in different combinations from whiteflies (Skaljac et al. 2010; Marubayashi et al. 2014). In populations of *Trialeurodes vaporariorum*, Skaljac et al. (2010) found co-occurrence of *Arsenophonus* with *Hamiltonella* (Enterobacteriales), whereas in *Bemisia tabaci* (Q biotype), from the same area, up to four endosymbiots co-occurred, in different combinations, the most common species being *Hamiltonella*, *Rickettsia*, *Wolbachia* (found only in bacteriocytes) and *Cardinium* (Bacteroidetes) (Parrella et al. 2014). Both host species also harbor *Portiera aleyrodidarum* in bacteriocytes as primary endosymbiont (Thao and Baumann 2004b; Gottlieb et al. 2008; Skaljac et al. 2010).

Replicated endosymbionts are not restricted to insects, as this process occurred also in other invertebrate taxa. The tick *Rhipicephalus turanicus* and a related species, harbor two symbionts, a *Coxiella*-like and a *Rickettsia* sp. (Lalzar et al. 2014). Several endosymbionts including *Wolbachia*, *Rickettsia* and *Spiroplasma* were also found together in linyphiid spiders (Goodacre et al. 2006).

In multiple associations, the integration and coexistence of endosymbionts rely on complementary contributions to the host metabolism, but interference may also occur. In some cases a process leading to the replacement of one symbiont by another species was observed, determined by a genetic drift modifying, through gene losses and consequential reduced functionality, the metabolic efficiency of an ancient endosymbiont. These processes erode the fitness of a "degenerated" guest bacterium, as was the case of a *Nardonella*, an ancestral symbiont of grain weevils (Curculionoidea), under replacement by the γ-proteobacterium *Sodalis glossinidius* (Moran 2002; Conord et al. 2008; McCutcheon and Moran 2012).

Dual or multiple endosymbionts, each located in a specific bacteriocyte type, can be found also in xylem sap feeder spittlebugs (Hemiptera). One of them, *Ca*. "Sulcia muelleri" (Bacteroidetes), was considered as the most ancestral. It provides, in conjunction with *Ca*. "Baumannia cicadellinicola" or *Ca*. "Zinderia insecticola", all the ten essential amino acids needed by their hosts (Koga et al. 2012). In a spittlebug lineage (tribe Philaenini), however, *Ca*. "Zinderia insecticola" was found under a selective pressure leading to its replacement by a newly recruited *Sodalis* sp. (Koga et al. 2013).

In nine *Drosophila* spp., apart of the *Wolbachia* endosymbiont, a secondary *Spiroplasma* sp. guest bacterium occurs. The *Spiroplasma* was shown to produce different outcomes, including male killing in some species, differing from other hosts in which no effect was exerted on their sex ratios. Sequencing and phylogenetic analyses showed that at least four phylogenetically distinct *Spiroplasma* haplotypes were produced in five distinct, ancestral host invasion events, through a series of vertical and horizontal transmissions. No evidence for an interference effect was found, however, in this association (Goto et al. 2006; Haselkorn et al. 2009). *Wolbachia* was also reported from the aphid *Cinara cedri*, in bacteriocytes hosting also the bacterium *Serratia symbiotica* (Gómez-Valero et al. 2004) and in the weevil *Sitophilus oryzae*, with its primary symbiont (Heddi et al. 1999).

Considering other invertebrate phyla and environments, the Bathymodiolinae mussels (*Bathymodiolus* spp.) – deep sea colonizers growing around hydrothermal vents – harbor two groups of sulphur- and methane-oxidizing symbionts. Cells of these bacteria occur in vacuoles present in specialized epithelial bacteriocytes and in the gill filament lateral zone (Fig. 3.3d). In *B. azoricus*, the dominant mussel species off the Azores Islands hydrothermal vents, sulfur-oxidizing bacteria occur with methane-oxidizing species (Fiala-Médioni et al. 2002). At the issue of their 16S rRNA analysis, up to six different endosymbionts were reported from *Bathymodiolus*, together with a further *Methylophaga* sp. capable of degrading toxic C1 byproducts but methane (Duperron et al. 2006, 2009; Petersen et al. 2011).

Hematophagous leeches, including the medicinal leech *Hirudo medicinalis* (Hirudinidae, Anellida), host in their intestine the facultative symbiont *Aeromonas veronii* biovar *sobria* (Graf 2005). Other studies showed that the acquisition of endosymbiotic bacteria occurred at least five times during the radiation of blood feeding leeches (Perkins et al. 2005). In the family Glossiphoniidae, two distincts endosymbionts belonging to the γ-Proteobacteria are hosted in specialized trophosomes (mycetomes) with different morphologies depending on their host species (*Placobdelloides jaegerskioeldi* and related species, or *Haementeria ghilianii*). The leech *Placobdella parasitica* and other related species harbor an endosymbiotic α-proteobacterium belonging to the genus *Reichenowia* (Fig. 3.4). This genus was described to accomodate the new species and appeared phylogenetically related either to *Sinorhizobium meliloti* (Rhizobiaceae), a nitrogen fixing bacterium endosymbiotic in the root nodules of leguminous plants, and to the animal pathogenic genera *Brucella* and *Bartonella* (Siddall et al. 2004; Perkins et al. 2005). Finally, a further, distinct Rickettsia-related and vertically transmitted endosymbiont was found in tissues of *Hemiclepsis marginata*, *Torix tagoi* and other aquatic leeches feeding on amphibians and fishes (Kikuchi et al. 2002; Kikuchi and Fukatsu 2005) (Fig. 3.4).

2.2.3 Evolutive Ties

As previously shown, phylogenetic approaches have been successfully applied to investigate the origins of symbionts and the subsequent evolution of these associations. Phylogenetic trees based on 16S rRNA or other gene sequences can provide visual clues on the evolutive patterns and distances separating endosymbionts, and have been routinely used to infer evolutionary paths characterizing species. Such

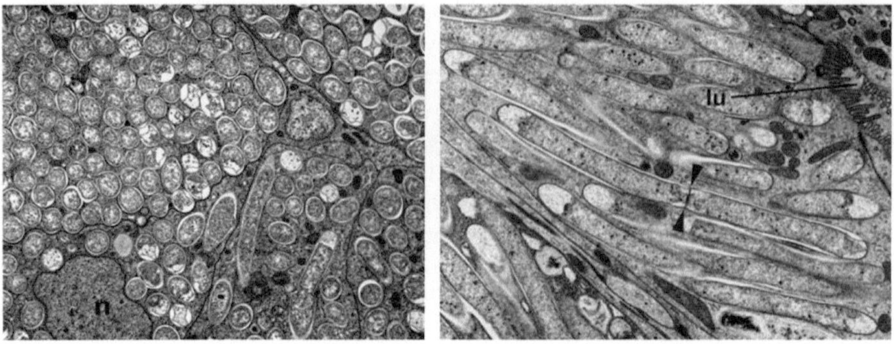

Fig. 3.4 TEM micrographs of transversal (*left*) and longitudinal (*right*) sections of an endosymbiotic α-proteobacterium from the leech *Desserobdella picta* (*Anellida, Clitellata*) (*n*, host cell nucleus; lu, trophosome lumen; arrowheads show septa) (Adapted from Siddall et al. 2004)

analyses may in fact uncover a number of events like host – symbiont congruent coevolution, host switch, extinction or horizontal transmission, elucidating complex mechanisms concerning association insurgence and evolution. Initial views of endosymbiont radiations considered in fact the bacteria as descendent from single ancestral infective agents through unique events, later maintained in intimate relationships with their hosts and refined throughout the subsequent cospeciation processes. This view has been in fact enriched by recent acquisitions originated by phylogenetic analyses offering new perspectives concerning coevolution, guest substitution, extinctions, HGT and/or co-occurrence of multiple lineages (Mitsuhashi et al. 2002; van Borm et al. 2003; Bordenstein and Wernegreen 2004; Bordenstein et al. 2009).

The search for congruent phylogenies and their evaluation can be considered as a valid approach to confirm or reject hypotheses concerning hosts and bacteria coevolution. This process implies the maintenance, on a many Myrs scale, of reciprocal effects driving evolution and gene changes of symbionts, maintaining their fitness and reinforcing existing functional links. These changes can be traced back when examining the properties of the symbionts phylogenetic trees, that can be defined as congruent when one dendrogram significantly mirrors that of the other lineage.

Congruent phylogenies occur within many host phyla and different kinds of endo or ectosymbioses, i.e. the flies of the Tephritinae subfamily (Diptera,Tephritidae) and their extracellular gut symbiont *Ca.* "Stammerula" spp. (Mazzon et al. 2010). However, if the analyses consider many distinct lineages, a broader spectrum of events can be observed. When pooling sequence data from different symbionts and hosts, a broader gene variability reveals the occurrence of multiple associations of distinct bacterial taxa within the same host lineages. This is the case i.e. of leeches and redworms (Anellida: Clitellata), which show associations with five evolutionarly distinct endosymbiont lineages. Other phyla, i.e. Nematoda, show association to four separated bacterial lineages. Many insect classes have been repeatedly interested by endosymbionts proceeding from a wide range of phylogenetically distant bacterial phyla. Viceversa, the colonization of hosts from different phyla by closely related bacterial lineages can also be observed (Fig. 3.5).

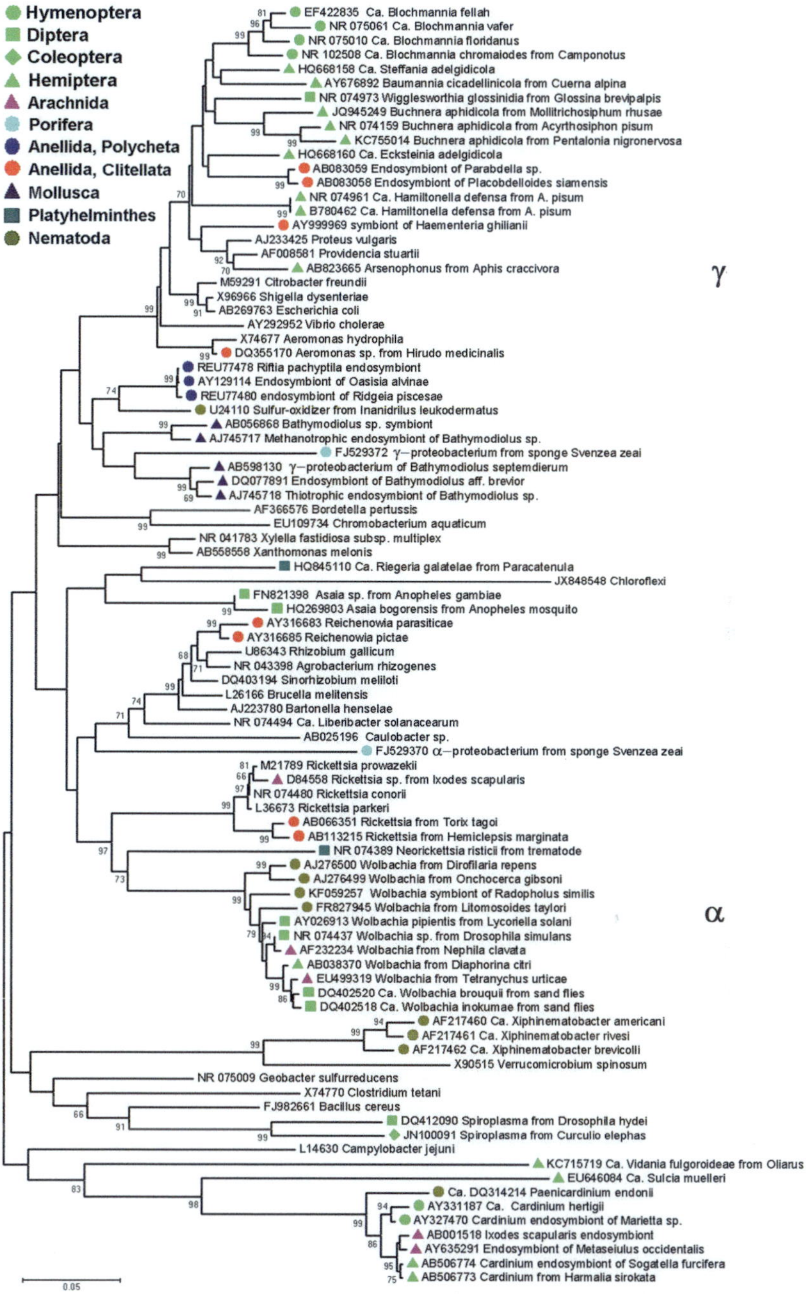

Fig. 3.5 Neighbour-Joining phylogenetic tree of invertebrate endosymbionts with other bacteria, based on aligned 16S rDNA sequences and the Jukes-Cantor method, with base substitutions per site (*scale bar*). *Color symbols* show host taxa (see legend). The percentage of trees (>70 %) including the same taxa in the bootstrap test (200 replicates) is shown next to branches. Constructed with ClustalW, BioEdit and MEGA6 (see Chap. 2 for references)

Apart of *Wolbachia*, present in hosts ranging from insects to nematodes and trematodes, many other associations of closely related bacterial lineages with hosts from distinct phyla reveal the occurrence of past events of host switching. This may be the case i.e. of the maternally transmitted endosymbiont of insects, *Cardinium* sp., which is also widespread among arachnids, including a wide array of spiders, scorpions and mites (Martin and Goodacre 2009). A polymorphic member of this lineage (*Ca.* "Paenicardinium endonii") was also discovered in the soybean cyst nematode *Heterodera glycines* (Noel and Atibalentja 2006).

Many phylogenetic applications mostly deal with vertically and maternally transmitted endosymbionts (or parasites, which may be considered as a special case of symbiosis). Statistical, inferential tools usually applied rely on 'molecular clocks', which are based on the number of substitutions per site and Myrs, possibly estimated or derived using available fossil records. Molecular clock applications were used to support congruent phylogenies linking the xylophagous cockroaches of the genus *Cryptocercus* spp. and the transovarially transmitted *Blattabacterium* endosymbionts. They also allowed the inference of the divergence time and evolution of the host species and elucidated eventual phylogeographic implications occurring during their speciation paths (Clark and Kambhampati 2003; Maekawa et al. 2005). Similarly, molecular clock phylogenies showed that the ant *Camponotus* and its primary, fast evolving *Blochmannia* endosymbiont established a stable evolutive relationship during tens of Myrs (Degnan et al. 2004; Adams et al. 2013).

Maternally inherited *Buchnera* endosymbionts of aphids showed high levels of phylogenetic congruence when their genes were used for comparison with the mitochondrial-based phylogenies of their hosts. Given the robustness of these relationships, the *Buchnera* phylogenies were applied to infer, at least in part, the main traits characterizing the evolution of their aphid hosts (Nováková et al. 2013).

This condition was not verified instead for a second, not strictly host-specific S-type enteric-like endosymbiont of *Buchnera*, which is vertically and possibly horizontally transmitted (Fukatsu et al. 2000). Phylogenetic analyses can also reveal an opposite situation, in which no evolutive congruent link exists between hosts and symbionts, as for instance in the case of bathymodiolin mussels (Bivalvia: Mytilidae) and associated sulfur-oxidizing bacteria (Won et al. 2008) or the bryozoans of the genus *Bugula* and associated *Ca.* "Endobugula" symbionts (Lim-Fong et al. 2008).

Similarly, analysis of the 16S rRNA sequence divergence may also be used to investigate presence/absence of evolutive links within the bacterial symbiont lineages, as applied to i.e. bathymodiolin ectosymbionts which did not show descendence from other host mussels endosymbionts (Duperron et al. 2009).

2.3 Ectosymbionts (Epibionts)

Several bacteria adapted to live symbiotically on the external body surface of invertebrates (i.e. the exoskeleton or other cuticular layers) or on the surface of body parts connected to the external environment (i.e. gills, gut or oral apertures). These associations, often specific, result from a strict functional relationship with their hosts. In this case the bacteria are called ectosymbionts or epibionts (from Ancient Greek επί = above and βιωντες = living). There are many examples of symbiosis of this kind as ectosymbiotic bacteria may be found in almost all invertebrate groups, being commonly encountered in insects, mites, crustaceans and others. Also in this process a wide range of environmental conditions are present, including extreme habitats (Childress et al. 1987). Many studies described the feeding of crustaceans on ectosymbiotic bacteria which are harbored on their body surfaces, a kind of ectosymbiosis often observed in reduced ecosystems. A first, well studied organism is *Rimicaris exoculata*, a decapod shrimp living in deep ocean floors of the Mid Atlantic Ridge, recorded down to 4080 m in depth (Guri et al. 2012). This species belongs to the family Alvinocarididae, which derived its name from Alvin, the popular deep sea manned submersible, instrumental in carrying out many explorative campaigns on life in deep ocean environments.

Colonies of *R. exoculata* were discovered in the surroundings of hydrothermal vents, where they are abundant (Martin and Haney 2005). The hydrothermal chimney environments of the Mid Atlantic Ridge are highly productive habitats, considered as real 'islands of life' in an otherwise relatively desert deep oceanic space. The shrimp represents the dominant member of a food chain not linked to photosynthesis but relying on primary chemoautotrophic bacteria (Polz and Cavanaugh 1996). The inhabiting communities depend on geothermal energy, reduced sulfur compounds and methane, which are released by the high-temperature and anoxic thermal springs originating in underground deeper layers. The areas around the hydrothermal vents (also described as black smokers and mounds) are covered by colonies of *R. exoculata*, which may reach enormous densities, up to $5 \cdot 10^4$ individuals per square meter. Differing from the deep Pacific hydrothermal invertebrate fauna of tubeworms and mussels, provided with their own endosymbiont communities (see previous section), shrimps like *R. exoculata* harbor filamentous ectosymbiotic bacteria, growing in very high densities on their extremities and on the inner surface of the carapace (Polz and Cavanaugh 1996). The epibionts fill an enlarged gill chamber, covering shrimps' filaments or setae (called bacteriophores) originating in the mouth. In spite of their pleomorphism, the majority of bacteria was initially found to belong to a single dominant species, a member of a monophyletic sulfur-oxidizing group of the ε-Proteobacteria division. The *R. exoculata* hosts colonize and occupy specific areas in which the vent reduced and the ocean oxygenated waters mix. This provides the bacteria with high amounts of energy and food, having access to both electron donors and acceptor substrates (Petersen et al. 2010). The shrimps feed in turn on the bacterial cells, which are ingested together with mineral sulfide particles (Polz and Cavanaugh 1996).

More recent studies showed a higher complexity of the shrimp epibiont community, as indicated by the presence of three metabolic signatures (for sulfur, iron and methane) (Zbinden et al. 2008) and by the occurrence of seven bacterial groups, two of which were the most represented (Petersen et al. 2010). A switch in epibionts species composition was observed when a second, dominant filamentous γ-proteobacterium ectosymbiont was found mostly associated to the eggs of *R. exoculata* (Guri et al. 2012). The diversity changes for the dominant cephalothorax epibionts – a type-I methanotrophic γ-proteobacterium morphologically distinct from the ε-proteobacterium – appeared to depend on specific recognition mechanisms and were followed in a chronological association throughout the host life-stages. When using 16S rRNA data, both bacterial groups appeared to dominate the epibiont community, the ε-proteobacteria sequences being associated to larvae and adults. These showed proximity to the 'hydrothermal invertebrates associated epibionts' (Marine Group-1) of sulfur-oxidizing chemolithoautotrophs. The bacteria were horizontally transmitted among the different oceanic sites examined, as shown by PCR amplification from sampled seawater (Guri et al. 2012).

The association of crustaceans with ectosymbiotic bacteria represents a successful evolutionary process, that occurred with similar traits also in other occasions. Bacteria phylogenetically related to those dominant in the *R. exoculata* community were also found on two crab species, *Kiwa hirsuta* and *K. puravida*, known as 'Yeti crabs' because of the dense layer of filamentous epibionts present on their clams, on which they evolved a grazing and feeding relationship (Thurber et al. 2011).

In mussels, extracellular thiotrophic symbionts have been reported from gill filaments of species living on sunken wood and whale decays. The bacterial cells are located extracellularly among surface microvilli (Fig. 3.6), in direct contact with seawater (Gros et al. 2007; Duperron et al. 2009). The symbionts possibly shelter the host cells from toxic reduced sulfur compounds, dissolved in the surrounding water. They are eventually endocytosed and digested in lysosomal cell structures (also referred to as bacteriocytes), possibly providing the host an additional source of nutrients (Gros et al. 2007). Extracellular thiotrophic ectosymbionts have been reported from Mytilidae collected at 2200 m depth from hydrothermal vents (McKiness et al. 2005) and from members of the bivalve family Thyasiridae (Dufour 2005; Dias Passos et al. 2007).

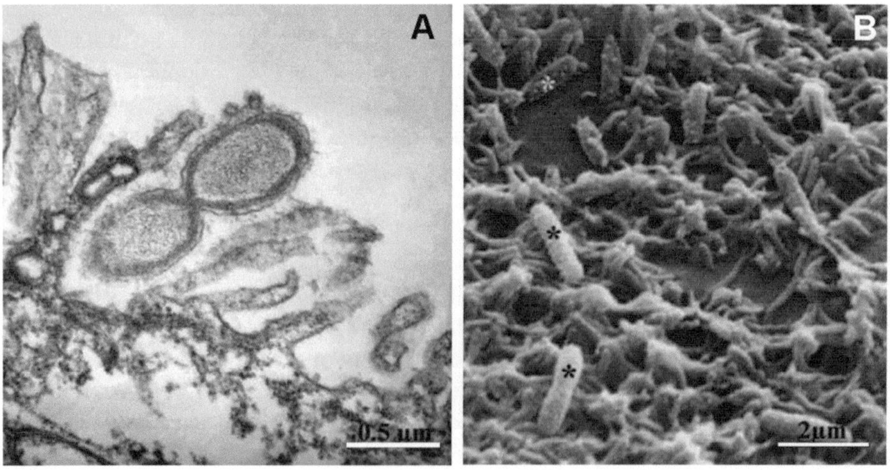

Fig. 3.6 TEM (**a**) and SEM (**b**) images of bacteria (*asterisks*) adhering to the microvilli on the external surface of the gill filament of the bivalve *Idas woodia* (Adapted from Gros et al. 2007)

Further examples of invertebrate-ectosymbionts associations include the deep sea tube-dwelling polychete *Alvinella pompejana* (Anellida), which harbors a cover on its dorsal epidermal surface made by a dense layer of epibiotic bacteria, mainly ε-proteobacteria (Campbell et al. 2003). Members of this bacterial community also include spirochetes (Campbell and Cary 2001). The polychete is highly thermotolerant and lives in extreme environments of the EPR, in proximity of chimneys formed around high-temperature hydrothermal vents. The hair-like filamentous or rod-shaped bacteria have a specific, obligate link with their *A. pompejana* host and are characterized by a chemolithoautotrophic metabolism. When studied with molecular methods, the bacterial community consistently showed the activation and functional expression of two genes – ATP citrate lyase and 2-oxoglutarate:acceptor oxidoreductase – of the reverse tricarboxylic acid (rTCA) cycle, involved in CO_2 fixation (Campbell et al. 2003). One member of the community was identified as the anaerobic, Gram-negative ε-proteobacterium *Nautilia profundicola* (Campbell et al. 2009).

Sponges show a wide array of bacterial associations, which elicited a particular interest in the scientific community due to the potential application in biotechnology and farmacology of a variety or biologically active compounds produced by their symbionts (Wagner and Horn 2006; Kennedy et al. 2007; Wang et al. 2013). In some cases, i.e. the so called "bacteriosponges", bacteria may account for around 40–50 % of the sponge biomass, with approximately 10^8–10^{10} cells per g of wet sponge tissue, a density several orders of magnitude higher than that usually found in the surrounding seawater. The taxa present in sponges are mainly unculturalble species from a wide array of phyla including *Acidobacteria*, *Bacteroidetes*, *Chloroflexi*, *Cyanobacteria*, *Nitrospira*, *Planctomycetes*, the *Candidate* phylum "Poribacteria", whose members are predominant among sponge symbionts, *Proteobacteria* and also Archaea (Vacelet and Donadey 1977; Webster et al. 2010). Apart of the production of food through the photosynthesis of *Cyanobacteria*, the functions attributable to sponge symbiotic species include waste disposal, detoxification and, in general, a defensive role through the production, mainly by *Actinobacteria*, of a wide range of metabolites with antibacterial, antifungal or cytotoxic properties (Chelossi et al. 2004; Kennedy et al. 2007; Wang et al. 2013).

2.3.1 Ectosymbionts Location

The host tissues and anatomical regions interested by ectosymbiosis are commonly related to the functional role of the activity charaterizing the guest bacterium and to the particular conditions of the extracellular environment in which most associations take place. For many marine invertebrates their ectosymbionts location is in fact directly related to the surrounding water conditions and to the availability of energy or food sources (see previous section).

Gut symbionts are often found in insects whose specialized feeding diets require the recruitment of organisms allowing the digestion of food with a poor nutritional quality, being rich in natural polymers normally difficult to digest, i.e. cellulose, wood or other similar materials. These associations may show a complex structure.

For example, the common wood-feeding termites (Isoptera) host hindgut microbial communities including symbiotic eukaryota (Protozoa) provided with externally attached ectosymbiotic spirochetes and hosting further, internal endosymbionts (Noda et al. 2003; Stingl et al. 2005). The unique phylogenetic identity of the protozoal endosymbiont allowed the erection of the new phylum "Endomicrobia" identified at the issue of 16S rRNA sequencing (Stingl et al. 2005). Termites, however, host a further cellulose degrading bacterial lineage, recognized as a subphylum of Fibrobacteres, in their intestinal content (Ransom-Jones et al. 2012).

Symbiotic bacteria of the gut and digestive tracts are indeed common among insect lineages, and in part underpin their evolutive success. The gut and other internal digestive structures are also an "hot spot" of bacterial diversity and can provide a precious source of new or rare taxa. This sheds light either on hosts evolution and on the genetic recombination of associated symbionts, taking place in a confined, niche environment (Dillon and Dillon 2004).

Among termites, *Reticulitermes* spp. harbor, on their hindgut wall, the filamentous and segmented cells of the genus *Arthromitus*, a monophyletic lineage of Gram-positive, spore-forming bacteria of the family *Lachnospiraceae*, and distantly related to *Clostridium piliforme* and *C. colinum*. The cell surface of *Arthromitus* is besides colonized by a further ectosymbiotic member of *Bacteroidales*. Apart of termites, members of these taxa are also found in the gut microbiota of cockroaches, scarab beetles and millipedes (Thompson et al. 2012).

Members of Rhizobiales and other groups of gut symbionts are commonly reported among ants, being likely functional to their herbivore feeding (Russell et al. 2009). Other examples include the symbiont of the gastric caeca of the stink bug *Nezara viridula*, vertically transmitted to the offsprings by females (Buchner 1965).

Many species-specific ectosymbionts are associated to earthworms. They are even considered as an experimental model system useful to the study of extracellular symbiosis (Lund et al. 2014). In *Eisenia fetida*, Gram-negative *Acidovorax*-like extracellular bacteria inhabit the earthworm excretory organs (nephridia) which are directly in connection with the outer body space. Earthworm nephridial bacteria include *Verminephrobacter* (β-Proteobacteria), a genus of common, species-specific and vertically transmitted ectosymbionts which have a long (>60 Myrs) history of association with their Lumbricidae hosts. The bacteria, located in the nephridial ampulla, provide the earthworm an osmoregulation and a nitrogenous waste excretory function. They are considered to sustain the host diet by providing vitamins and other nutrient factors or by playing a defensive role (Lund et al. 2014).

Other ectosymbiont associations are based on more complex microbial communities. For example, by applying pyrosequencing genome analyses, the microbial community present in the gut of the medicinal leech, *Hirudo verbana* (Anellida) revealed a complex structure based on 36 bacterial taxa, including the two most dominant symbionts – *Aeromonas veronii* and a *Rikenella*-like bacterium – plus members of *Proteus*, *Clostridium* and other less represented lineages such as *Eryspelothrix*, *Desulfovibrio* and *Fusobacterium* (Maltz et al. 2014).

Several chemolithoautotrophic sulfur-oxidizing (thiotrophic) bacteria have been reported from marine invertebrates, including members of the phyla Ciliophora, Polychaeta, Crustacea and Nematoda. On Stilbonematinae nematodes, one or more unculturable bacteria may be found as colonizers of the external cuticle body surface. The bacterial community is often composed by a single, dominant species, i.e. the γ-proteobacterium of *Laxus* sp., which is horizontally transmitted. Its cell adhere to the host by means of a dense extracellular matrix bearing D-mannose and L-rhamnose residues (Nussbaumer et al. 2004).

As a general rule, the host body structure affects the location of ectosymbionts. In sponges this location reflects the host organization, characterized by a sessile, filter feeder organism, in which water flows through specialized cells called porocytes, carrying food from the external environment (Fig. 3.7). The internal sponge

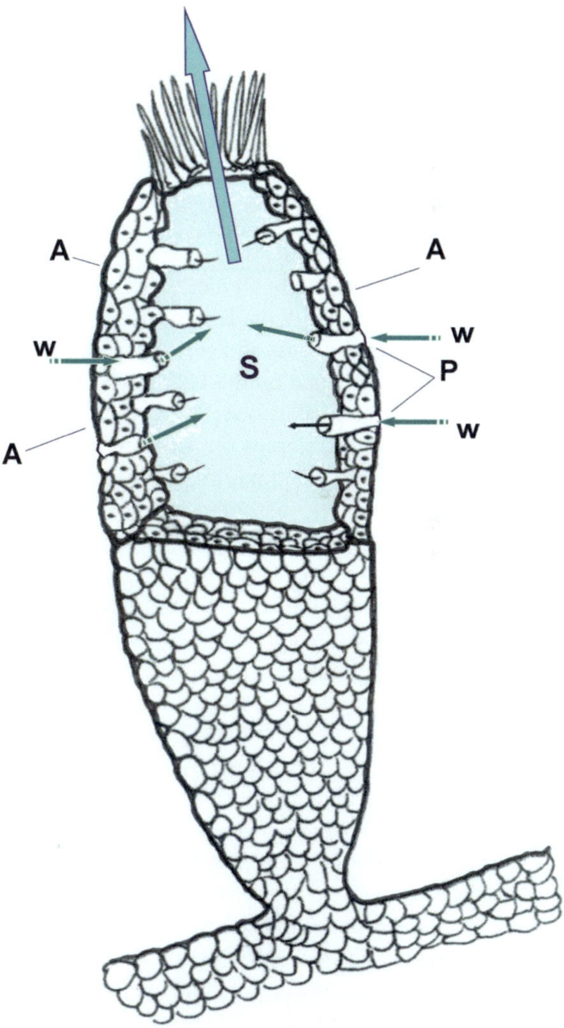

Fig. 3.7 A schematic drawing of a sponge structure, showing the water flow from the external environment (w, *small arrows*) reaching, through the porocytes (P), the internal space (mesohyl, S) in which the archaeocytes (A) perform bacteria phagocytosis. Water is propelled outward (*larger arrow*) by flagellated cells through the apical, opercolated aperture (Re-drawn from Kennedy et al. 2007)

cavity, the mesohyl, is a region characterized by bacteria phagocytosis, performed by a layer of cells called the archaeocytes. On this cell layer many species may become established, especially in larger sponges, remaining either externally in the mesohyl matrix, or internally in bacteriocytes. In many sponges, photosynthetic bacteria may also be found on the outer body surface exposed to light, whereas the mesohyl harbors mainly autotrophic or heterotrophic species (Kennedy et al. 2007).

3 Functionality of Symbiosis

Webster (2014) listed four fundamental factors that characterize the persistent, evolutionary success of a symbiotic relationship, namely: (1) the host and symbiont specificity, (2) the mechanisms of acquisition, development and maintenance, (3) the presence of functional mechanisms eventually yielding one or more benefits to the symbionts and (4) the host response. The first three factors will be treated in this section, the latter, given its structural nature, will be discussed in Sect. 4.

3.1 Specificity

There is a general agreement about the origin of many endosymbionts from invasive bacteria. This hypothesis considers the evolution of a strict host relationship arising from ancient parasitic events and is supported, for vertically transmitted bacteria, by their lifecycle dependence on the host body environment. Some symbiont lineages evolved to get established as beneficial or mutualistic organisms, developing a variety of multiplication strategies, in some cases highly sofisticated like those involving i.e. host cytoplasm incompatibility (CI) (Weeks et al. 2007). Other routes leading to specialized symbiotic associations possibly originated from phagocytosis or peculiar host feeding behaviours. In many cases the specialization and host specificity adaptations that occurred several Myrs ago can still be putatively reconstructed using indirect evidence from processes like CI, bacterial genome resizing or presence of specific biochemical signatures.

Although cospeciation and vertical transmission do not necessarily imply a high degree of host specificity and specialization, as shown by processes like host switching, multiple hosts or symbiont substitutions, specificity and specialization traits are common among invertebrate endosymbiosis. Specificity is intimately linked to the induction of direct effects on the host biology, as shown by i.e. the host reproductive manipulation (see Sect. 4), or may arise from and/or sustain the host adaptation to new ecological niches. It may reinforce symbiotic relationship when hosts face significant changes concerning their ecology and evolution, i.e. when adapting to new habitats or behaviours. For example, a new clade of γ-Proteobacteria found in bat flies of the genus *Nycterophilia* (Diptera: Streblidae) showed an ancient, obligate relationship with their hosts, as indicated by high prevalence, presence in pupae and phylogenetic host codivergence with duplications and losses. Specificity allowed the endosymbiont to adapt to the blood feeding bat flies when these switched from cave habitats with a broad temperature range required for life-cycle completion, to a narrower range of hot caves and roosts. Analyses of different endosymbiont clades showed their secondary adaptation to the bat flies, when these followed the bats in new hot cave habitats (Morse et al. 2012).

There is evidence that specialization and specificity may provide higher host benefits. To demonstrate that selective benefits were associated to specialization, comparative experimental assays were performed on sponges and symbiotic cyanobacteria, by controlling in situ the bacterial photosynthesis through shading. Data on the sponges biomass showed that more specialized symbionts, like the cyanobacterium *Oscillatoria spongeliae* from the marine sponge *Lamellodysidea chlorea*, provided a greater benefit, when compared to the generalist cyanobacterium *Synechococcus spongiarum*, hosted in the sponge *Xestospongia exigua*. Furthermore, the *O. spongeliae* abundance was not affected by shading, indicating that the increased level of host benefit was linked to the bacterium reduced dispersal and higher specificity (Thacker 2005).

The correlation between specificity and benefits of symbiosis is indicative of an evolutive path in which the relationship is mutually reinforced and plays a fundamental role in the host ecology. Insects obligate symbioses provide many examples of host specificity and bacterial specialization, conferring or reinforcing key fitness traits. Using microinjection, Moran and Yun (2015) were able to transfer cells of a heat-resistant *Buchnera* endosymbiont in a *A. pisum* matriline hosting heat-sensitive bacterial mutants. Prior to the transfer, the latter symbionts had been removed from the recipient aphid line through repeated exposure for a few hours to lethal temperatures (34–35 °C). Some endosymbionts could be introgressed in the survivors' recipient embryos and could be inherited. Their persistance in the following generations provided a remarkable experimental demonstration of a symbiont's impact on the hosts ecology, as they conferred a higher tolerance to heat.

The link between benefits and specificity does not characterize, however, all endosymbiotic associations, being evident in vertically transmitted but less common among horizontally transmitted symbioses. The type of ecological contribution to the host fitness provided by symbionts is fundamental at this regard, since a lack of specificity may reflect a different host and symbiont ecology. An example is given by deep sea bone worms of the genus *Osedax* (Polychaeta: Siboglinidae). These invertebrates use some particular anatomical structures (also known as "roots") to colonize and degrade sunken bones from whale carcasses or other floor organic materials. *Osedax* spp. lack mouth and digestive tracts and environmentally acquire primary and secondary endosymbionts. The diversity of the primary heterotrophic *Neptunomonas* (γ-Proteobacteria: Oceanospirillales), from bacteriocytes located in the tissues around the ovisac and proliferative roots, showed that they are acquired through a non-specific recruitment process, known only in part (Salathé and Vrijenhoek 2012).

3.2 Acquisition

Different mechanisms characterize the acquisition of symbiotic bacteria and their persistence through the host generations, reflecting the mode of transmission, the ecology of either the hosts and bacteria, the symbionts location inside the host organism as well as their functional role. Three main modes of acquisition are recognized: vertical or horizontal transmission and horizontal-lateral acquisition from another host. Further MMT (mixed mode) mechanism have been also recognized as widespread among invertebrates (Ebert 2013; Szafranski et al. 2014). MMT has been reported for *Hamiltonella defensa* and *Regiella insecticola* (Enterobacteriaceae) in the aphid *A. pisum*, *Paenibacillus larvae* in *Apis mellifera Wolbachia* sp. in parasiotic wasps, *Rickettsia* spp. in many arthropods or *Borrelia duttoni* in ticks (for a review see Ebert 2013). MMT may have dramatic effects on the ecology and epidemiology of the hosts, increasing the chances of environmental persistance and spread, given the links of horizontal and vertical transmission modes with the host densities and fecundity, respectively. Both processes also interact with the host evolution (Ebert 2013).

In insects with bacteriocytes, vertical transmission shows a variety of processes that take place mostly during oocyte or embryo development. In *A. pisum* the transmission of *B. aphidicola* cells proceeds through the release of maternal bacteriocytes from the ovaria into the extracellular space, in which they are later endocytosed by the posterior syncytial cytoplasm of the developing blastula (Koga et al. 2012).

The persistance of male killing bacteria depends entirely on the manipulation of the host reproductive biology (Hurst and Jiggins 2000). Members of *Wolbachia* are maternally inherited as cytoplasmic endosymbionts widespread in many arthropod lineages, including insects. One of the most remarkable traits of *Wolbachia* endosymbiosis is its capability to interfere with the host reproductive biology. In many lineages these bacteria are localized within the host reproductive tissues or cells. As other reproductive parasites, *Wolbachia* evolved sophisticated strategies to infect the host reproductive tissues and mainly colonize the ovary and oocytes, reaching the egg germ pole to increase its persistance though germline colonization (White et al. 2013). The interference of *Wolbachia* with the host reproduction is characterized by a variety of outcomes related to the host, including the induction of parthenogenesis (Hymenoptera), male feminization (Isopoda) or male killing, as well as the CI insurgence, often observed in insects (Table 3.2) (Rokas 2000; Saridaki and Bourtzis 2010; White et al. 2013). Horizontal transmission and adaptation to host populations were also reported for *Wolbachia* infection of terrestrial isopods (Rigaud et al. 2001).

Transmission by feeding was experimentally demonstrated for *Ca.* "Hepatoplasma crinochetorum", endosymbiotic in the hepatopancreas of the freshwater isopod *Asellus aquaticus* and environmentally transmitted to the progeny (Wang et al. 2007). Insect endosymbionts like *Rickettsia* and *Spiroplasma* require a mix of vertical and horizontal transmission mechanisms, the latter based on acquisition when the host feeds on an intermediate vertebrate or plant.

Table 3.2 Mechanisms of reproductive manipulation by endosymbionts

Endosymbionts	Mechanism
Wolbachia, Cardinium	Cytoplasm incompatibility between gametes and post-zygotic sterility
Wolbachia, Rickettsia, Spiroplasma, Flavobacteria, Arsenophonus	Male killing through targeted death of male progeny
Wolbachia, Cardinium	Feminization, converting genetic males into functional females
Wolbachia, Cardinium, Rickettsia	Parthenogenesis, by production of asexual daughters

Oviposition is a further mechanism through which vertical transmission may take place. Pentatomid stinkbugs females harbor endosymbiotic γ-Proteobacteria in their midgut caecum. In *Sibaria englemani*, newly hatched first instar nymphs aggregate on the egg masses, probing the egg chorion to acquire the bacterial symbiont, through maternal transmission at oviposition. Experimental assays showed that the removal of nymphs soon after hatching interfered with transmission and halted bacterial acquisition. The assay also showed that symbiont-deprived insects lost some benefits of symbiosis, with a prolonged second instar stage and reduced growth rates, coupled to other morphological changes. The smearing of eggs with other bacteria also proved the possibilty of symbiont contamination and substitution mechanisms (Bistolas et al. 2014). Further examples of acquisition via oviposition include the symbiont of the gastric caeca of the stink bug *N. viridula*, which are vertically transmitted to the offsprings by females that smear the bacterial cells on the egg after release, for later acquisition by the hatching larvae when probing the egg surface (Buchner 1965).

In the earthworm *E. fetida* a more complex transmission mechanism occurs at oviposition. The *Acidovorax*-like extracellular symbionts are acquired during the early developmental stage, being deposited into the cocoon, an external capsule enclosing the eggs. The cocoons are formed after mating, by a polymerized secretion originated from a tube-like organ, the clitellum. The symbionts colonize the nephridia of juveniles before hatching, through specific transient embryonic ducts (not through the nephridiopores) formed during nephridia development. During maturation of the earthworm segments, the bacterial cells get established through their flagellar and twitching motility. Subsequently, the recruitment canal is closed, whereas the nephridiopore opens (Davidson and Stahl 2008; Lund et al. 2014).

Bacterial acquisition in marine invertebrates may occur through horizontal transmission from surrounding waters, as shown for the squid *E. scolopes* and its bioluminescent symbiont *V. fischeri*, acquired from seawater at the early larval stage (Visick and Mcfall-Ngai 2000). Similarly, also the γ-proteobacterium symbiont adhering to the external cuticle surface of the nematode *Laxus* sp. is horizontally acquired from the surrounding marine environment (Nussbaumer et al. 2004). A variety of transmission modes was also observed in other lineages, including bivalves in which the symbiont acquisition mechanism are family-dependent. The Lucinidae, for example, show horizontal acquisition from surrounding waters, whereas vertical transmission mechanisms have been documented in *Isorropodon bigoti* (Vesicomyidae). In this clam, sulfur-oxidizing symbiotic bacteria are hosted within gill epithelial cells and occur intracellularly in female gametes throughout all maturation stages, that transfer the symbiont to the embryos and larval stages (Szafranski et al. 2014). Other mechanisms of transmission include ingestion, as shown for the bivalve larvae of *Solemya reidi* (Solemyidae) (Gustafson and Reid 1988; Cavanaugh et al. 2013).

3.3 *Benefits of Symbiosis*

A basic framework in the study of the symbiosis evolution is that such partnerships have been driven and favoured by one or more benefits improving fitness and/or yielding selective advantages for at least one of the organisms involved. The benefits linking hosts and guests may be classified in categories mostly related (but not limited) to nutritional and defensive functions. The discoveries concerning the biochemical and molecular traits of the nutritional basis of symbiosis exerted a broader impact on science. The consolidated mechanisms of selection and evolution have been enriched by the observation of new processes like gene losses and genome reductionism (Moran 2002; Gil et al. 2003). Many studies reinforced current views about the effects of evolutive convergence and finally provided new perspectives in the study of molecular and cellular communication (Webster 2014).

The benefits ascribed to symbiosis include many examples, like the possibility to colonize or spread within new ecological niches or environments, or the adaptation to strict, poor trophic conditions. It is also accepted that in many evolutionary paths the initial symbiotic relationships may have changed from one type to another, i.e. shifting from parasitism to commensalism or mutualistic associations, in some cases indispensable to warrant the host survival.

3.3.1 Nutrition and Metabolism

Enriched trophism is one of the most studied benefits recognized among those yielded by symbiosis. During the evolutionary history of many invertebrate lineages, symbiosis has provided a successful key for the colonization of previously unaccessible environments by improving host nutrition and fitness. In these relationships bacteria conferred their hosts one or more metabolic advantages allowing i.e. the integration of a deprived diet, the use of poor substrates (for example cellulose) or the progressive colonization of extreme habitats, characterized by particular chemical compositions.

In terrestrial habitats, many insects integrate their strict and incomplete diets with products synthesized by their endosymbionts. The symbiotic relationship is often fundamental for the survival of the species involved, and any interference may lead to fatal outcomes for the host. This strict dependence was revealed by the deleterious effects, leading to infertility or even death, observed after treating insects with antibiotics which eliminated their endosymbiotic bacteria. Similarly, ectosymbionts provide their host direct or indirect nutritional benefits, either as food, when hosts graze on the bacterial colonies they harbor, or by transforming a substrate to make it available for the host metabolism, i.e. in the case of termites. In deep sea hydrothermal vents, both ecto- and endosymbiotic relationships are equally effective in sustaining primary productions from chemautotrophic metabolism, relying on the reduction of sulfur, iron, methane or hydrogen.

Nutrition is one of the most important functions recognized to bacteriocytes, in which endosymbionts provide essential amino acids or vitamins, necessary to integrate unbalanced host diets. This relationship is fundamental and can be found among many distinct insect lineages. These include not only plant pests (i.e. aphids and other phloem suckers), but also wood consumers, insects of public health importance (cockroaches), blood suckers (bed bugs), vectors of human diseases (i.e. the tsetse fly *Glossina*) or ticks (Douglas 1998).

Many symbiotic associations yielded functional, coevolutive adaptations steering the trophism and specialization of the species involved. For a number of primary symbionts of insects (γ-Proteobacteria) these processes are often characterized by the lack of molecular products or biochemical patterns. This effect is shown by the loss of functional genes or of entire chromosome parts, as revealed by the sequenced genomes (Pérez-Brocal et al. 2006; Dale and Moran 2006). This trend is often flanked by the production of specific metabolites, i.e. essential amino acids used by the host and/or by the complete dependence on the host for other metabolic needs (Nakabachi et al. 2005; Douglas 2014a). Metabolites exchanged with hosts include vitamin B, needed to integrate the diets of blood feeding insects (Douglas 2014b).

Studies on bacteriocytes transcriptome shed light on their metabolic role and on the peculiar evolutive adaptations characterizing these structures. For example, the relationship between *Buchnera* and its aphid host is very strict, both organisms depending on each other for survival (mutualistic symbiosis). During its evolutive history in the aphid lineage the bacterium lost several key biochemical pathways, including genes coding for non-essential amino acids or phospholipid biosynthesis. This genomic shrinkage limits its ability to synthesize cell membranes, depending on phospholipids acquired by the host through the bacteriocyte vesicular transport system. The cells of *Buchnera* are in fact surrounded by a membrane of host origin, which acts as an interface along which metabolic exchanges occur (Douglas 1998).

The genes expressed in the *Buchnera* bacteriocytes of a parthenogenetic clone of the pea aphid *A. pisum* included transcripts related to amino acids metabolism, as well as genes coding for the biosynthesis of amino acids that the bacterium cannot produce. Experimental data showed that some genes coding for essential amino acids and those related to the synthesis of nonessential ones were highly expressed in the bacteriocyte, confirming the *Buchnera* involvement in metabolic exchanges between the host and its endosymbiont. Other expressed genes were related to transport and bacterial cell wall degradation or to intermediate metabolites in the amino acid synthesis. Mitochondria-rich bacteriocytes also showed a high level of mitochondrial activity, always related to the synthesis of amino acids, a process requiring high energy levels. Transcriptome analyses confirmed that the bacterium activity is mainly, but not only, centered on the production of essential amino acids which are absent or scarce in the phloem sap assumed by the aphids when feeding on plants. *Buchnera* in turn receives back from the host some nonessential amino acids and other fundamental compounds, necessary for its survival (Nakabachi et al. 2005; Douglas 2014b).

Using chemically defined diets with or without essential amino acids and treatments with antibiotics, Feldhaar et al. (2007) could experimentally demonstrate a fundamental nutritional role for *Blochmannia*, a genus of primary symbionts from the omnivorous carpenter ant *C. floridanus*. Also in this case, in spite of its reduced genome, the bacterium provided the host with essential amino acids, with an additional role in recycling N-based metabolites through the production of functional urease.

The advantage deriving by a direct diet enrichment is not, however, the unique mechanism of trophic benefit that can be achieved by invertebrates through symbiosis. The mutualistic Gram-negative *Xenorhabdus* and *Photorhabdus* spp. (γ-Proteobacteria) have a complex relationship with their entomopathogenic host nematodes (*Steinernema* and *Heterorhabditis* spp., respectively). The bacteria play an indirect nutritional role as food source in the life cycle of their hosts that inoculate them into target insect preys. The septicemia originated by the infection and the eventual bacterial multiplication in the preys is followed by a nutrition phase of the nematodes, which complete their life cycle feeding on the bacterial cells filling the insect cadavers (Forst et al. 1997; see also Chap. 6).

Nutritional benefits of symbioses may also derive from complex bacterial communities integrating members of other Kingdoms. The rich gut microbiota of termites include eukaryotes like several species of flagellates and hundreds of unculturable bacterial and archaeal species. Studies with advanced molecular techniques, including genomic and transcriptomic approaches, revealed a main contribution of gut bacteria in higher termites (Termitidae) concerning cellulose and hemicellulose digestion. This family of wood-feeding termites lacks cellulolytic gut flagellates and shows extremely high pH levels in midgut and hindgut compartments with varying oxygen contents. In the hindgut environments cellulolytic activity is attributed to spirochetes of the genus *Treponema* and to uncultured Fibrobacteres, involved in the hydrolysis of lignocellulose, together with other pentosidase, carbohydrate and chitin catalytic activities.

Members of *Treponema* are also the main bacterial producers of H_2 which is in turn removed by further reductive *Treponema* acetogenic metabolism and by methanogenic archaea, i.e. *Methanobrevibacter*. Further activities of *Treponema* include N_2 fixation. This reaction is also carried out by the bacterial symbionts of gut flagellates, that are mainly responsible for the anaerobic digestion of lignocellulose in the other six families of phylogenetically basal lower termites (Hongoh 2011).

In marine environments, the dual sulphur and methane metabolism of mussel endosymbionts provides an example of trophic benefit acquired through chemiosynthetic symbionts. The advantage is linked to the variable amounts of substrates available in the deep sea environments as shown by *Bathymodiolus* spp., that rely either on C conversion from methane and energy derivation from sulphur oxidation (Duperron et al. 2009).

In the marine oligochaete *Olavius algarvensis*, more complex bacterial communities show the effect of the host ecology on the evolution of symbiosis. Metagenomic analyses showed the dependence of the host (lacking mouth, gut and anus, and with a reduced nephridial system) on four distinct extracellular bacterial endosymbionts (two γ-Proteobacteria and δ-Proteobacteria) located below its cuticle. Metagenomic analyses showed that the symbionts are involved in autotrophic CO_2 fixation through sulphur-oxidizing or heterotrophic sulphate-reducing metabolism. Coupled to the nitrate reduction under anoxic conditions and use of sulphur compounds or hydrogen as energy sources, the lysed symbiont cells yield the food sources needed by the host to acquire its nutrients under different circumstances, as it changes location between oxidized and reduced sediment environments (Woyke et al. 2006).

3.3.2 Host Defense

A fundamental category of functional benefits provided by symbionts concerns host defensive effects which are ubiquitous among invertebrates, based on a variety of interactions like the antagonism towards parasitoids or predators, lower prevalence of pathogens or parasites and higher fitness or resistance levels in response to environmental stress.

Maternally inherited endosymbionts like *Wolbachia* and *Spiroplasma* have been studied in *Drosophila* for their defensive role. Both bacteria are known for the manipulation of the host sex ratio that increases their persistance among offsprings. Lineages not interfering with host reproduction revealed a different mechanism of persistance, related to the insect protection from natural enemies. *Spiroplasma* are extracellular endosymbionts that occupy the host hemolymph. In *Drosophila melanogaster* the presence of this symbiont was correlated to an increased survival to attacks from a parasitoid wasp (Xie et al. 2010). *Spiroplasma* was also observed to restore the host fertility in *D. neotestacea* after infection by a nematode parasite, reducing subsequent transmission (Jaenike et al. 2010).

In *Drosophila* spp. *Wolbachia* showed inhibition of viral replication and higher resistance to infection with reduced mortality rates, although limited to RNA viruses (Hedges et al. 2008; Hamilton and Perlman 2013). *Wolbachia* also reduced infection by the dengue virus in *A. albopictus* cell lines (Moreira et al. 2009; Frentiu et al. 2010). The *Wolbachia* strain *wMel* was shown to block transmission of the dengue serotype 2 and invaded *A. aegypti* populations in cage assays, suggesting a potential for disease management (Walker et al. 2011).

Insect protection from parasites has also been reported in other bacterial lineages, i.e. Enterobacteriaceae. As facultative secondary endosymbiont of aphids, *Hamiltonella defensa* protects *A. pisum* from the parasitoid wasp *Aphidius ervi*. Experimental data from parasitism bioassays showed that the wasp is killed before completion of its lifecycle by a toxin-encoding and lysogenic lambda-like bacteriophage. The symbiont reduced wasp infection by 90 % and allowed host survival and reproduction. This mechanism appeared likely present also in other aphid species. It allowed a multiplication advantage and selective spread in natural populations either for the symbiont and host, in spite of processes inducing secondary symbiont loss that also occur in nature (Oliver et al. 2009).

Two morphologically distinct clades of the secondary symbiont *Ca.* "Serratia symbiotica" were found in members of the aphid subfamily Lachninae as primary or secondary symbionts (Lamelas et al. 2008). As primary endosymbionts they are involved in a nutritional provision to hosts through the synthesis of amino acids. As secondary endosymbiont the bacterium is located extracellularly in a separate bacteriocyte distinct from that of the primary endosymbiont *Buchnera*. In *A. pisum* the symbiont protects the host from the wasp *A. ervi* (Oliver et al. 2003, 2005) and also confers higher tolerance to thermal stress (Montllor et al. 2002).

Also a specific isolate of *Regiella insecticola* showed protection of the peach-potato aphid *Myzus persicae* from parasitic wasps (Vorburger et al. 2010). Host protection exerted by this bacterium is not limited, however, to wasps. Other *R. insecticola* strains protect *A. pisum* from the specialized fungal pathogens *Pandora neoaphidis* and *Zoophthora occidentalis*, but not from the parasitic hyphomycete *Beauveria bassiana* (Scarborough et al. 2005; Parker et al. 2013).

The release of toxic compounds provides a direct protection of hosts, as in the carnivore beetle *Paederus fuscipes*, which protects its progeny by transferring the toxin pederin into the eggs. Pederin is released in the hemolymph by an endosymbiotic bacterium related to *Pseudomonas aeruginosa*. It is effective in deterring predating wolf spiders and is produced by the bacterium through a gene cluster acquired via HGT (Kellner and Dettner 1996; Piel et al. 2004).

Chemical defensive processes are not limited to insects but also occur in other Arthropoda, i.e. the shrimp *Palaemon macrodactylus* and the lobster *Homarus americanus*. Both crustaceans are protected from the pathogenic aquatic fungus *Lagenidium callinectes* by means of antifungal metabolites released by their bacterial epibiotic symbionts (Gil-Turnes et al. 1989; Gil-Turnes and Fenical 1992). *Santia* spp. (Isopoda) settle in suitable habitats in which they cultivate and consume environmentally acquired ectosymbiotic cyanobacteria that render them unpalatable, thus allowing protection from fish predation (Lindquist et al. 2005). A further example of predators avoidance is given by the bioluminescence of *Vibrio fischeri* that helps its host squid *E. scolopes* to escape predation (see Sect. 2.2.1).

Benfits yielded by endosymbionts also concern effects induced on the host life-stage and survival. A *Spiroplasma* sp. protects the fly *Drosophila neotestacea* from the castrating effects induced by infection of the roundworm *Howardula aoronymphium* (Jaenike et al. 2010). In the filarial nematode *Onchocerca ochengi* a higher longevity is induced by a *Wolbachia* endosymbiont by diverting the potentially lethal eosinophils produced in the immune response of the nematode parasitized animals (Hansen et al. 2011).

3.3.3 Detoxification and Other Functional Benefits

Blochmannia spp. mutualist in Camponotini ants are located in midgut specialized cells and play a direct, nutritional role. They are also involved in the detoxification of urea that is recycled to glutamine through the activity of two encoded enzymes, urease and glutamine synthetase. The latter gene was, however, absent in the sequenced genome of *B. vafer*, suggesting alternative mechanisms of ammonia detoxification including assimilation by the host though a Class II glutamine synthetase or its use as N donor in the carbamoyl phosphate synthase pathway, leading to arginine biosynthesis by the host cells (Williams and Wernegreen 2010).

Other examples of benefits related to detoxification processes include the gut bacterial communities of the bark beetle *Dendroctonus ponderosae*. They include members of *Pseudomonas*, *Serratia*, *Rahnella* and *Burkholderia* which express genes active in the complementary degradation of a variety of toxic terpenes, released by the conifers during beetles attack and wood penetration (Adams et al. 2013; Six 2013).

Wolbachia was observed to interfere with iron metabolism in insects, lowering concentrations levels damaging cells through oxidative stress and apoptosis (Kremer et al. 2009). Members of this lineage also induced higher fecundity in *Trichogramma bourarchae* (Vavre et al. 1999), increased fecundity and longevity in *A. albopictus* (Dobson et al. 2002) and higher lifespan in *D. melanogaster* (Fry and Rand 2002).

Host waste disposal is performed by the specialized extracellular symbionts of the marine oligochaete *O. algarvensis*, which provide to the uptake and recycling of ammonium, urea and other host fermentation waste products such as dicarboxylate succinate, monocarboxylate acetate and propionate (Woyke et al. 2006).

4 Adaptive Processes

4.1 Reproductive Manipulation

Endosymbionts that uniquely multiply through the female reproductive system (egg cytoplasm) induce a shift in the offsprings sex ratios leading to higher frequencies of females or even to parthenogenesis. From an evolutionary perspective this shift is advantageous for the symbionts since the host manipulation allows higher reproductive rates. It has, however, dramatic effects on the host biology even leading to the insurgence of new speciation patterns. Many examples of this behaviour can be observed among Arthropoda or Nematoda, involving a variety of bacterial lineages (Vandekerckhove et al. 2000). The most studied and widespread is *Wolbachia*, observed in a broad host range of insects, crustaceans, spiders and mites (Breeuwer and Jacobs 1996; Johanowicz and Hoy 1996; Tsagkarakou et al. 1996; van Borm et al. 2003; Cordaux et al. 2011; White et al. 2013). Apart of CI, in which infected males cross with uninfected females, reproductive manipulation mechanisms include male killing, feminization of genetic males and parthenogenesis (Table 3.2).

Members of the *Wolbachia* group have been found in diverse invertebrate phyla. In Nematoda, they were found in the animal parasitic family Onchocercidae and in the plant parasitic nematode *Radopholus similis* (Lo et al. 2007; Werren et al. 2008; Haegemann et al. 2009). Their role as reproductive parasites of nematodes has not been, however, clarified yet.

The adaptation of *Wolbachia* to insects is the result of a complex and diversified evolution and involves many types of associations, ranging from parasitism to mutualism (Dedeine et al. 2001). In some cases the bacterium is also associated to the presence of other additional symbiotic bacteria. As for many other symbionts, it is generally assumed that *Wolbachia*, at the issue of the different types of associations, receives a benefit or reproductive gain by its activity inside the host. The different mechanisms characterizing the wide range of relationships and the resulting outcomes, however, may vary from host to host, and detailed knowledge is far from being complete. Similar considerations hold for the phylogenetic status of this complex of species, which in some associations has also been considered as horizontally transmitted (Plantard et al. 2012; Frost et al. 2014).

The taxonomy of *Wolbachia* is debated, some authors considering the bacterium as a single species (*W. pipientis*), while others suggest a polyphyletic complex of species (Lo et al. 2007; Pfarr et al. 2007). Molecular phylogenetic studies showed the presence of several "supergroups" in the *Wolbachia* evolutive radiation, partially identified by their associations to different host taxa (Bordenstein et al. 2004). The occurrence of abundant and widespread bacteriophages, mediating intracellular HGT was also demonstrated (Bordenstein and Wernegreen 2004).

Drosophila spp. have been actively investigated to understand the effects of *Wolbachia* sp. on reproduction and related mechanisms of action. In many cases the bacterium acts at the cytoplasmic level as a biological reproductive barrier among distinct host populations or between closely related species that recently diverged (Rousset and Solignac 1995; Rokas 2000).

Other cases of insects whose reproductive biology is affected by *Wolbachia* spp. through induced parthenogenesis include *E. formosa*. This parasitoid is thelytokous,[3] and is a suitable biocontrol agent that attacks the homopteran white fly *Trialeurodes vaporariorum*. In *E. formosa*, which is characterized by few occasional, sterile or non-functional males, treatments with the antibiotic tetracycline eliminated the *Wolbachia* symbionts, inducing the appearance of males whose insemination, however, was not successful (Zchori-Fein et al. 1992). A similar reaction had been previously observed also in the parasitoid wasps *Trichogramma* spp., in which antibiotic treatments induced the appearance of males, restoring in this case a permanent bisexual arrhenotokous[4] reproduction (Stouthamer et al. 1990). In both cases, the causal agent responsible for the absence of males were demonstrated to be the *Wolbachia* endosymbionts, which in *Trichogramma* were found to act through dominant nuclear effects (Russell and Stouthamer 2011). Although the induction of a partenogenetic life-cycle may have a positive impact on the rearing of parasitoids for exploitation in biological control and/or reduce the time spent by females in

[3] *Thelitoky* = a form of parthenogenesis in which female offsprings are produced from unfertilized eggs.

[4] *Arrhenotoky* = sexual reproduction in which haploid males emerge from unfertilized eggs.

mate searching (Stouthamer 1993), the absence of males may also represent a drawback, reducing host sexual recombination and fitness (Zchori-Fein et al. 1992).

A further example of interference with the host reproduction is given by the obligatory and highly specific mutualistic association of a *Wolbachia* sp. to the hymenopteran parasitic wasp *Asobara tabida*. In this case the presence of the bacterium appeared necessary for the successful completion of the host oogenesis, as demonstrated by antibiotic-based assays (Dedeine et al. 2001).

4.2 Host Adaptive Changes

Many endosymbiotic bacteria acted as direct evolutionary forces, affecting the host anatomy or speciation, often in an irreversible way. In some associations the endosymbiotic bacteria induced the evolution of particular anatomical outcomes in their recipient organisms, i.e. the changes observed in the marine invertebrate *Astomonema* (Nematoda) species. Members of this taxon are characterized by adults deprived of oral and anal apertures, provided with a reduced, non functional gut. Their nutrition depends on the metabolic activity of endosymbiotic sulfur-oxidizing bacteria (γ-Proteobacteria, Chromatiaceae) that fill their body (Musat et al. 2007).

Many anatomical adaptations shown by associations of this kind can also be found in other, distinct invertebrate radiations, leading to similar functional outcomes. At the issue of a convergent evolutive process, chemoautotrophic, sulfur-oxidizing bacteria adapted to live as endosymbionts in the gutless oligochaete *I. leukodermatus*, which also host a further, smaller endosymbiotic bacterial species (Krieger et al. 2000) (Fig. 3.8).

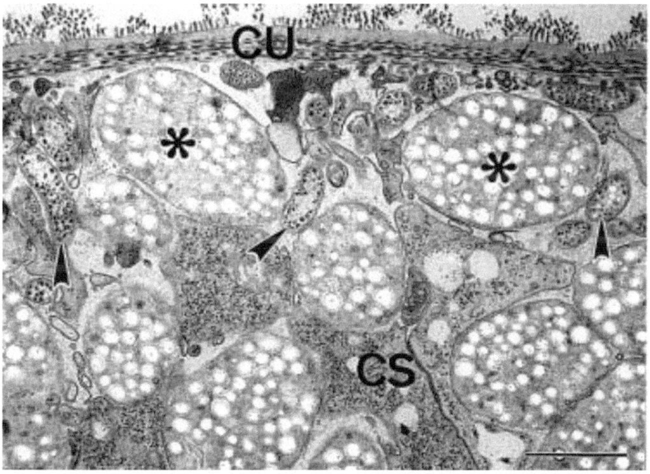

Fig. 3.8 Transmission electron micrograph of a cross section of the annelid Inanidrilus *leukodermatus* showing large (*asterisks*) and small (*arrowheads*) endosymbiotic bacteria within the subcuticular (CU) cytoplasm (CS) (*Scale bar* = 1 μm; adapted from Krieger et al. 2000)

A similar, convergent evolutive outcome was observed also in marine platyhelmints of the genus *Paracatenula*, that are deprived of mouth, pharynx and gut. In these organisms, that live in shallow-water subtidal sands, the majority of body cells are bacteriocytes that adapted to harbor intracellular bacterial symbionts, originating a tissue called trophosome (Fig. 3.3). The nematode α-proteobacterium symbiont, Ca. "*Riegeria*", is a thiotrophic, sulfur-oxidizing species that provides all needed nutrients to the *Paracatenula* host, gaining from this association a stable and sheltered environment, suitable for its trophic demand (Leisch et al. 2011).

The light organ of the squid *E. scolopes* has a complex anatomy, which evolved at the issue of its association with the bioluminescent *V. fischeri* that it contains. This organ develops after host colonization by the bacterium from surrounding waters. The bacteria are extracellular but in intimate contact with host microvilli in crypts aligned with columnar epithelial cells. The organ is completed by a reflector tissue and is provided with a lens and ink sac, in order to modulate the light emission under host control (McFall-Ngai and Montgomery 1990; Visick and McFall-Ngai 2000).

4.3 Genome Shrinking

One of the most relevant evolutive traits shown by endosymbiotic bacteria is the genome size reduction, resulting in a functional dependence on recipient cells metabolism (Canbäck et al. 2004; Delaye and Moya 2010). This process is often characterized by a nucleotidic composition different from closest non-symbiotic siblings and by higher rates of evolution (Moran and Wernegreen 2000) or even by differences from close endosymbiotic species, affecting the bacterium-host by products (Williams and Wernegreen 2010). Reduced genomes also display, when compared to free living non-symbiotic bacteria from the same lineages, a more conservative evolution with a lower number of insertion or deletions and lower frequencies of genetic rearrangements related to HGT or nonsynonymous nucleotide substitutions (Tamas et al. 2002; Canbäck et al. 2004).

The complementarity of the host and guest metabolism is reflected in the eventual gene losses and genome resizing. Host environment, alimentary habits with food source and host ecology are main drivers affecting gene retention and losses (Rio et al. 2003). Gene losses and genome reductions show the adaptation to new metabolic and ecological niches, as well as the consolidation of novel evolutive paths, including peculiar life condition on minimal gene pools, required to keep the endosymbiont active and functional within the host cell environment (Canbäck et al. 2004; Delaye and Moya 2010).

References

Adams, A. S., et al. (2013). Mountain pine beetles colonizing historical and naïve host trees are associated with a bacterial community highly enriched in genes contributing to terpene metabolism. *Applied Environmental Microbiology, 79*, 3468–3475.

Attardo, G. M., et al. (2008). Analysis of milk gland structure and function in *Glossina morsitans*: Milk protein production, symbiont populations and fecundity. *Journal of Insect Physiology, 54*, 1236–1242.

Bandi, C., et al. (1995). The establishment of intracellular symbiosis in an ancestor of cockroaches and termites. *Proceedings of the Royal Society B: Biological Sciences, 259*, 293–299.

Baumann, P. (2005). Biology of bacteriocyte-associated endosymbionts of plant sap-sucking insects. *Annual Reviews in Microbiology, 59*, 155–189.

Baumann, P., Moran, N. A., & Baumann, L. (2006). Bacteriocyte-associated endosymbionts of insects. In M. Dworkin, S. Falkow, E. Rosenberg, K. H. Schleifer, & E. Stackebrandt (Eds.), *The Prokaryotes. Vol. 1: Symbiotic associations, biotechnology, applied microbiology* (pp. 465–496). Dordrecht: Springer.

Bian, G., Xu, Y., Lu, P., Xie, Y., & Xi, Z. (2010). The endosymbiotic bacterium *Wolbachia* induces resistance to dengue virus in *Aedes aegypti. PLoS Pathogens, 6*, e1000833.

Bistolas, K. S. I., Sakamoto, R. I., Fernandes, J. A. M., & Goffredi, S. K. (2014). Symbiont polyphyly, co-evolution, and necessity in pentatomid stinkbugs from Costa Rica. *Frontiers in Microbiology, 5*, 349.

Bordenstein, S. R., & Wernegreen, J. J. (2004). Bacteriophage flux in endosymbionts (*Wolbachia*): Infection frequency, lateral transfer, and recombination rates. *Molecular Biology and Evolution, 21*, 1981–1991.

Bordenstein, S. R., et al. (2009). Parasitism and mutualism in *Wolbachia*: What the phylogenomic trees can and cannot say. *Molecular Biology and Evolution, 26*, 231–241.

Breeuwer, J. A., & Jacobs, G. (1996). *Wolbachia*: Intracellular manipulators of mite reproduction. *Experimental Applied Acarology, 20*, 421–434.

Bright, M., & Bulgheresi, S. (2010). A complex journey: Transmission of microbial symbionts. *Nature Review Microbiology, 8*, 218–230.

Buchner, P. (1965). Symbiosis in animals which suck plant juices. In *Endosymbiosis of animals with plant microorganisms* (pp. 210–432). New York: Interscience.

Burke, G. R., Normark, B. B., Favret, C., & Moran, N. A. (2009). Evolution and diversity of facultative symbionts from the aphid subfamily Lachninae. *Applied and Environmental Microbiology, 75*, 5328–5335.

Campbell, B. J., & Cary, S. C. (2001). Characterization of a novel spirochete associated with the hydrothermal vent polychaete annelid, *Alvinella pompejana. Applied and Environmental Microbiology, 67*, 110–117.

Campbell, B. J., Stein, J. L., & Cary, S. C. (2003). Evidence of chemolithoautotrophy in the bacterial community associated with *Alvinella pompejana*, a hydrothermal vent polychaete. *Applied and Environmental Microbiology, 69*, 5070–5078.

Campbell, B. J., et al. (2009). Adaptations to submarine hydrothermal environments exemplified by the genome of *Nautilia profundicola. PLoS Genetics, 5*, e1000362.

Canbäck, B., Tamas, I., & Andersson, S. G. E. (2004). A phylogenomic study of endosymbiotic bacteria. *Molecular Biology and Evolution, 21*, 1110–1122.

Cavanaugh, C. M., McKiness, Z. P., Newton, I. L. G., & Stewart, F. J. (2013). Marine chemosynthetic symbioses. In E. Rosenberg et al. (Eds.), *The Prokaryotes; Prokaryotic biology and symbiotic associations* (pp. 579–607). Berlin/Heidelberg: Springer.

Chaves, S., Neto, M., & Tenreiro, R. (2009). Insect-symbiont systems: From complex relationships to biotechnological applications. *Biotechnology Journal, 4*, 1753–1765.

Chelossi, E., Milanese, M., Milano, A., Pronzato, R., & Riccardi, G. (2004). Characterisation and antimicrobial activity of epibiotic bacteria from *Petrosia ficiformis* (Porifera, Demospongiae). *Journal of Experimental Marine Biology and Ecology, 309*, 21–33.

Chen, X., Li, S., & Aksoy, S. (1999). Concordant evolution of a symbiont with its host insect species: Molecular phylogeny of genus *Glossina* and its bacteriome-associated endosymbiont, *Wigglesworthia glossinidia. Journal of Molecular Evolution, 48*, 49–58.
Childress, J. J., Felbeck, H., & Somero, G. N. (1987). Symbiosis in the deep sea. *Scientific American, 256*, 115–120.
Clark, J. W., & Kambhampati, S. (2003). Phylogenetic analysis of *Blattabacterium*, endosymbiotic bacteria from the wood roach, *Cryptocercus* (Blattodea: Cryptocercidae), including a description of three new species. *Molecular Phylogenetics and Evolution, 26*, 82–88.
Clark, M. E., Veneti, Z., Bourtzis, K., & Karr, T. L. (2002). The distribution and proliferation of the intracellular bacteria *Wolbachia* during spermatogenesis in *Drosophila. Mechanisms of Development, 111*, 3–15.
Conord, C., et al. (2008). Long-term evolutionary stability of bacterial endosymbiosis in Curculionoidea: Additional evidence of symbiont replacement in the Dryophthoridae family. *Molecular Biology and Evolution, 25*, 859–868.
Cordaux, R., Bouchon, D., & Grève, P. (2011). The impact of endosymbionts on the evolution of host sex-determination mechanisms. *Trends in Genetics, 27*, 332–341.
Dale, C., & Moran, N. A. (2006). Molecular interactions between bacterial symbionts and their hosts. *Cell, 126*, 453–465.
Davidson, S. K., & Stahl, D. A. (2008). Selective recruitment of bacteria during embryogenesis of an earthworm. *The ISME Journal, 2*, 510–518.
Dedeine, F., et al. (2001). Removing symbiotic *Wolbachia* bacteria specifically inhibits oogenesis in a parasitic wasp. *Proceedings of the National Academy of Sciences, USA, 98*, 6247–6252.
Degnan, P. H., Lazarus, A. B., Brock, C. D., & Wernegreen, J. J. (2004). Host-symbiont stability and fast evolutionary rates in an ant-bacterium association: Cospeciation of *Camponotus* species and their endosymbionts, *Candidatus* Blochmannia. *Systematic Biology, 53*, 95–110.
Delaye, L., & Moya, A. (2010). Evolution of reduced prokaryotic genomes and the minimal cell concept: Variations on a theme. *BioEssays, 32*, 281–287.
Dias Passos, F., de Lima Curi Meserani, G., & Gros, O. (2007). Structural and ultrastructural analysis of the gills in the bacterial-bearing species *Thyasira falklandica* (Bivalvia, Mollusca). *Zoomorphology, 126*, 153–162.
Dillon, R. J., & Dillon, V. M. (2004). The gut bacteria of insects: Nonpathogenic interactions. *Annual Review of Entomology, 49*, 71–92.
Dobson, S., Marsland, E., & Rattanadechakul, W. (2002). Mutualistic *Wolbachia* infection in *Aedes albopictus*: Accelerating cytoplasmic drive. *Genetics, 160*, 1087–1094.
Douglas, A. (1994). *Symbiotic interactions* (156 pp). Oxford: Oxford University Press.
Douglas, A. (1998). Nutritional interactions in insect-microbial symbioses: Aphids and their symbiotic bacteria B*uchnera. Annual Review of Entomology, 43*, 17–37.
Douglas, A. E. (2014a). Symbiosis as a general principle in eukaryotic evolution. In P. Keeling, & E. Koonin (Eds.), *Origin and evolution of Eukaryotes* (Cold Spring Harbor Perspectives in Biology 6, a016113). New York: Cold Spring Harbor Laboratory Press.
Douglas, A. E. (2014b). The molecular basis of bacterial–insect symbiosis. *Journal of Molecular Biology, 426*, 3830–3837.
Douglas, A. E., & Dixon, A. F. G. (1987). The mycetocyte symbiosis of aphids: Variation with age and morph in virginoparae of *Megoura viciae* and *Acyrthosiphon pisum. Journal of Insect Physiology, 33*, 109–113.
Dubilier, N., Giere, O., Distel, D. L., & Cavanaugh, C. M. (1995). Characterization of chemoautotrophic bacterial symbionts in a gutless marine worm (Oligochaeta, Annelida) by phylogenetic 16S rRNA sequence analysis and in situ hybridization. *Applied Environmental Microbiology, 61*, 2346–2350.
Dubilier, N., Bergin, C., & Lott, C. (2008). Symbiotic diversity in marine animals: The art of harnessing chemosynthesis. *Nature Review Microbiology, 6*, 725–740.
Dufour, S. C. (2005). Gill anatomy and relationship to chemoautotrophic symbiont presence in the bivalve family Thyasiridae. *Biological Bulletin, 208*, 200–212.

Duperron, S., et al. (2006). A dual symbiosis shared by two mussel species, *Bathymodiolus azoricus* and *Bathymodiolus puteoserpentis* (Bivalvia: Mytilidae), from hydrothermal vents along the northern Mid-Atlantic Ridge. *Environmental Microbiology, 8*, 1441–1447.

Duperron, S., Lorion, J., Samadi, S., Gros, O., & Gaill, F. (2009). Symbioses between deep-sea mussels (Mytilidae: Bathymodiolinae) and chemosynthetic bacteria: Diversity, function and evolution. *Comptes Rendus Biologies, 332*, 298–310.

Duron, O., et al. (2014). Origin, acquisition and diversification of heritable bacterial endosymbionts in louse flies and bat flies. *Molecular Ecology, 23*, 2105–2117.

Ebert, D. (2013). The epidemiology and evolution of symbionts with mixed-mode transmission. *Annual Review of Ecology, Evolution, and Systematics, 44*, 623–643.

Engel, M. S., & Grimaldi, D. A. (2004). New light shed on the oldest insect. *Nature, 427*, 627–630.

Estes, A. M., Hearn, D. J., Bronstein, J. L., & Pierson, E. A. (2009). The olive fly endosymbiont, "Candidatus Erwinia dacicola", switches from an intracellular existence to an extracellular existence during host insect development. *Applied Environmental Microbiology, 75*, 7097–7106.

Feldhaar, H., et al. (2007). Nutritional upgrading for omnivorous carpenter ants by the endosymbiont *Blochmannia*. *BMC Biology, 5*, 48.

Ferree, P. M., et al. (2005). *Wolbachia* utilizes host microtubules and dynein for anterior localization in the *Drosophila* oocyte. *PLoS Pathogens, 1*, 111–124.

Ferri, E., et al. (2011). New Insights into the evolution of *Wolbachia* infections in filarial nematodes inferred from a large range of screened species. *PloS One, 6*, e20843.

Fiala-Médioni, A., & Métivier, C. (1986). Ultrastructure of the gill of the hydrothermal vent bivalve *Calyptogena magnifica*, with a discussion of its nutrition. *Marine Biology, 90*, 215–222.

Fiala-Médioni, A., et al. (2002). Ultrastructural, biochemical, and immunological characterization of two populations of the mytilid mussel *Bathymodiolus azoricus* from the Mid-Atlantic Ridge: Evidence for a dual symbiosis. *Marine Biology, 141*, 1035–1043.

Forst, S., Dowds, B., Boemare, N., & Stackebrandt, E. (1997). *Xenorhabdus* and *Photorhabdus* spp.: Bugs that kill bugs. *Annual Review of Microbiology, 51*, 47–72.

Frenkiel, L., & Mouëza, M. (1995). Gill ultrastructure and symbiotic bacteria in *Codakia orbicularis* (Bivalvia, Lucinidae). *Zoomorphology, 115*, 51–61.

Frentiu, F. D., Robinson, J., Young, P. R., McGraw, E. A., & O'Neill, S. L. (2010). *Wolbachia* mediated resistance to dengue virus infection and death at the cellular level. *PloS One, 5*, e13398.

Frost, C. L., Pollock, S. W., Smith, J. E., & Hughes, W. O. H. (2014). *Wolbachia* in the flesh: Symbiont Intensities in germ-line and somatic tissues challenge the conventional view of *Wolbachia* transmission routes. *PloS One, 9*, e95122.

Fry, A., & Rand, D. (2002). *Wolbachia* interactions that determine *Drosophila melanogaster* survival. *Evolution, 56*, 1976–1981.

Fukatsu, T., Nikoh, N., Kawai, R., & Koga, R. (2000). The secondary endosymbiotic bacterium of the pea aphid *Acyrthosiphon pisum* (Insecta: Homoptera). *Applied and Environmental Microbiology, 66*, 2748–2758.

Gardebrecht, A., et al. (2012). Physiological homogeneity among the endosymbionts of Riftia pachyptila and Tevnia jerichonana revealed by proteogenomics. *The ISME Journal, 6*, 766–776.

Gehrer, L., & Vorburger, C. (2012). Parasitoids as vectors of facultative bacterial endosymbionts in aphids. *Biology Letters, 8*, 613–615.

Gil, R., et al. (2003). The genome sequence of *Blochmannia floridanus*: Comparative analysis of reduced genomes. *Proceedings of the National Academy of Sciences, USA, 100*, 9388–9393.

Gil-Turnes, M. S., & Fenical, W. (1992). Embryos of *Homarus americanus* are protected by epibiotic bacteria. *Biological Bulletin, 182*, 105–108.

Gil-Turnes, M. S., Hay, M. E., & Fenical, W. (1989). Symbiotic mariner bacteria chemically defend Crustancean embryos from a pathogenic fungus. *Science, 246*, 116–118.

Gómez-Valero, L., et al. (2004). Coexistence of *Wolbachia* with *Buchnera aphidicola* and a secondary symbiont in the aphid *Cinara cedri. Journal of Bacteriology, 186*, 6626–6633.

Gonella, E., et al. (2012). Horizontal transmission of the symbiotic bacterium *Asaia* sp. in the leafhopper *Scaphoideus titanus* Ball (Hemiptera: Cicadellidae). *BMC Microbiology, 12*(Suppl 1), S4.

Goodacre, S. L., & Martin, O. Y. (2013). Endosymbiont infections in spiders. In *Spider ecophysiology* (pp. 93–105), New York: Springer.

Goodacre, S. L., Martin, O. Y., Thomas, C. F., & Hewitt, G. M. (2006). *Wolbachia* and other endosymbiont infections in spiders. *Molecular Ecology, 15*, 517–527.

Goto, S., Anbutsu, H., & Fukatsu, T. (2006). Asymmetrical interactions between Wolbachia and Spiroplasma endosymbionts coexisting in the same insect host. *Applied and Environmental Microbiology, 72*, 4805–4810.

Gottlieb, Y., et al. (2008). Inherited intracellular ecosystem: Symbiotic bacteria share bacteriocytes in whiteflies. *FASEB Journal, 22*, 2591–2599.

Graf, J. (2005). Molecular requirements for the colonization of *Hirudo medicinalis* by *Aeromonas veronii*. In J. Overmann (Ed.), *Progress in molecular and subcellular biology* (Molecular Basis of Symbiosis, pp. 291–303). Berlin: Springer.

Gros, O., Guibert, J., & Gaill, F. (2007). Gill-symbiosis in mytilidae associated with wood fall environments. *Zoomorphology, 126*, 163–172.

Gruber-Vodicka, H. R., et al. (2011). *Paracatenula*, an ancient symbiosis between thiotrophic Alphaproteobacteria and catenulid flatworms. *Proceedings of the National Academy of Sciences, USA, 108*, 12078–12083.

Gruwell, M. E., Morse, G. E., & Normark, B. B. (2007). Phylogenetic congruence of armored scale insects (Hemiptera: Diaspididae) and their primary endosymbionts from the phylum Bacteroidetes. *Molecular Phylogenetics and Evolution, 44*, 267–280.

Guri, M., et al. (2012). Acquisition of epibiotic bacteria along the life cycle of the hydrothermal shrimp *Rimicaris exoculata. The ISME Journal, 6*, 597–609.

Gustafson, R. G., & Reid, R. G. B. (1988). Association of bacteria with larvae of the gutless protobranch bivalve *Solemya reidi* (Cryptodonta, Solemyidae). *Marine Biology, 97*, 389–401.

Haegemann, A., et al. (2009). An endosymbiotic bacterium in a plant-parasitic nematode: Member of a new *Wolbachia* supergroup. *International Journal for Parasitology, 39*, 1045–1054.

Hamilton, P. T., & Perlman, S. J. (2013). Host defense via symbiosis in *Drosophila. PLoS Pathogens, 9*, e1003808.

Hansen, R. D. E., et al. (2011). A worm's best friend: Recruitment of neutrophils by *Wolbachia* confounds eosinophil degranulation against the filarial nematode *Onchocerca ochengi. Proceedings of the Royal Society B, 278*, 2293–2302.

Harmer, T. L., et al. (2008). Free-living tube worm endosymbionts found at deep-sea vents. *Applied and Environmental Microbiology, 74*, 3895–3898.

Haselkorn, T. S., Markow, T. A., & Moran, N. A. (2009). Multiple introductions of the *Spiroplasma* bacterial endosymbiont into *Drosophila. Molecular Ecology, 18*, 1294–1305.

Haygood, M. G., Schmidt, E. W., Davidson, S. K., & Faulkner, D. J. (1999). Microbial symbionts of marine invertebrates: Opportunities for microbial biotechnology. *Journal of Molecular Microbiology and Biotechnology, 1*, 33–43.

Haynes, S., et al. (2003). The diversity of bacteria associated with natural aphid populations. *Applied and Environmental Microbiology, 69*, 7216–7223.

Heddi, A., Grenier, A. M., Khatchadourian, C., Charles, H., & Nardon, P. (1999). Four intracellular genomes direct weevil biology: Nuclear, mitochondrial, principal endosymbiont, and *Wolbachia. Proceedings of the National Academy of Sciences, USA, 96*, 6814–6819.

Hedges, L. M., Brownlie, J. C., O'Neill, S. L., & Johnson, K. N. (2008). *Wolbachia* and virus protection in insects. *Science, 322*, 702.

Henry, L. M., et al. (2013). Horizontally transmitted symbionts and host colonization of ecological niches. *Current Biology, 23*, 1713–1717.

Hertig, M., & Wolbach, S. B. (1924). Studies on *Rickettsia*-like microorganisms in insects. *Journal of Medical Research, 44*, 329–374.

Hilgenboecker, K., Hammerstein, P., Schlattmann, P., Telschow, A., & Werren, J. H. (2008). How many species are infected with *Wolbachia*? A statistical analysis of current data. *FEMS Microbiology Letters, 281*, 215–220.

Hongoh, Y. (2011). Toward the functional analysis of uncultivable, symbiotic microorganisms in the termite gut. *Cellular and Molecular Life Sciences, 68*, 1311–1325.

Hosokawa, T., Koga, R., Kikuchi, Y., Meng, X. Y., & Fukatsu, T. (2010). *Wolbachia* as a bacteriocyte-associated nutritional mutualist. *Proceedings of the National Academy of Sciences, USA, 107*, 769–774.

Hubert, J., et al. (2012). Detection and identification of species-specific bacteria associated with synanthropic mites. *Microbial Ecology, 63*, 919–928.

Hurst, G. D. D., & Jiggins, F. M. (2000). Male-killing bacteria in insects: Mechanisms, incidence and implications. *Emerging Infectious Diseases, 6*, 329–336.

Husnik, F., et al. (2013). Horizontal gene transfer from diverse bacteria to an insect genome enables a tripartite nested mealybug symbiosis. *Cell, 153*, 1567–1578.

Ijichi, N., et al. (2002). Internal spatiotemporal population dynamics of infection with three *Wolbachia* strains in the adzuki bean beetle, *Callosobruchus chinensis* (Coleoptera: Bruchidae). *Applied and Environmental Microbiology, 68*, 4074–4080.

Jaenike, J., Unckless, R., Cockburn, S. N., Boelio, L. M., & Perlman, S. J. (2010). Adaptation via symbiosis: Recent spread of a *Drosophila* defensive symbiont. *Science, 329*, 212–215.

Johanowicz, D. L., & Hoy, M. A. (1996). *Wolbachia* in a predator prey system: 16S ribosomal RNA analysis of two phytoseiids (Acari: Phytoseiidae) and their prey (Acari: Tetranychidae). *Annals of the Entomological Society of America, 89*, 435–441.

Kaltenpoth, M., et al. (2006). '*Candidatus* Streptomyces philanthi', an endosymbiotic streptomycete in the antennae of *Philanthus* digger wasps. *International Journal of Systematic and Evolutionary Microbiology, 56*, 1403–1411.

Kellner, R. L. L., & Dettner, K. (1996). Differential efficacy of toxic pederin in deterring potential arthropod predators of Paederus (Coleoptera: Staphylinidae) offspring. *Oecologia, 107*, 293–300.

Kelly, M. S., & McKenzie, J. D. (1995). Survey of the occurrence and morphology of sub-cuticular bacteria in shelf echinoderms from the north-east Atlantic Ocean. *Marine Biology, 123*, 741–756.

Kennedy, J., Marchesi, J. R., & Dobson, A. D. W. (2007). Metagenomic approaches to exploit the biotechnological potential of the microbial consortia of marine sponges. *Applied Microbiology and Biotechnology, 75*, 11–20.

Kikuchi, Y., & Fukatsu, T. (2005). *Rickettsia* infection in natural leech populations. *Microbial Ecology, 49*, 265–271.

Kikuchi, Y., Sameshima, S., Kitade, O., Kojima, J., & Fukatsu, T. (2002). Novel clade of *Rickettsia* spp. from leeches. *Applied and Environmental Microbiology, 68*, 999–1004.

Kochevar, R. E., Childress, J. J., Fisher, C. R., & Minnich, E. (1992). The methane mussel: Roles of symbiont and host in the metabolic utilization of methane. *Marine Biology, 112*, 389–401.

Koga, R., Meng, X. Y., Tsuchida, T., & Fukatsu, T. (2012). Cellular mechanism for selective vertical transmission of an obligate insect symbiont at the bacteriocyte–embryo interface. *Proceedings of the National Academy of Sciences, USA, 109*, E1230–E1237.

Koga, R., Bennett, G. M., Cryan, J. R., & Moran, N. A. (2013). Evolutionary replacement of obligate symbionts in an ancient and diverse insect lineage. *Environmental Microbiology, 15*, 2073–2081.

Kölsch, G., & Pedersen, B. V. (2010). Can the tight co-speciation between reed beetles (Col., Chrysomelidae, Donaciinae) and their bacterial endosymbionts, which provide cocoon material, clarify the deeper phylogeny of the hosts? *Molecular Phylogenetics and Evolution, 54*, 810–821.

Kramer, L. H., Passeri, B., Corona, S., Simoncini, L., & Casiraghi, M. (2003). Immunohistochemical/immunogold detection and distribution of the endosymbiont *Wolbachia* of *Dirofilaria immitis* and *Brugia pahangi* using a polyclonal antiserum raised against WSP (Wolbachia surface protein). *Parasitology Research, 89*, 381–386.

Kremer, N., et al. (2009). *Wolbachia* interferes with ferritin expression and iron metabolism in insects. *PLoS Pathogens, 5*, e1000630.

Krieger, J., Giere, O., & Dubilier, N. (2000). Localization of RubisCO and sulfur in endosymbiotic bacteria of the gutless marine oligochaete *Inanidrilus luekodermatus* (Anellida). *Marine Biology, 137*, 239–244.

Lalzar, I., Friedmann, Y., & Gottlieb, Y. (2014). Tissue tropism and vertical transmission of *Coxiella* in *Rhipicephalus sanguineus* and *Rhipicephalus turanicus* ticks. *Environmental Microbiology*. doi:10.1111/1462-2920.12455.

Lamelas, A., et al. (2008). Evolution of the secondary symbiont "*Candidatus* Serratia symbiotica" in Aphid species of the subfamily Lachninae. *Applied and Environmental Microbiology, 74*, 4236–4240.

Landmann, F., Foster, J. M., Slatko, B., & Sullivan, W. (2010). Asymmetric Wolbachia segregation during early *Brugia malayi* embryogenesis determines its distribution in adult host tissues. *PLoS Neglected Tropical Diseases, 4*, e758.

Lee, O. O., Chui, P. Y., Wong, Y. H., Pawlik, J. R., & Qian, P. Y. (2009). Evidence for vertical transmission of bacterial symbionts from adult to embryo in the caribbean sponge *Svenzea zeai. Applied and Environmental Microbiology, 75*, 6147–6156.

Leisch, N., et al. (2011). Microanatomy of the trophosome region of *Paracatenula* cf. *polyhymnia* (Catenulida, Platyhelminthes) and its intracellular symbionts. *Zoomorphology, 130*, 261–271.

Lim-Fong, G. E., Regali, L. A., & Haygood, M. G. (2008). Evolutionary relationships of "Candidatus Endobugula" bacterial symbionts and their Bugula bryozoan hosts. *Applied Environmental Microbiology, 74*, 3605–3609.

Lindquist, N., Barber, P. H., & Weisz, J. B. (2005). Episymbiotic microbes as food and defence for marine isopods: Unique symbioses in a hostile environment. *Proceedings of the Royal Society B, 272*, 1209–1216.

Lo, N., et al. (2007). Taxonomic status of the intracellular bacterium *Wolbachia pipientis. International Journal of Systematic and Evolutionary Microbiology, 57*, 654–657.

Lund, M. B., Kjeldsen, K. U., & Schramm, A. (2014). The earthworm—*Verminephrobacter* symbiosis: An emerging experimental system to study extracellular symbiosis. *Frontiers in Microbiology, 5*, 128.

Maekawa, K., Park, Y. C., & Lo, N. (2005). Phylogeny of endosymbiont bacteria harbored by the woodroach *Cryptocercus* spp. (Cryptocercidae: Blattaria): Molecular clock evidence for a late Cretaceous – Early Tertiary split of Asian and American lineages. *Molecular Phylogenetics and Evolution, 36*, 728–733.

Maltz, M. A., et al. (2014). Metagenomic analysis of the medicinal leech gut microbiota. *Frontiers in Microbiology, 5*, 151.

Martin, O. Y., & Goodacre, S. L. (2009). Widespread infections by the bacterial endosymbiont *Cardinium* in arachnids. *The Journal of Arachnology, 37*, 106–108.

Martin, J. W., & Haney, T. A. (2005). Decapod crustaceans from hydrothermal vents and cold seeps: A review through 2005. *Zoological Journal of the Linnean Society, 145*, 445–522.

Marubayashi, J. M., et al. (2014). Diversity and localization of bacterial endosymbionts from whitefly species collected in Brazil. *PloS One, 9*, e108363.

Mazzon, L., et al. (2010). Phylogenetic relationships between flies of the Tephritinae subfamily (Diptera, Tephritidae) and their symbiotic bacteria. *Molecular Phylogenetics and Evolution, 56*, 312–326.

McCutcheon, J. P., & Moran, N. A. (2012). Extreme genome reduction in symbiotic bacteria. *Nature Reviews Microbiology, 10*, 13–26.

McCutcheon, J. P., & von Dohlen, C. D. (2011). An interdependent metabolic patchwork in the nested symbiosis of mealybugs. *Current Biology, 21*, 1366–1372.

McFall-Ngai, M. J. (1999). Consequences of evolving with bacterial symbionts: Lessons from the squid-vibrio associations. *Annual Review of Ecology and Systematics, 30*, 235–256.

McFall-Ngai, M., & Montgomery, M. K. (1990). The anatomy and morphology of the adult bacterial light organ of *Euprymna scolopes* Berry (Cephalopoda:Sepiolidae). *Biological Bulletin, 179*, 332–339.

McKiness, Z. P., McMullin, E. R., Fisher, C. R., & Cavanaugh, C. M. (2005). A new bathymiodoline mussel symbiosis at the Juan de Fuca hydrothermal vents. *Marine Biology, 148*, 109–116.

Min, K. T., & Benzer, S. (1997). *Wolbachia*, normally a symbiont of *Drosophila*, can be virulent, causing degeneration and early death. *Proceedings of the National Academy of Sciences, USA, 94*, 10792–10796.

Mitsuhashi, W., Saiki, T., Wei, W., Kawakita, H., & Sato, M. (2002). Two novel strains of *Wolbachia* coexisting in both species of mulberry leafhoppers. *Insect Molecular Biology, 11*, 577–584.

Montllor, C. B., Maxmen, A., & Purcell, A. H. (2002). Facultative bacterial endosymbionts benefit pea aphids *Acyrthosiphon pisum* under heat stress. *Ecological Entomology, 27*, 189–195.

Moran, N. A. (2002). Microbial minimalism: Genome reduction in bacterial pathogens. *Cell, 108*, 583–586.

Moran, N. A., & Wernegreen, J. J. (2000). Lifestyle evolution in symbiotic bacteria: Insights from genomics. *TREE, 15*, 321–326.

Moran, N. A., & Yun, Y. (2015). Experimental replacement of an obligate insect symbiont. *Proceedings of the National Academy of Sciences, USA, 112*, 2093–2096.

Moran, N. A., Tran, P., & Gerardo, N. M. (2005). Symbiosis and insect diversification: An ancient symbiont of sap-feeding insects from the bacterial phylum Bacteroidetes. *Applied and Environmental Microbiology, 71*, 8802–8810.

Moreira, L. A., et al. (2009). A *Wolbachia* symbiont in *Aedes aegypti* limits infection with Dengue, Chikungunya, and *Plasmodium. Cell, 139*, 1268–1278.

Morse, S. F., Dick, C. W., Patterson, B. D., & Dittmar, K. (2012). Some like it hot: Evolution and ecology of novel endosymbionts in bat flies of cave-roosting bats (Hippoboscoidea, Nycterophiliinae). *Applied and Environmental Microbiology, 78*, 8639–8649.

Müller, W. E. G., et al. (2004). Sustainable production of bioactive compounds by sponges – Cell culture and gene cluster approach: A review. *Marine Biotechnology, 6*, 105–117.

Musat, N., Giere, O., Gieseke, A., Thiermann, F., Amann, R., & Dubilier, N. (2007). Molecular and morphological characterization of the association between bacterial endosymbionts and the marine nematode *Astomonema* sp. from the Bahamas. *Environmental Microbiology, 9*, 1345–1353.

Nakabachi, A., et al. (2005). Transcriptome analysis of the aphid bacteriocyte, the symbiotic host cell that harbors an endocellular mutualistic bacterium, *Buchnera. Proceedings of the National Academy of Sciences, USA, 102*, 5477–5482.

Noda, S., Ohkuma, M., Yamada, A., Hongoh, Y., & Kudo, T. (2003). Phylogenetic position and in situ identification of ectosymbiotic spirochetes on protists in the termite gut. *Applied and Environmental Microbiology, 69*, 625–633.

Noel, G. R., & Atibalentja, N. (2006). '*Candidatus* Paenicardinium endonii', an endosymbiont of the plant-parasitic nematode *Heterodera glycines* (Nemata: Tylenchida), affiliated to the phylum Bacteroidetes. *International Journal of Systematic and Evolutionary Microbiology, 56*, 1697–1702.

Nováková, E., et al. (2013). Reconstructing the phylogeny of aphids (Hemiptera: Aphididae) using DNA of the obligate symbiont *Buchnera aphidicola. Molecular Phylogenetics and Evolution, 68*, 42–54.

Nussbaumer, A. D., Bright, M., Baranyi, C., Beisser, C. J., & Ott, J. A. (2004). Attachment mechanism in a highly specific association between ectosymbiotic bacteria and marine nematodes. *Aquatic Microbial Ecology, 34*, 239–246.

Oliver, K. M., Russell, J. A., Moran, N. A., & Hunter, M. S. (2003). Facultative bacterial symbionts in aphids confer resistance to parasitic wasps. *Proceedings of the National Academy of Sciences, USA, 100*, 803–1807.

Oliver, K. M., Moran, N. A., & Hunter, M. S. (2005). Variation in resistance to parasitism in aphids is due to symbionts not host genotype. *Proceedings of the National Academy of Sciences, USA, 102*, 12795–12800.

Oliver, K. M., Degnan, P. H., Hunter, M. S., & Moran, N. A. (2009). Bacteriophages encode factors required for protection in a symbiotic mutualism. *Science, 325*, 992–994.

Oliver, K. M., Degnan, P. H., Burke, G. R., & Moran, N. A. (2010). Facultative symbionts in aphids and the horizontal transfer of ecologically important traits. *Annual Review of Entomology, 55*, 247–266.

Osborne, S. E., Iturbe-Ormaetxe, I., Brownlie, J. C., O'Neill, S. L., & Johnson, K. N. (2012). Antiviral protection and the importance of *Wolbachia* density and tissue tropism in *Drosophila simulans. Applied and Environmental Microbiology, 78*, 6922–6929.

Parker, B. J., Spragg, C. J., Altincicek, B., & Gerardo, N. M. (2013). Symbiont-mediated protection against fungal pathogens in pea aphids: A role for pathogen specificity? *Applied Environmental Microbiology, 79*, 2455–2458.

Parrella, G., Nappo, A. G., Manco, E., Greco, B., & Giorgini, M. (2014). Invasion of the Q2 mitochondrial variant of Mediterranean *Bemisia tabaci* in southern Italy: Possible role of bacterial endosymbionts. *Pest Management Science, 70*, 1514–1523.

Pérez-Brocal, V., et al. (2006). A small microbial genome: The end of a long symbiotic relationship? *Science, 314*, 312–313.

Perkins, S. L., Budinoff, R. B., & Siddall, M. E. (2005). New gammaproteobacteria associated with blood-feeding leeches and a broad phylogenetic analysis of leech endosymbionts. *Applied and Environmental Microbiology, 71*, 5219–5224.

Perlman, S. J., Kelly, S. E., Zchori-Fein, E., & Hunter, M. S. (2006). Cytoplasmic incompatibility and multiple symbiont infection in the ash whiteXy parasitoid, *Encarsia inaron. Biological Control, 39*, 474–480.

Petersen, J. M., et al. (2010). Dual symbiosis of the vent shrimp *Rimicaris exoculata* with filamentous gamma- and epsilonproteobacteria at four Mid-Atlantic Ridge hydrothermal vent fields. *Environmental Microbiology, 12*, 2204–2218.

Petersen, J. M., et al. (2011). Hydrogen is an energy source for hydrothermal vent symbioses. *Nature, 476*, 176–180.

Pfarr, K., Foster, J., Slatko, B., Hoerauf, A., & Eisen, J. A. (2007). On the taxonomic status of the intracellular bacterium *Wolbachia pipientis*: Should this species name include the intracellular bacteria of filarial nematodes? *International Journal of Systematic and Evolutionary Microbiology, 57*, 1677–1678.

Piel, J., Hofer, I., & Hui, D. (2004). Evidence for a symbiosis island involved in horizontal acquisition of pederin biosynthetic capabilities by the bacterial symbiont of *Paederus fuscipes* beetles. *Journal of Bacteriology, 186*, 1280–1286.

Plantard, O., et al. (2012). Detection of *Wolbachia* in the tick *Ixodes ricinus* is due to the presence of the Hymenoptera endoparasitoid *Ixodiphagus hookeri. PloS One, 7*, e30692.

Polz, M. F., & Cavanaugh, C. M. (1996). The ecology of ectosymbiosis at a Mid-Atlantic Ridge hydrothermal vent site. In F. Ublein, J. Ott, & M. Stachowtisch (Eds), *Deep-sea and extreme shallow-water habitats: Affinities and adaptations* (Biosystematics and ecology series, Vol. 11, pp. 337–352). Wien: Österreichische Akademie der Wissenschaften.

Ransom-Jones, E., Jones, D. L., McCarthy, A. J., & McDonald, J. E. (2012). The Fibrobacteres: An important Phylum of cellulose-degrading bacteria. *Microbial Ecology, 63*, 267–281.

Rigaud, T., Pennings, P. S., & Juchault, P. (2001). *Wolbachia* bacteria effects after experimental interspecific transfers in terrestrial isopods. *Journal of Invertebrate Pathology, 77*, 251–257.

Rio, R. V. M., Lefevre, C., Heddi, A., & Aksoy, S. (2003). Comparative genomics of insect-symbiotic bacteria: Influence of host environment on microbial genome composition. *Applied and Environmental Microbiology, 69*, 6825–6832.

Robidart, J. C., et al. (2008). Metabolic versatility of the *Riftia pachyptila* endosymbiont revealed through metagenomics. *Environmental Microbiology, 10*, 727–737.

Rokas, A. (2000). *Wolbachia* as a speciation agent. *Tree, 15*, 45–46.

Rousset, F., & Solignac, M. (1995). Evolution of single and double *Wolbachia* symbioses during speciation in the *Drosophila simulans* complex. *Proceedings of the National Academy of Sciences, USA, 92*, 6389–6393.

Russell, J. E., & Stouthamer, R. (2011). The genetics and evolution of obligate reproductive parasitism in *Trichogramma pretiosum* infected with parthenogenesis-inducing *Wolbachia. Heredity, 106*, 58–67.

Russell, J. A., Moreau, C. S., Goldman-Huertas, B., Fujiwara, M., Lohman, D. J., & Pierce, N. E. (2009). Bacterial gut symbionts are tightly linked with the evolution of herbivory in ants. *Proceedings of the National Academy of Sciences, USA, 106*, 21236–21241.

Sachs, J. L., Skophammer, R. G., & Regus, J. U. (2011). Evolutionary transitions in bacterial symbiosis. *Proceedings of the National Academy of Sciences, USA, 108*, 10800–10807.

Saha, S., et al. (2012). Survey of endosymbionts in the *Diaphorina citri* metagenome and assembly of a *Wolbachia* wDi draft genome. *PloS One, 7*(11), e50067. doi:10.1371/journal.pone.0050067.

Salathé, R. M., & Vrijenhoek, R. C. (2012). Temporal variation and lack of host specificity among bacterial endosymbionts of *Osedax* bone worms (Polychaeta: Siboglinidae). *BMC Evolutionary Biology, 12*, 189.

Saridaki, A., & Bourtzis, K. (2010). *Wolbachia*: More than just a bug in insects genitals. *Current Opinion in Microbiology, 13*, 67–72.

Sauer, C., Stackebrandt, E., Gadau, J., Hölldobler, B., & Gross, R. (2000). Systematic relationships and cospeciation of bacterial endosymbionts and their carpenter ant host species: Proposal of the new taxon *Candidatus* Blochmannia gen. nov. *International Journal of Systematic and Evolutionary Microbiology, 50*, 1877–1886.

Scarborough, C. L., Ferrari, J., & Godfray, H. C. J. (2005). Bacterial endosymbiont increases aphid inclusive fitness after pathogen attack. *Science, 310*, 1781.

Schuett, C., Doepke, H., Gratho, A., & Gedde, M. (2007). Bacterial aggregates in the tentacles of the sea anemone *Metridium senile. Helgoland Marine Research, 61*, 211–216.

Sebastien, A., Gruber, M. A. M., & Lester, P. J. (2012). Prevalence and genetic diversity of three bacterial endosymbionts (*Wolbachia*, *Arsenophonus*, and *Rhizobiales*) associated with the invasive yellow crazy ant (*Anoplolepis gracilipes*). *Insectes Sociaux, 59*, 33–40.

Shigenobu, S., & Wilson, A. C. C. (2011). Genomic revelations of a mutualism: The pea aphid and its obligate bacterial symbiont. *Cellular and Molecular Life Sciences, 68*, 1297–1309.

Siddall, M. E., Perkins, S. L., & Desser, S. S. (2004). Leech mycetomes endosymbionts are a new lineage of alphaproteobacteria related to the Rhizobiaceae. *Molecular Phylogenetics and Evolution, 30*, 178–186.

Six, D. L. (2013). The bark beetle holobiont: Why microbes matter. *Journal of Chemical Ecology, 39*, 989–1002.

Skaljac, M., Zanic, K., Goreta Ban, S., Kontsedalov, S., & Ghanim, M. (2010). Co-infection and localization of secondary symbionts in two whitefly species. *BMC Microbiology, 10*, 142.

Steindler, L., Huchon, D., Avni, A., & Ilan, M. (2005). 16S rRNA pylogeny of sponge-associated Cyanobacteria. *Applied and Environmental Microbiology, 71*, 4127–4131.

Stewart, F. J., & Cavanaugh, C. M. (2006). Symbiosis of thioautotrophic bacteria with *Riftia pachyptila*. In J. Overmann, & W. E. G. Müller (Eds.), *Molecular basis of symbiosis* (Progress in molecular and subcellular biology series, pp. 197–225). Berlin: Springer.

Stingl, U., Radek, R., Yang, H., & Brune, A. (2005). *Endomicrobia*: Cytoplasmic symbionts of termite gut protozoa form a separate Phylum of Prokaryotes. *Applied and Environmental Microbiology, 71*, 1473–1479.

Stoll, S., Feldhaar, H., Fraunholz, M. J., & Gross, R. (2010). Bacteriocyte dynamics during development of a holometabolous insect, the carpenter ant *Camponotus floridanus*. *BMC Microbiology, 10*, 308.

Stouthamer, R. (1993). The use of sexual versus asexual wasps in biological control. *Entomophaga, 38*, 3–6.

Stouthamer, R., Luck, R. F., & Hamilton, W. D. (1990). Antibiotics cause parthenogenetic *Trichogramma* (Hymenoptera, Trichogrammatidae) to revert to sex. *Proceedings of the National Academy of Sciences, USA, 87*, 2424–2427.

Szafranski, K. M., Gaudron, S. M., & Duperron, S. (2014). Direct evidence for maternal inheritance of bacterial symbionts in small deep-sea clams (Bivalvia: Vesicomyidae). *Naturwissenschaften, 101*, 373–383.

Tamas, I., et al. (2002). 50 million years of genomic stasis in endosymbiotic bacteria. *Science, 296*, 2376–2379.

Telschow, A., Flor, M., Kobayashi, Y., Hammerstein, P., & Werren, J. H. (2007). *Wolbachia*-induced unidirectional cytoplasmic incompatibility and speciation: Mainland-island model. *PloS One, 2*, e701.

Thacker, R. W. (2005). Impacts of shading on sponge-Cyanobacteria symbioses: A comparison between host-specific and generalist associations. *Integrative and Comparative Biology, 45*, 369–376.

Thao, M. L., & Baumann, P. (2004a). Evidence for multiple acquisition of *Arsenophonus* by whitefly species (Sternorrhyncha: Aleyrodidae). *Current Microbiology, 48*, 140–144.

Thao, M. L., & Baumann, P. (2004b). Evolutionary relationships of primary prokaryotic endosymbionts of whiteflies and their hosts. *Applied Environmental Microbiology, 70*, 3401–3406.

Thao, M. L., et al. (2000). Secondary endosymbionts of psyllids have been acquired multiple times. *Current Microbiology, 41*, 300–304.

Thompson, C. L., Vier, R., Mikaelyan, A., Wienemann, T., & Brune, A. (2012). '*Candidatus* Arthromitus' revised: Segmented filamentous bacteria in arthropod guts are members of *Lachnospiraceae. Environmental Microbiology, 14*, 1454–1465.

Thurber, A. R., Jones, W. J., & Schnabel, K. (2011). Dancing for food in the deep sea: Bacterial farming by a new species of Yeti crab. *PloS One, 6*(11), e26243.

Toenshoff, E. R., et al. (2012). Bacteriocyte-associated gammaproteobacterial symbionts of the *Adelges nordmannianae/piceae* complex (Hemiptera: Adelgidae). *The ISME Journal, 6*, 384–396.

Tsagkarakou, A., Guillemaud, T., Rousset, F., & Navajas, M. (1996). Molecular identification of a *Wolbachia* endosymbiont in a *Tetranychus urticae* strain. *Insect Molecular Biology, 5*, 217–221.

Urban, J. M., & Cryan, J. R. (2012). Two ancient bacterial endosymbionts have coevolved with the planthoppers (Insecta: Hemiptera: Fulgoroidea). *BMC Evolutionary Biology, 12*, 87.

Vacelet, J., & Donadey, C. (1977). Electron microscope study of the association between some sponges and bacteria. *Journal of Experimental Marine Biology and Ecology, 30*, 301–314.

van Borm, S., Wenseleers, T., Billen, J., & Boomsma, J. J. (2003). Cloning and sequencing of wsp encoding gene fragments reveals a diversity of co-infecting *Wolbachia* strains in *Acromyrmex* leafcutter ants. *Molecular Phylogenetics and Evolution, 26*, 102–109.

Vandekerckhove, T. T. M., Willems, A., Gillis, M., & Coomans, A. (2000). Occurrence of novel Verrucomicrobial species, endosymbiotic and associated with parthenogenesis in *Xiphinema americanum* group species (Nematoda, Longidoridae). *International Journal of Systematic and Evolutionary Microbiology, 50*, 2197–2205.

Vavre, F., Girin, C., & Bouletreau, M. (1999). Phylogenetic status of a fecundity-enhancing *Wolbachia* that does not induce thelytoky in *Trichogramma. Molecular Biology and Evolution, 8*, 67–72.

Veneti, Z., Clark, M. E., Karr, T. L., Savakis, C., & Bourtzis, K. (2004). Heads or tails: Host parasite interactions in the *Drosophila-Wolbachia* system. *Applied and Environmental Microbiology, 70*, 5366–5372.

Visick, K. L., & Mcfall-Ngai, M. J. (2000). An exclusive contract: Specificity in the *Vibrio fischeri-Euprymna scolopes* partnership. *Journal of Bacteriology, 182*, 1779–1787.

von Dohlen, C. D., Kohler, S., Alsop, S. T., & McManus, W. R. (2001). Mealybug b-proteobacterial endosymbionts contain g-proteobacterial symbionts. *Nature, 412*, 433–436.

Vorburger, C., Gehrer, L., & Rodriguez, P. (2010). A strain of the bacterial symbiont *Regiella insecticola* protects aphids against parasitoids. *Biology Letters, 6*, 109–111.

Wagner, M., & Horn, M. (2006). The *Planctomycetes*, *Verrucomicrobia*, *Chlamydiae* and sister phyla comprise a superphylum with biotechnological and medical relevance. *Current Opinion in Biotechnology, 17*, 241–249.

Walker, T., et al. (2011). The *wMel Wolbachia* strain blocks dengue and invades caged *Aedes aegypti* populations. *Nature, 476*, 450–453.

Wang, Y., Brune, A., & Zimmer, M. (2007). Bacterial symbionts in the hepatopancreas of isopods: Diversity and environmental transmission. *FEMS Microbiology Ecology*, 61, 141–152.

Wang, J. B., & Cheung, W. W. K. (1997). Electron microscopy studies on the a-bacteroids in the fat bodies of the lantern bug *Pyrops candelaria* Linn (Homoptera: Fulgoridae). *Parasitology Research, 83*, 499–503.

Wang, J. B., & Cheung, W. W. K. (1998). Multiple bacteroids in the bacteriome of the lantern bug *Pyrops candelaria* Linn. (Homoptera: Fulgoridae). *Parasitology Research, 84*, 741–745.

Wang, X., et al. (2013). Molecular cross-talk between sponge host and associated microbes. *Phytochemistry Reviews, 12*, 369–390.

Webster, N. (2014). Cooperation, communication, and co-evolution: Grand challenges in microbial symbiosis research. *Frontiers in Microbiology*. doi:10.3389/fmicb.2014.00164

Webster, N. S., et al. (2010). Deep sequencing reveals exceptional diversity and modes of transmission for bacterial sponge symbionts. *Environmental Microbiology, 12*, 2070–2082.

Weeks, A. R., Turelli, M., Harcombe, W. R., Reynolds, K. T., & Hoffmann, A. A. (2007). From parasite to mutualist: Rapid evolution of *Wolbachia* in natural populations of *Drosophila. PLoS Biology, 5*, 997–1005.

Werren, J. H., Baldo, L., & Clark, M. E. (2008). *Wolbachia*: Master manipulators of invertebrate biology. *Nature Review Microbiology, 6*, 741–751.

White, J. A., Kelly, S. E., Cockburn, S. N., Perlman, S. J., & Hunter, M. S. (2011). Endosymbiont costs and benefits in a parasitoid infected with both *Wolbachia* and *Cardinium. Heredity, 106*, 585–591.

White, J. A., Giorgini, M., Strand, M. R., & Pennacchio, F. (2013). Arthropod endosymbiosis and evolution. In A. Minelli et al. (Eds.), *Arthropod biology and evolution* (pp. 441–477). Berlin: Springer.

Wilkinson, C. R. (1984). Immunological evidence for the Precambrian origin of bacterial symbioses in marine sponges. *Proceedings of the Royal Society of London B, 220*, 509–517.

Wilkinson, D. (2001). At cross purposes. *Nature, 412*, 485.

Williams, L. E., & Wernegreen, J. J. (2010). Unprecedented loss of ammonia assimilation capability in a urease-encoding bacterial mutualist. *BMC Genomics, 11*, 687.

Won, Y. J., Jones, W. J., & Vrijenhoek, R. C. (2008). Absence of cospeciation between deepsea mytilids and their thiotrophic endosymbionts. *Journal of Shellfish Research, 27*, 129–138.

Woyke, T., et al. (2006). Symbiosis insights through metagenomic analysis of a microbial consortium. *Nature, 443*, 950–955.

Xie, J., Vilchez, I., & Mateos, M. (2010). *Spiroplasma* bacteria enhance survival of *Drosophila hydei* attacked by the parasitic wasp *Leptopilina heterotoma. PloS One, 5*, e12149.

Yin, L., Nordin, J. H., Lucches, P., & Giorgi, F. (2001). Cysteine proprotease colocalizes with vitellogenin in compound granules of the cockroach fat body. *Cell and Tissue Research, 304*, 391–399.

Zbinden, M., et al. (2008). New insights on the metabolic diversity among the epibiotic microbial community of the hydrothermal shrimp *Rimicaris exoculata. Journal of Experimental Marine Biology and Ecology, 359*, 131–140.

Zchori-Fein, E., Roush, R. T., & Hunter, M. S. (1992). Male production by antibiotic treatment in *Encarsia formosa* (Hymenoptera: Aphelinidae), an asexual species. *Experientia, 48*, 102–105.

Chapter 4
Parasitic Endosymbiosis

Abstract Processes accounting for the insurgence of parasitism are reviewed, including evolutionary mechanisms and Red Queen Model effects. Genetic races and host-parasite coevolution involve virulence and infectivity, with a concomitant role of environmental factors. Selective processes are active within host-parasite interactions, and their effects include the selection of different mechanisms of resistance, host immune response and tolerance. Host cuticle is the first physical barrier to pathogens invasion protecting many invertebrates and integrated by other resistance mechanisms. Several bacteria, including *Bacillus thuringiensis* and other taxa, are reviewed for the production and properties of a wide array of toxins. Factors related to pathogenicity, virulence and specificity of diseases are discussed. A number of invertebrate diseases are described from different groups involved in primary productions, including crustaceans, molluscs, insects and nematodes.

Keywords *Bacillus* • Cuticle • Defense • Diseases • Endospores • Host pathogenicity • Parasitism • *Pasteuria* • Red Queen model • Resistance • Toxins • *Vibrio* • Virulence

1 Introduction

No living organism is free from diseases and invertebrates do not escape this rule. They are the target of a large array of pathogens, including parasitic bacteria characterized by varying degrees of virulence and specificity. Outcomes may range from lethal pathologies to minor, asympthomatic conditions.

When compared to the unique phylum Vertebrata, the more than 30 invertebrate phyla include all together a number of species accounting to more than 90 % of all animals present on earth today. By applying a simple inductive estimation, the number of related infective agents, of which bacteria represent only a fraction, is consequently extremely high, including many species which are thus far unknown or not yet recognized as new taxa (Klaphake 2009). It has been estimated that the number of insects that appeared on earth during their evolutionary time is around 10^{18}, with similar numbers for copepoda and other crustaceans. Actual counts of species placed their number at more than one million (Foottit and Adler 2009). Considering

A. Ciancio, *Invertebrate Bacteriology*, DOI 10.1007/978-94-024-0884-3_4

that each species is (or has been) the host of at least one pathogenic bacterium, the toll of pathogenic associations achieves an ashtonishing level. In nature, indeed, these numbers are often much more higher, being frequent the co-occurrence of various bacterial species per infected individual.

Parasites (from Greek παράσιτο = eating, living at someone's side or expenses, Liddell and Scott 1940) represent a particular case of symbionts, that can be placed at a position opposite to that of bacteria specialized in beneficial mutualism or commensalism (see previous Chapter). However, due to practical implications, not last the regulation of pests or the epidemics in husbandries (i.e. in shrimp aquaculture or other industries) the topic is treated separately in this Chapter, whereas the aspects related to the ecology and regulation of pathogens in production cycles like i.e. agriculture will be treated in Chap. 10.

The relationships between parasitic bacteria and their hosts is in many cases very strict. Like for endosymbionts, some evolutive processes may be tracked back to the early origins of the bacterial associations. Several ecological factors are also involved in the insurgence of host specificity and long term maintenance, which represent important fields of study.

A variety of strategies has been deployed by invertebrate bacterial parasites, including infective mechanisms, the occurrence of genome insertions, yet unsuspected thus far, or the production of toxins. These strategies have been mirrored by host reactions like i.e. the evolution of the innate immune system (see Chap. 5), and by adaptations like the use of elusive behaviours or the selection of detoxification mechanisms. In this Chapter some of these processes are reviewed, with attention to parasite-host evolution and pathogenicity.

2 Parasitism and Evolution

2.1 Red Queen Model

Since the initial theory proposed by Haldane (1949) parasitism is recognized as a fundamental evolutive force leading to the insurgence, on evolutionary time scales, of sexuality and genetic recombination mechanisms. This hypothesis, which represents a milestone in evolutionary biology, was based on the observation that parasites exert a selective pressure on their target populations by attacking the hosts whose traits made them susceptible to parasitism. Within the host population, those individuals carrying different character(s) for the same allele(s) were instead capable to escape infection (or at least had a higher probability to do so). Higher reproductive rates of immune or less suitable hosts may thus establish a population with lower parasite prevalence, at a given time delay. This changes induce subsequently a new selective pressure on the parasite population that favors those mutants provided with novel virulent or adaptive characters, helpful for parasitism. These sequel of serial adaptations yields a so called "arms race", determining the insurgence on a long time scale of an evolutive path.

The hypothesis based on the occurrence of reciprocal selective pressures working as an evolutive force is known as the Red Queen model, as proposed and validated by Van Valen (1973) when studying the probability of extinction that, for many taxa, is a constant parameter over time. The term has been later successfully adopted by the scientific community, also due to its coming from the famous novel by Lewis Carrol.[1] Parasitism exerts a selective pressure that originates recombination cycles in both antagonistic populations (host and parasite), giving rise to—and maintaining in time—a genetic arms race that also counteracts other negative effects, such as those linked to genetic drifts (Hamilton 1980; Howard and Lively 1998).

This process may occur on populations spatially dispersed or separated, and largely contributes to the stability of a range of diverse alleles and, ultimately, to the speciation processes of both organisms (Bérénos et al. 2010). It also explains the persistance of sexual reproduction among animals and plants, in spite of the "cost of males", when compared to self-fertilization or asexual reproduction (Morran et al. 2011). Haldane's law and subsequent views hold for hosts and parasites of different Kingdoms and have been experimentally verified for invertebrates and bacteria. This theoretical framework has been enriched by the addition of genetic mutations as a flanking force supporting sexuality. The latter may confer higher population resistance and fitness, counteracting negative, cumulative trends increasing mutation rates in absence of sexual reproduction, also known as "Muller's rachet" (Muller 1964). Further integrations concern the effects of the immune host reactions and of sex on selection (Masri et al. 2013).

At present, either parasitism and mutations are recognized as two forces interacting in sinergism, responsible for the evolution of sexuality in Eukaryotes (Howard and Lively 1998; Buckling and Rainey 2002; Brockhurst et al. 2004). When present simultaneously, parasitism may also reinforce the strength of selection against parasitized individuals, that bear one or more deleterious mutations (Young et al. 2009).

2.2 *Genetic Races and Coevolution*

There is evidence that parasitism increases host population diversity and that the latter is linked to higher resistance among individuals (Altermatt and Ebert 2008; Wolinska and Spaak 2009). For invertebrates, the hypotheses underpinning the Red Queen model have been proved using the crustacean waterflea *Daphnia magna* and its bacterial parasite *Pasteuria ramosa* (Bacillaceae). This host-parasite association has been extensively studied in experimental assays to investigate the mechanisms of parasitism evolution (Ebert 2008). In a seminal experiment, dormant eggs of *Daphnia* were used with resting infective propagules of *P. ramosa*, collected from

[1] In "*Through the Looking-Glass*" the Red Queen game antagonist tells Alice to run, if she wants to maintain her position ("*in this place it takes all the running you can do, to keep in the same place*").

pond sediment layers accounting for a time series of four decades. Infection data with parasites of different ages showed significant changes in host fecundity and *P. ramosa* infections, evident when the host was challenged with propagules of different years and accounting for a range of host or parasite delays. The changes reflected the effects of the host and *P. ramosa* genetic race, based on the selection of a single allele determining the success of parasitism (Decaestecker et al. 2007; Luijckx et al. 2012, 2013). Red Queen evolutionary trends were also documented to occur in the wild for *Daphnia* spp. and associated bacterial parasites (Wolinska and Spaak 2009).

The evolution of host traits induced by parasitism and of related effects on the parasite population shows complex interactions (Schulte et al. 2010). Population-dependent evolutive patterns could be detected, however, in coevolving hosts and parasites, in comparison to independent populations that were not subject to reciprocal evolutive pressures (Schulte et al. 2011). In experimental assays involving pathogenic strains of *Bacillus thuringiensis* (Bt) and the free living nematode *Caenorhabditis elegans* a number of effects could be observed in the reciprocal evolutionary interactions. The nematode is characterized by two genders, hermaphrodites and males, which are present in equal sex ratios, with reproduction that favors the male sperms. The exposure of mixed nematode populations to parasitism by Bt isolates including nematocidal or not-pathogenic strains, showed an effect on male frequencies in the offspring generations. Males were reduced due to lower survival and resistance, higher susceptibility to the pathogenic bacteria and decreased sexual activity and mating ability. However, males were not completely eliminated and remained stable at a 10–20 % level, indicative of male-dependent benefits of the Red Queen type, likely linked to the maintenance of higher levels of heterozygosis among the offsprings (Masri et al. 2013).

Effects due to reciprocal selection (such as increasing bacterial virulence and host resistance) were identified in another experimental coevolution study on Bt and *C. elegans*. The assays were based on observations carried out during 48 host generations in controlled conditions (Schulte et al. 2010). The rapid changes found concerned characters directly involved in parasitism, with a direct implication on the biology and ecology of the pathogen, due to its shorter generation time and higher population density. Processes like recombination and mutation also competed with or reinforced the alleles related to virulence (pathogen) and resistance (host). A reduced reproduction fitness was observed in Bt and *C. elegans* due to the cost of reciprocal adaptation. Changes concerned the bacterial genes producing toxins, with higher genetic diversity and reassortement during the adaptive process, and higher host nematode mutation rates (Schulte et al. 2010).

Host-parasite coevolution may also affect behavioural traits which may appear in the host population as a direct outcome of parasitism. A behavioural response to parasitism based on avoidance of the Bt cells—as an alternative to a costly immune system activation—was observed in a *C. elegans* populations that coevolved with pathogenic bacterial lines (Hasshoff et al. 2007; Schulenburg and Ewbank 2007; Schulte et al. 2012).

The Red Queen model was also experimentally proved in a study with *C. elegans* populations characterized by different reproduction types (obligate selfing, outcrossing and wild types either reproducing by selfing or outcrossing). The populations were confronted with a lethal *Serratia marcescens* isolate. Coevolution assays showed extinction of the nematode selfing populations within 20 generations in presence of coevolving pathogens, maintaining outcrossing in the *C. elegans* wild type, whereas the bacteria evolved greater infectivity (Morran et al. 2011, 2012).

An increased complexity of the selective pressures induced by parasitism may be achieved if the reciprocal selective steps are affected by an interference due to other ecological factors, depending on (*i*) the environment in which the host-parasite associations take place, including spatial dispersion, (*ii*) the likelihood of host switch or (*iii*) the presence of multiple parasites and symbionts (Auld et al. 2012). Horizontal transmission may also restore high prevalence levels and increase infectivity if there is an exchange of parasites among spatially separated populations, i.e. due to host migration and recolonization, or passive mobility of the pathogen propagules. This process may lead to the reinsurgence of recessive traits "conserved" in diploid genomes and re-entering the reciprocal selection race when a parasite type reappears and colonizes again the host population.

Increased virulence and the infectivity of parasitic bacteria are traits involved in genotype (G) × genotype (G) evolutionary races. They may be also affected by the co-occurrence of multiple parasites and by the different strategies they deploy, i.e. ranging from competition for host exploitation to cooperation (Bose and Schulte 2014). Among the phenotypic outcomes of these interactions, virulence traits may induce structural adaptations of the parasite infective propagules, i.e. increasing their durability, resistance and/or infectivity.

Durable propagules represent a costly survival mechanism for pathogens and may depend, in specialized parasites, on the need to start a new cycle only when suitable hosts are available in their environment or microcosm. Structural adaptation of propagules like the endospores of *Bacillus* and related lineages likely involve proteins present on their surface coat. Endospores are produced by several bacterial parasites characterized by durable, highly resistant propagules, i.e. *Pasteuria* or *Bacillus* spp. It is likely that the durable *Pasteuria* endospores evolved as an adaptation to the likelihood of host extinction and/or lack of suitable, receptive hosts. Their disappearance may result from reduced infectivity, as traits selected through a genetic recombination allow hosts to escape infection, and/or because of density changes due to parasitism and host aggregation, like i.e. in plant parasitic nematodes. In this particular case the evolution of durable bacterial propagules also favors the parasite environmental and passive spread to other host populations. A further host-related factor underpinning the complexity of these relationships and involved in selection is represented by the likelihood, after a given time delay, of parasitic events targeting new susceptible generations of the host, that reappear in a population or that newly colonize the microcosm.

3 Host-Parasite Interactions

3.1 Environmental Factors

Many factors interfere with the coevolutive processes that link invertebrate hosts to their bacterial parasites and may affect, on the short term, the levels of prevalence. They are mainly related to the host population ecology and the environment in which the interactions take place, as well as to the presence of additional biotic factors, including i.e. the presence of further concomitant parasites or predators.

The environment has been recently re-evaluated as a fundamental factor significantly contributing to the coevolutive races. Considering that the majority of invertebrates are ectothermes, temperature has been identified as one of the main drivers controlling the fitness of the *parasite genotype* (G_P) × *host genotype* (G_H) × *environment* (E) interactions. Temperature affects these relationships by acting on variables like the induction and/or length of latency periods, the time required for host recovery, the parasite mortality and reproduction, as well as the levels of virulence or resistance achieved (Mitchell et al. 2005). Depending on the differences in thermal sensitivity of hosts and parasites, non-linear responses can dramatically affect their population dynamics with varying outcomes, as often observed in pest biocontrol attempts (Thomas and Blanford 2003).

Environmental parameters alter the selective processes active in the host-parasite interactions, but these effects received to date only a marginal attention. Genetic variations can be maintained in a population due to the fitness adaptive trade-off of the G_P and/or G_H × E interactions. These occur as the results of challenges due to the environmental variability at the level of the host population or of their innate immune system (Byers 2005; Lazzaro and Little 2009; Wolinska and King 2009). These considerations may have a practical consequence when i.e. choosing a bacterial biocontrol agent for application in a new environment, different from its original one, or when selecting bacterial isolates only in controlled conditions, among members of a collection of types that naturally occur in a given environment.

Other factors like diet and host body environment also play a role in host-parasite adaptive races. Experimental assays using *P. ramosa* in *Daphnia* spp. showed that poor invertebrate host diets and low availability of key nutritional elements like P reduced the host prevalence, by decreasing the developmental rates of the parasitic bacteria (Frost et al. 2008). Past food conditions also affected fitness of the parasite, as shown for *P. ramosa* propagules that were more infective when proceeding from well fed *D. magna* hosts (Little et al. 2007). In the same pathosystem, Mitchell and Read (2005) experimentally showed that the maternal exposure to favourable (rich) food levels negatively affected the resistance capacity of their first generation progeny. It is worth to note at this regard that one or more epigenetic[2]

[2] *Epigenetics*: it indicates the transmission to the progeny of a trait which is not coded by genes, but is based on different conditions, like DNA methylation or presence of small, non-coding RNA molecules affecting the gene expression patterns.

signals might play strong and previously unsuspected effects in these interactions, active along generations, which may be worth further investigations.

The presence of one or more additional bacteria or parasite(s) is also considered to exert an effect on the evolutionary G_P and/or $G_H \times E$ interactions. Some examples are given by the higher host resistance levels induced by mutualistic bacteria in insects, or by the coinfections with other parasites' genotypes or species found in the interactions between nematode parasitic Bt and *C. elegans* (Bose and Schulte 2014).

3.2 *Resistance*

The gut and its epithelium represent one of the most common interfaces between some invertebrate groups and their external environment. At this level, different mechanisms of resistance (through an immune response) and tolerance (by the activation of pathways responsible for damage repair and homeostasis recovery) take place. Gut regions are also common horizontal transmission entries exploited by pathogenic bacteria that eventually proliferate in the host body (Milutinović et al. 2013). The main alternative route is the body exoskeleton or cuticle, whose lysis or degradation allows the bacterium penetration into the host. In successful infections the pathogenic bacteria are capable to overcome some host immune response, and thus disrupt the gut homeostasis (Buchon et al. 2013).

3.2.1 The Cuticle Barrier

In some invertebrate phyla like Arthropoda and Nematoda, the cuticle forms a rigid and elastic cover, originated by layers of chitin, collagens and other proteins which yield an exoskeleton, to which muscles are attached for movements. This is an evolutionary successful structure that allowed the advent of locomotion in early invertebrates (and of flight later on), providing at the same time a set of protective and defensive functions. As an extra-cellular matrix, the cuticle covers the entire body in arthropods or nematodes, including respiratory tracheae (when present) or reproductive, gut and excretory openings and ducts. In other lineages, i.e. Mollusca, the cuticle can be limited to specific organs, like the beaks of squids and octopuses.

The cuticle is assembled in many layers in which chitin is linked to other proteins with different functions, including collagens and cuticlins (Johnstone 1994). Chitin is one of the main components of the cuticle. It is one of the most common natural polymer present on earth, in particular in marine environments, being found in the zooplankton. Chitin is also present in the cell wall of fungi. It is a resistant and flexible polymer, whose monomer is N-acetylglucosamine. Chitin has also been found in the cuticle of molluscs, as well as in the tegumentum, periostracum and radulae.

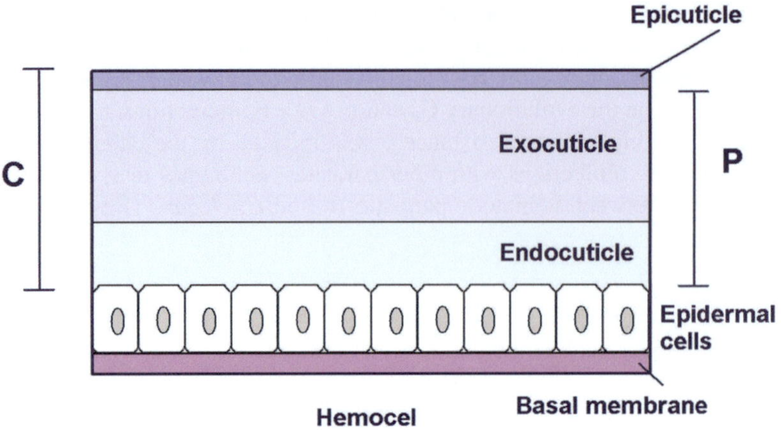

Fig. 4.1 Schematic structure of arthropod cuticle (*C*) and procuticle (*P*) layers, covering the hemocel layers (Re-drawn from Moret and Moreau 2012)

It has also been found in calcified fossil shell layers, suggesting that ancestral cuticle consisted already of proteins, mucopolysaccharides and chitin (Peters 1972).

The cuticle acts as a physical barrier opposing resistance to invasion through the adaptation of its architectural organization, and by means of its chemical composition and thickness. It is also a site of attachment for bacterial symbionts, and the ultimate barrier halting the body penetration process developed by the pathogenic bacteria present on its surface. This goal is also achieved by means of moulting (Moret and Moreau 2012).

In Arthropoda, the cuticle is a layered structure originated by a basal membrane or cell lamina that produces the epidermis, which secretes the outer layers and their various extracellular components. It is usually stratified in a layered procuticle covered by a waxy epicuticle. The procuticle (differentiated into a harder external exocuticle and a softer internal endocuticle, Fig. 4.1) includes proteins, lipids and chitin cross-linked at varying degree, providing elasticity and hardness. Crustaceans and miriapods show an intense carbonate calcification due to the presence of high amounts of $CaCO_3$ increasing the cuticular strength (Moret and Moreau 2012).

The cuticle is reformed at each moulting phase under hormonal control, through a process initiated by the hormone ecdysone. This molecule stimulates the epidermal growth, with the enzymatic digestion of the endocuticle layer and rupture of the exocuticle. During the hardening of the new exocuticle (sclerotization) the outer procuticle proteins covalently bound each other. Enzymatic synthesis of quinones, that polymerize as melanin, confer the cuticle its color and crosslinks with proteins and chitin, for final hardening (Moret and Moreau 2012).

Despite its robustness and thickness, bacterial pathogens produce various enzymes (chitinases and proteases) that can degrade the cuticle at the onset of infection, in spite of the toxicity and antimicrobial properties of i.e. the melanin therein present. Intense melanization and color in insects at moulting have been found associated to their high density, a condition increasing the risk of infection by one or more pathogens (Wilson et al. 2001). Moulting is also an effective process reducing the load of parasites attached to the cuticle, thus contributing to evade infection (Moret and Moreau 2012).

A further defense pathway active at the cuticle level involves the PO zymogens, which eventually leads to the formation of melanized spots in response to invading pathogens. Similarly, various AMPs like cecropins are present, produced by the underlying epidermal cells (Brey et al. 1993).

The cuticle is also found in nematodes and other invertebrates, in which hydrostatic and elastic, rather than rigid, skeletons basically provide similar protective and mechanical functions. It is involved in defense, locomotion (through the internally attached muscles) giving shape to the body by sustaining its turgor, and acting as an hydrostatic skeleton.

In nematodes, the cuticle is present in all stages, including the eggshell and post-embryonic stage and is visible and fully functional already in emerging second stage juveniles. It covers the body anal, genital and stomatal apertures, originates lip structures like hooks, bristles, papillae, probolae, stomatal rhabdions or stylets, and further body appendages. The nematode cuticle is organized in an inner syncitial cell layer called hypodermis, followed by basal, median and cortical layers (Fig. 4.2), with a further external epicuticle covered by a thin surface coat. This is formed by mucins and other proteins, carbohydrates and lipids. The basal layer shows fine networks of collagen fibrils, whereas the median layer is rich in collagens and soluble proteins. The cortical layer has variable thickness depending on the species, and shows collagens and insoluble cuticlins. The surface coat shows antibody domains specific for lectins and other proteins, involved in many interactions with plants or pathogenic microorgansims, with a variable turn over of secreted proteins that are later released in the nematode environment (Page and Johnstone 2007; Curtis et al. 2011).

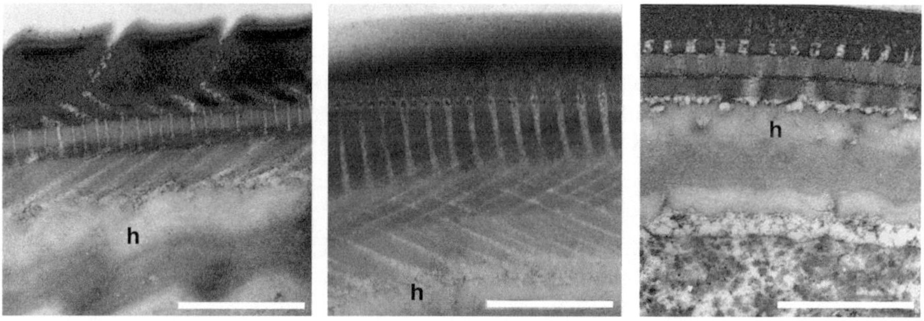

Fig. 4.2 TEM of cuticle from the nematode *Discolaimus major*, showing the hypodermis (*h*) and the network of collagen bands, as visible in longitudinal (*left*) semi-transversal (*center*) and transversal body sections (*right*) (*Scale bars*: 1 μm; images by M. Cermola, A. Ciancio and R. Favre, CNR)

The nematode cutile is the first barrier opposed to invasive bacteria like i.e. the endoparasites of the genus *Pasteuria*, which adhere to their hosts through a highly specific process. The endospore subsequently germinates and the germ peg penetrates their bodies, with a minimal loss in body turgor pressure. The endospore and central core size measured among several members of these bacteria were correlated to the thickness of the cuticle and hypodermis of the corresponding hosts, (Ciancio 1995). This was a re-discovery of a rule defined by Harrison in 1915, relating the size of parasites to the dimension of their hosts (Johnson et al. 2005).

3.2.2 Other Resistance Mechanisms

The gut epithelium and associated membranes represent critical tissues for the selection of resistance traits to bacterial pathogens, being ingestion a common route of entry for bacteria, either as vegetative cells or resting spores. Various insect lines have been selected in the laboratory for resistance to Bt, based on mechanisms like changes of the midgut digestive proteases or ABC transporters, decreased peritrophic membrane permeability, more effective immune responses or esterase production. A further mechanism was found in the cabbage looper, *Trichoplusia ni*, through the reduced binding of Cry Bt toxin due the differential regulation of two midgut aminopeptidase N (APN). An APN1, responsible for susceptibility, was downregulated conferring a resistance trait through a parallel increase in upregulation of a APN6, compensating for the loss of the APN1 functions (Tiewsiri and Wang 2011).

Mosquito control programmes rely on the application of spores of *Lysinibacillus sphaericus* to control susceptible *Culex* populations. In spite of the potential risk for insurgence of resistance among the target populations, due to the use of single bacterial lines and related toxins, host resistance traits showed to be recessive, allowing a reliable use of the spores applied (Rodcharoen and Mulla 1997; Mulla et al. 2003; Oliveira et al. 2003; Berry 2012).

Reported mechanisms of resistance include changes of the bacterium toxin receptors present in the *Culex* gut, through mutations that remove an anchor sequence from the insect receptor protein (Nielsen-Leroux et al. 1995; Oliveira et al. 2004). Other mechanisms involve the deletion of internal receptor domains, necessary for the toxin anchoring, through the insertion of transposable elements (Darboux et al. 2002, 2007). *Culex* populations that have been discovered with resistance traits to most binary (Bin-) toxins are still susceptible, however, to other types of toxins and bacterial strains (Berry 2012).

The identification of the various mechanisms conferring resistance represents a relatively recent field of study, worth the effort. These mechanisms have to be known in detail for practical applications of the bacteria for insect management and control. Resistance may affect indeed not only the choice of the strains to be applied in a given insect control programme, but also the selection of genes variants to introduce in plants or those to be used. GM crops produce in fact a strong field pressure capable to select populations of insects resistant to the Bt genes introgressed (Bravo and Soberón 2008; Tabashnik et al. 2013).

Dietary immunostimulants have been studied in shrimp aquaculture. Products included fragments of bacterial cell wall, with LPS or by β-glucans from fungal or algal species. Exposure to glucans from *Schizophyllum commune* increased resistance of the shrimp *Penaeus monodon* to infection by the white spot syndrome virus (WSSV) (Chang et al. 1999) and of *P. japonicus* and *Litopenaeus vannamei* to *Vibrio* sp. (Itami et al. 1994; Burgents et al. 2004).

3.3 *Toxins*

Many pathogenic bacteria produce a wide array of toxins, which play a central role in the host infection process and its subsequent killing. Several are pore-forming toxins, disrupting the continuity of lipid membranes in the epithelium of i.e. the insect host gut, after bacteria ingestion.

Some of the most studied toxins have implications in the use of one or more bacterial strains in programmes aiming at the control of important insect pests and/or vectors. Other toxins have a role in the infection of invertebrates used as human food, and may have implications also in human health. Together with their bacterial producers some of these toxins are reviewed in the following paragraphs.

3.3.1 Bt Toxins

Several Bt subspecies produce proteinaceous toxins which are mainly responsible for their selective activity against targeted invertebrate hosts. These metabolites are classified as: Cry and Cyt toxins (also known as δ-endotoxins), vegetative insecticidal proteins Vip, secreted insecticidal proteins Sip and parasporins, with further crystal toxins not yet characterized (Estruch et al. 1996; Bravo et al. 2007; Ye et al. 2012; Palma et al. 2014). The toxicity of insect parasitic Bt is initiated after ingestion, being linked to the host feeding activity, and takes place in the midgut and digestive apparatus.

Cry proteins are among the most studied Bt toxins. They are parasporal crystals characterized by varying insecticidal properties and different morphologies. The Bt toxins are produced during the bacterium stationary growth phase and in the sporulation process, and are located as parasporal crystalline inclusions of the endospore (Federici et al. 2006). Cry proteins form a family including several distant types of toxins, thus far accounting for more than 586 amino acid sequences, classified in 73 classes (*cry1* to *cry73* genes)[3] (Shin et al. 1995; Bravo et al. 2007; Ye et al. 2012; Palma et al. 2014;

[3] The Cry toxin classes were initially indicated by roman numbers, accounting for the target insect groups (i.e. I for lepidopterans; II for lepidopterans and dipterans; III for coleopterans or IV for dipterans). This classification has been abandoned in favor of a sequence-based system, due to the increased number of variants with dual targets, the lack of toxicity shown by some proteins and the number of specificity assays required for description of any new toxin.

Crickmore et al. 2016). The coding genes *cry* are mainly located in the Bt plasmid, with only a few strains showing a bacterial chromosome location (Kronstad et al. 1983; Ye et al. 2012).

Cyt toxins are parasporal inclusion proteins similar to Cry, and are produced by Bt lineages characterized by a cytolitic activity. Both Cry and Cyt are pore-forming toxins (PFT), a class of water soluble bacterial proteins that are activated by host proteases. Cry proteins show an α-helical structure leading to the formation of a transmembrane pore, typical of a first group of PFTs. Cyt toxins belong to a second PFT group with a β-barrel conformation and are inserted into the host cell membrane through a number of β-sheet hairpin monomers (Parker and Feil 2005). The numerous toxins forming these classes are distinguished and classified through the pairwise identity of their primary amino acid sequences with those of other known proteins (Crickmore et al. 1998; Bravo et al. 2007).

Both Cry and Cyt toxin types are secreted as protoxins. Cry proteins have a three-domain structure and after secretion are digested by host proteases in smaller protease-resistant polypeptides. These recognize transmembrane glycoproteins (i.e. cadherins and amino peptidases) as specific receptor epitopes on the membranes of the host midgut cells. After binding, they undergo a number of changes that affect their tri-dimensional conformation and reactivity, resulting in the toxin insertion in the epithelial cell membrane. These changes are triggered by a reduction of pH and are eventually followed by a process leading to the formation of pores. These are permeable to inorganic ions, amino acids and sugars, producing the cell lysis and host death. Although this mechanism is typical of the three-domain Cry toxins, the class also includes other unrelated protein types, with different modes of action. Further host death mechanisms also include the Bt proliferation in the host hemolymph, with production of a generalized septicemia (Bravo et al. 2007; Palma et al. 2014; Soberón et al. 2009; Van Frankenhuyzen 2009).

Protoxins of the two Cyt types (Cyt1 and Cyt2) are activated after removal of their -C and -N termini (Li et al. 1996; Bravo et al. 2007). Cyt toxins may also act synergistically with other Bt proteins (Yu et al. 2012).

Vip and Sip toxins are expressed and secreted during the vegetative growth in a number of Bt strains. They are divided into four classes of proteins, lethal for Lepidoptera, Coleoptera or Homoptera (Palma et al. 2014).

Parasporins (Ps) are a further class of Bt toxins showing high similarities to Cry, including the parasporal location of the crystals. However, they are deprived of any insecticidal activity. Strong cytocidal effects against human cancer cells have been reported for this group of proteins (Ohba et al. 2009).

Different shapes of Bt crystals have been reported. Their organizations vary from bi-pyramidal or diamond-shape (measuring ca 0.5–0.7 × 1.0–1.5 μm), to smaller rhomboidal to oval, spindle, spherical or cubic forms (Feng et al. 2001). Bi-piramidal and oval crystals are usually found in Bt *kurstaki* (Tyrell et al. 1981), whereas Bt *israelensis* is characterized by a wider range of cubic-rhomboid crystal shapes, associated to the presence of a net-coat (Abdel-Hameed et al. 1990).

The Bt crystals are formed during the sporulation process. Sporulation in *Bacillus* is governed by a series of biochemical signals exchanged between the two cells arising from first division, through the expression of several *spo* genes and is concluded by the release of an endospore covered by the exosporium, derived by the original cell membrane. As for many *Bacillus* spp. this process can be divided in seven phases starting with the cell asymmetric division and characterized by the subsequent formation of the spore cortex, internally to the mother cell. An axial filament is formed in stage I, followed by the formation of a forespore septum (stage II), of a cell engulfment (stage III), and parasporal crystals and forespore formation (stage IV–VI). These stages are followed by the formation of the exosporium, primordial cell wall, cortex and spore coats with transformation of spore nucleoid and (stage VII) by the spore maturation and sporangial lysis (Fig. 4.3) (Ibrahim et al. 2010).

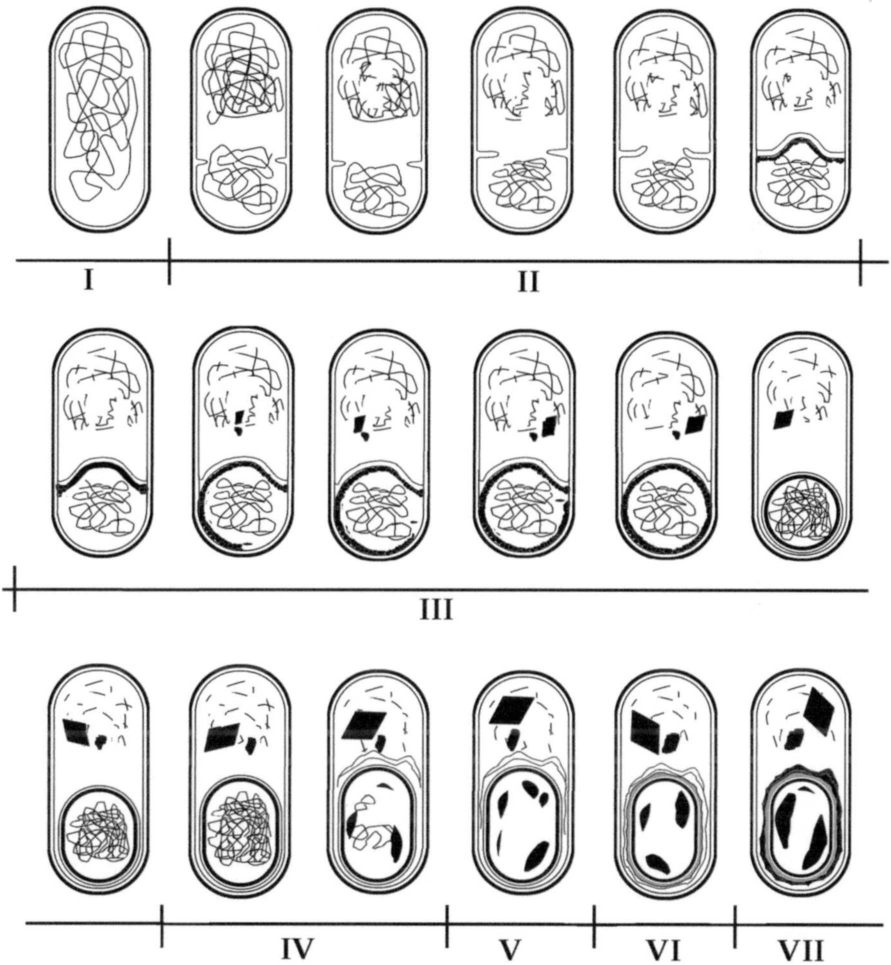

Fig. 4.3 A schematic drawing of the seven phases of the Bt sporulation process, showing the development of two toxin parasporal crystals (Re-drawn with modifications from Bechtel and Bulla 1976)

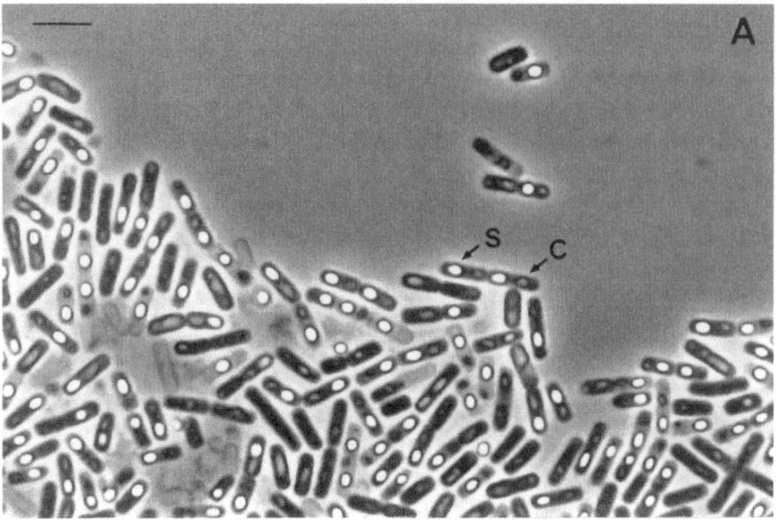

Fig. 4.4 Phase-contrast image of Bt serovar *dakota* cells showing spores (*S*) and crystals (*C*) (Adapted from Kim et al. 2003)

Bacillus endospores are extremely resistant to high temperatures (in some cases they can withstand up to 120 °C for 30 min) and can maintain their vitality after a long "suspended" life, as dormant propagules and in dry or extreme conditions. In optimal growth conditions Bt sporulation can be completed within 10–20 h (Bechtel and Bulla 1976).

The spores can be recognized by their refractivity when examined in phase-contrast light microscopy (Fig. 4.4). Crystals are mostly found outside the exosporium, in a position opposite to the forespore and become separated from the spore at the end of the sporulation process. However, in Bt ssp. *finitimus* and Bt ssp. *oyamesis* the parasporal crystal proteins are located between the exosporium and the spore coat, and adhere to the spore after sporulation (Aronson et al. 1976; Debro et al. 1986; Wojciechowska et al. 1999; Zhu et al. 2011).

Bt toxins have been extensively studied due to the bacterium importance as a biopesticide for management of coleopteran, lepidopteran and dipteran pests. A number of isolates also showed activity against molluscs like the intermediate host of the trematode *Schistosoma japonicum*, or against nematodes like *C. elegans* or plant parasitic species, with a few strains also characterized by antibacterial effects (Palma et al. 2014).

In the last decades the Bt toxins raised a great interest because of the insertion of their coding genes in the genome of genetically modified (GM) plants. The expression of the introduced Bt genes producing the toxins is induced in all tissues and during the entire plant life by the 35S promoter of the Cauliflower Mosaic Virus (Perlak et al. 1991; Li et al. 2003), or may be activated in pollen or green tissues until photosynthesis is active (by PEP, carboxylase promoter). However, due to the high A + T content of the *cry* sequence (non optimal for plants), for an efficient expression of the Bt toxin in GM plants the nucleotidic coding sequence (but not the traduced amino acid) had to be modified to increase its G + C content, and thus become more suitable for expression in green tissues (Murray et al. 1991; Sanchis and Bourguet 2008). By this way a potent and efficient system protecting the plant by the pest attack and killing the larvae is activated (Meiyalaghan et al. 2006). Benefits include a more effective and low cost pest management strategy, that ultimately reduces the amount of chemicals that otherwise had to be required for crop protection, in a conventional management approach.

Bt toxins are associated to the specificity of the producing isolate towards its target insect lineage, and are used in the classification of the bacterium strains. These originate a subspecific ordering and are often referred to as "serovars", "subspecies" or "varieties". The Bt classification was initially based on serology and phenotypic characters, which also appeared correlated to esterase data (Norris 1964). This approach was later integrated by the use of characters like flagellar antigens (De Barjac and Frachon 1990).

In the last years, a growing number of toxins has been discovered and added to the list of those produced by Bt. They are classified by their pairwise amino acid identity with toxins that are already known. Actually, whole genome sequencing and gene and/or toxin amino acid sequence data have been made available in public databases (GenBank, EMBL) for several hundred isolates. The amino acid sequences are used for more stringent and functional toxin nomenclature and classification (Crickmore et al. 1998; Lecadet et al. 1999; Palma et al. 2014). For Cry, a list of 306 holotype toxins is given in the online database developed by the "*Bacillus thuringiensis* δ-endotoxin Nomenclature Committee" which started its activity in 1993 (see http://www.btnomenclature.info/).

Cry toxins represent an heterogeneous group of Bt proteins, whose name arose because of the *crystal* nature of the parasporal inclusions. The group is formed by a number of independent lineages from distinct protein families, the best known of which are the three-domain Cry proteins, Bin- and Etx_Mtx2-like toxins, the latter also produced by the close species *Lysinibacillus sphaericus* (Palma et al. 2014).

The quaternary ranking code adopted for nomenclature of Cry has been adopted also for Cyt and secretable (Vip and Sip) Bt toxins. It shows the protein position in reference to a given class, with 45 %, 78 % and 95 % amino acid pairwise identity as rank borders. The ranking uses the toxin class (i.e. Cry or Vip), followed by a first arabic number, which is different for proteins with <45 % of amino acid identity (i.e. Cry1 and Cry2). The secondary rank is an uppercase letter (different for proteins showing <78 % identity), followed by the tertiary rank (a lowercase letter for <95 % identity) and by the quaternary one, a further arabic number differentiating proteins having pairwise identity >95 % (Palma et al. 2014). This approach simplifies the identification and grouping of new toxins, reducing the need for bioassays when studying their properties and activity (see Table 4.1 for some examples of Cry codes and diversity of targeted groups).

Table 4.1 Examples of Bt sub-species, target hosts and cry toxins produced[a]

Bt ssp.	Hosts	Toxins	References
aizawai	Lepidoptera	Cry1Aa Cry1Ba Cry1Ab7 Cry1Ab8 Cry1Ab9 Cry1Ad1 Cry1Ca2 Cry1Ca3 Cry1Ca5 Cry1Da1 Cry1Eb1 Cry1Fa1 Cry1Fa2	Shimizu et al. (1988), Haider and Ellar (1988), Oeda et al. (1987), Chak and Chen (1993), Feitelson (1993), Sanchis et al. (1989), Höfte et al. (1990), and Chambers et al. (1991)
darmstadiensis	Myceliophagous and plant parasitic nematodes	Cry5Aa1 Cry5Ab1	Narva et al. (1991)
entomocidus	Lepidoptera (*Diatraea saccharalis*, *Plutella xylostella*)	Cry58Aa1 Cry1Ib1	Noguera and Ibarra (2010) and Shin et al. (1995)
finitimus	Lepidoptera	Cry26Aa1 Cry28Aa1 Cry28Aa2	Wojciechowska et al. (1999)
galleriae	Lepidoptera (*Galleria mellonella*)	Cry9Aa1	Smulevitch et al. (1991) and Shevelev et al. (1993)
israelensis	Diptera (mosquitoes: *Aedes aegypti*, *Culex* spp., *Anopheles* spp.)	Cry4Aa Cry4Ba5 Cry11Aa3	Margalith and Ben-Dow (2000) and Berry et al. (2002)
japonensis	Coleoptera (scarabaeid beetles)	Cry8Ca1 Cry9Bb1	Sato et al. (1994) and Silva-Werneck and Ellar (2008)
jegathesan	Diptera (mosquitoes, *Aedes aegypti*)	Cry11Ba1Cry19Aa1	Delécluse et al. (1995), Rosso and Delécluse (1997), and Likitvivatanavong et al. (2010)
kenyae	Lepidoptera (*Spodoptera exigua*)	Cry1Ea1 Cry1Ea2 Cry1Ea4	Visser et al. (1990), Bossé et al. (1990), and Barboza-Corona et al. (1998)
kim	Lepidoptera	Cry57Aa1 Cry59Aa1	Noguera and Ibarra (2010)
kumamotoensis	Coleoptera (*Leptinotarsa decemlineata*)	Cry8Aa1 Cry8Ba1 Cry7Ab2	Zeigler (1999)
kurstaki	Lepidoptera	Cry1Aa1 Cry1Aa6 Cry1Ac6 Cry1Ab2 Cry1Ab6 Cry1Ab12	Schnepf et al. (1985), Masson et al. (1994), Thorne et al. (1986), Hefford et al. (1987), and Silva-Werneck et al. (1999)

(continued)

Table 4.1 (continued)

Bt ssp.	Hosts	Toxins	References
medellin	Diptera (Culicidae)	Cry29Aa1	Delécluse et al. (2000)
		Cry30Aa1	
morrisoni	Lepidoptera (*Artogeia rapae*)	Cry1Ka1	Koo et al. (1995)
san diego	Coleoptera	Cry3Aa	Herrnstadt et al. (1987)
sotto	Lepidoptera	Cry1Aa13	Zhong et al. (2004)
tenebrionis	Coleoptera	Cry3Aa4	McPherson et al. (1988)
thompsoni	Diptera (Muscidae)	Cry15Aa1	Brown and Whiteley (1992)
thuringiensis	Diptera, Coleoptera, Lepidoptera	CryA4, Cry1Ba1	Brizzard and Whiteley (1988), and Zhong et al. (2000)
tolworthi	Coleoptera	Cry3ba1	Sick et al. (1990)

[a]For the full updated list of Bt δ-endotoxins and available NCBI entries see Crickmore et al. (2016), *Bacillus thuringiensis* toxin nomenclature, *http://www.btnomenclature.info/*

3.3.2 *Lysinibacillus sphaericus* Toxins

The genus *Lysinibacillus* (Bacillales) is characterized by the inability to ferment sugars and utilize polysaccharides, by the presence of a peptidoglycan with lysine and aspartic acid, by the resistance to the antibiotics chloramphenicol, streptomycin and tetracycline and by the ability to use arginine as a sole C source (Ahmed et al. 2007; Berry 2012).

Lysinibacillus sphaericus has attracted attention for the potential of mosquitocidal strains in biological control and management of some important vectors of human diseases, with a satisfactory level of biosafety for other superior organisms (Singer 1987; Berry 2012). Insecticidal activity is mainly due to the toxins produced by these strains. They are: a sporulation binary toxin (Bin proteins), a vegetative Mtx toxins, and a Cry toxin with two components (Cry48 and Cry49). The toxins are lethal to the mosquito larvae of the genus *Culex*, that ingest the bacterial cells during their feeding activities (Myers and Yousten 1980; Jones et al. 2007; Berry 2012). Further experimental assays also showed toxic effects on eggs of the nematode *Trichostrongylus colubriformis*, the grass shrimp *Palaemonetes pugio*, and activity of the Mtx1 toxin against larvae of *Chironomus riparius*.

In mosquitocidal strains, the BinA and BinB binary toxins are deposited as parasporal crystals formed by two proteins expressed at the end of the exponential growth phase and at the beginning of sporulation (Ahmed et al. 1995). The toxins are solubilized in the alkaline environment of the insect gut, and undergo a lytic process yielding two products of lower molecular weights. These are necessary to produce a highly toxic activity in *Culex* or *Anopheles* gut tissues, with additional damages induced on nerves and muscles (Berry 2012). After binding to receptors (a GPI anchored maltase, Cpm1) in the caecum and posterior midgut in *Culex* and in other regions in *Anopheles*, the two toxins induce formation of a membrane pore and increase cell vacuolation after introgression in the cytoplasm and/or transfer to other cells. These steps are followed by the subsequent cell death and damage to the internal tissues, but the details of these mechanisms are not yet completely understood (Berry 2012).

Two major classes of Mtx toxins are produced by *L. sphaericus* during the vegetative growth, categorized in Mtx1 and Mtx2 (Berry 2012). Mtx1 is a 100 kDa protein with an N-terminus signal sequence. Its removal yields a 97 kDa toxin, whose lysis produces a 27 kDa product with an ADP-ribosyl transferase sequence. The second product of ca. 70 kDa shows internal repeat sequences showing lectin-like motifs (Schirmer et al. 2002). The toxin crystal structure shows four ricin B-like lectin domains surrounding the ADP-ribosyl transferase, causing inhibition of the enzyme activity (Treiber et al. 2008). The proteolytic cleavage of the complex allows the enzyme uptake into the target cell by means of the lectin domain. The C-terminal 70 kDa region of Mtx1 with the lectin repeats causes morphological alterations in *Culex quinquefasciatus* and *Aedes aegypti* cells. Either the 27 kDa and the 70 kDa domains are necessary for toxicity to mosquito larvae. Cytotoxicity has also been reported for the ADP-ribosylating 27 kDa domain, on HeLa cells (Berry 2012).

Mtx2 is the second class of the *L. sphaericus* toxin gene family. Further toxins (Mtx3 and Mtx4) are also present, likely originated by gene duplication events. Mtx2 are pore-forming toxins produced during the vegetative phase and show mosquitocidal activity. One Mtx2 property is its specificity towards different mosquitoes species, as determined by the amino acid at position 224. In case of lysine the toxin activity is targeted towards *C. quinquefasciatus*, whereas threonine is required for activity against *A. aegypti* (Chan et al. 1996). Synergic coexpression of Mtx1 and Mtx2 coding genes showed higher toxicity on larvae of *A. aegypti* (Rungrod et al. 2009).

The *L. sphaericus* Cry toxin is a complex of two proteins, Cry48/Cry49 which were discovered in some strains capable to kill insects that were already resistant to other strains (Jones et al. 2007). Cry48 is related to the Bt Cry toxins, whereas Cry49 is a toxin of the Bin type. The two toxins have a high toxicity towards *Culex* larvae but only when both are present, not being able to kill the larvae when alone. Since resistance to both Bin and Cry toxins appeared in artificial conditions as a low frequency trait, use of bacterial strains expressing both toxins represents a promising approach for insect control (Berry 2012).

Further *L. sphaericus* toxins include a sphaericolysin, deriving from the cleavage of its precursor, a 53 kDa secreted protein, toxic to eukaryotic cells. It belongs to the Thiol cytolysin superfamily (also including perfringolysin O and the Bt alveolysin). Its insecticidal activity is significantly reduced by co-administration with cholesterol and consists in the production of pores, approx. 35 nm wide, in the insect erythrocytes. It was found in a *L. sphaericus* strain proceeding from the pit-making, ant predator *Myrmeleon bore* (Neuroptera), an insect that injects its gut fluids into its preys to kill it. This behaviour suggests a sort of cooperation between the predator and the bacterium (Nishiwaki et al. 2007). Sphaericolysin showed activity on the cockroach *Blattella germanica* and the common cutworm *Spodoptera litura*, induced by injection (Nishiwaki et al. 2007; Berry 2012). Further subtoxic effects were also observed on a mosquito predator, the emipteran *Laccotrephes griseus* and other aquatic organisms (Mathavan et al. 1987).

3.3.3 *Brevibacillus laterosporus*

A further *Bacillus*-related bacterium producing insecticidal toxins is *Brevibacillus laterosporus*. The genus *Brevibacillus* was separated from *Bacillus* by Shida et al. (1996), and placed in the *Brevibacillus brevis* cluster on the basis of 16S rRNA ribosomal sequences and phylogenetic analyses. Further TEM studies confirmed the peculiar trait of the spores, that Laubach already identified in 1916, when providing the first description of the bacterium. This trait originated its name: a sort of canoe-shaped parasporal body adiacent to one side of the spore (Fig. 4.5a–c), determining its lateral position within the sporangium (Fig. 4.5d) (Ruiu 2013).

In comparison to Bt, a 100-fold lower toxicity was reported for *B. laterosporus* toxins, with a low frequency of strains producing crystalline deposits. Two isolates tested against mosquitoes, however, were toxic for larvae of *A. aegypti* and *Anopheles stephensi* and, at a minor extent, to *C. pipiens* fed with 2.4–4.0 $\cdot$ 10^7 spores $\cdot$ ml^{-1} of feed soup. Mosquitocidal activities were associated to sporulated cells, and considered due to irregularly squared crystals produced by the bacteria (Orlova et al. 1998). Larvicidal activities of crystal-producing strains was confirmed by Zubasheva et al. (2010), who showed that the purified tetragonal to romboid crystals were proteins of 68–130 kDa.

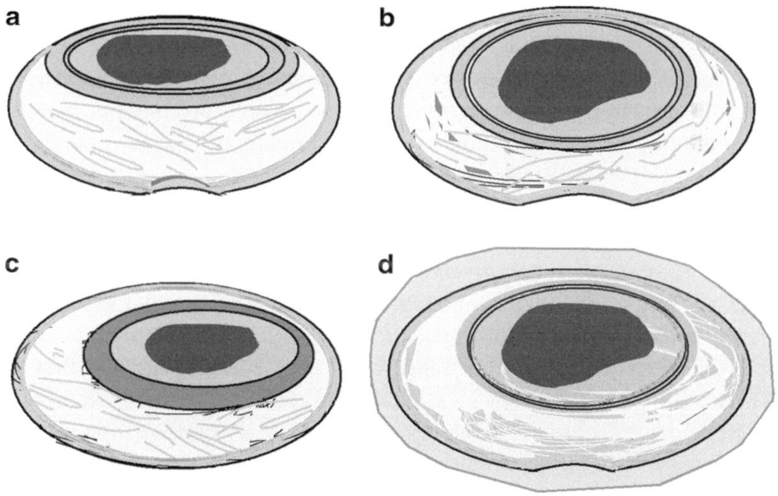

Fig. 4.5 A schemating drawing of the *Brevibacillus laterosporus* typical canoe-shaped resting spore, as seen from lateral (**a**) semi-lateral (**b**) and top view (**c**), or contained within the sporangium (**d**)

Assays and applications of *B. laterosporus* isolates aimed at testing their activity against Coleoptera, Diptera or Lepidoptera. Further applications concerned plant pathogenic bacteria and fungi, due to toxins and antimicrobial or antibiotic properties observed for some strains, also revealed by their sequenced genomes (Djukic et al. 2011; Sharma et al. 2012; Ruiu 2013). The toxins are secreted as insecticidal proteins with homology to some Bt Vip toxins (then classified following their nomenclature). They are produced by isolates with a specific toxicity towards their target insects (Coleptera or Lepidoptera). The proteins are active only upon ingestion, and need complimentary products to become toxic (Ruiu 2013). Among Diptera, several strains showed pathogenicity towards larvae of Culicidae (*Culex quinquefasciatus* and *A. aegypti*) and Simuliidae (the black fly *Simulium vittatum*), but they were 1000 times less effective than Bt *israelensis*. Other strains from Russia, however, showed crystalline inclusions whose insecticidal activity against *A. aegypti* and *A. stephensi* was comparable to Bt (Orlova et al. 1998; Ruiu 2013).

Sporulated cultures of *B. laterosporus* were also toxic to larvae and adults of the house fly *Musca domestica*. Toxicity was mainly associated to the canoe-shaped parasporal body, obtained through cell disruption. Dissected insects showed several types of histological damages with a deteriorated midgut epithelium, together with damages to muscles and connective tissues (Ruiu et al. 2012). A further species of *Brevibacillus*, *B. formosus*, also showed molluscicidal activity towards the snail vector of schistosomiasis, *Biomphalaria glabrata* (Zeigler 2013).

3.3.4 *Paenibacillus* spp.

The genus *Paenibacillus* includes Gram-positive, rod-shaped bacteria pathogenic to insects like *P. popilliae* (former *Bacillus popilliae*, causal agent of the "milky disease of insects", Pettersson et al. 1999), *P. lentimorbus* and *P. larvae*, causing "foulbrood", a severe epizootic disease of honeybee larvae, endemic worldwide (see Sect. 4.3.3). Extracts from cultures of *P. larvae* showed production, with release into the medium, of low-molecular-weight compounds with a potent toxicity towards honeybee larvae (Schild et al. 2014).

Two binary AB toxins, Plx1 and Plx2, are produced by *P. larvae* and act as virulence factors (Fünfhaus et al. 2013). The first one is a member of a toxin family including the *L. sphaericus* Mtx1, whereas the second consists of two subunits. The AB toxins are formed by two components (the A and B subunits) which bind to a cell surface T receptor, for subsequent translocation of the A subunit within the cytosol. Here it acts on a different range of targets. In *P. larvae*, the first subunit (Plx2A) has similarity to a mono-ADP-ribosyltransferase C3 of *P. dendritiformis* and a ADP-ribosyltransferase of *Bacillus cereus*. The sequence of the second subunit (Plx2B) has similarity to the C2 actin-ADP-ribosylating binary toxin of *Clostridium botulinum* and the CDT toxin of *C. difficile*. Comparative genome sequencing of two *P. larvae* strains (genotypes ERIC I and II) also showed the presence of genes coding for further functional toxins (Djukic et al. 2014).

In silico analysis of the sequenced genome of *P. larvae* showed also the presence of C3larvin, a further toxin that appeared highly cytotoxic once cloned and expressed in a yeast (Krska et al. 2015). Toxic factors were also involved in the molluscicidal activity of some isolates of a further species, *P. alvei*, towards *B. glabrata* and the zebra mussel *Dreissena polymorpha*. A nematocidal activity was also shown against the soybean cyst nematode, *Heterodera glycines* (Singer 1996).

3.3.5 Toxins of Other Entomopathogens

Among Gram-negative bacteria, the *Drosophila* pathogen *Pseudomonas entomophila* produces monalysin, a pore-forming toxin. Coupled to an excess of reactive oxygen species (ROS, see Chap. 6) produced by the host during its defense response, monalysin originates gut damages preventing tissue repair. By this way it allows the introgression of bacterial metabolites into the insect hemolymph (Opota et al. 2011).

The sequenced genome of *P. entomophila* showed genes coding for "toxin complex" (*tc*), of the TccC-type (Vodovar et al. 2006). These are insecticidal toxins found only in the two rod-shaped, Gram-negative, entomopathogenic bacteria, *Photorhabdus luminescens* and *Xenorhabdus nematophilus* (γ-Proteobacteria: Enterobacteriaceae). These are two genera of facultative, anaerobic endosymbionts that are stored into specialized intestinal pockets of entomopathogenic nematodes (EPN) of the genera *Heterorhabditis* and *Steinernema*, respectively. EPNs regurgitate the bacteria and introduce them into their target insects. The bacteria release a number of toxic compounds (toxins, drugs and proteases) that overcome the insect immune defense. The sepsis eventually originated kills the hosts, the multiplying bacterial cells subsequently serving as food for the bacteriovorous nematodes (ffrench-Constant et al. 2007; Hinchliffe et al. 2010; Dieppois et al. 2015).

The relationships between EPNs and their endosymbionts is complex and involves the expression of various *tc* genes. *Photorhabdus luminescens* shows four distinct *tc* loci (*tca*, *tcb*, *tcc* and *tcd*), each one with multiple genes of different homology types. These are named A, B and C, the highest toxicities being given by A and B+C. The coded proteins display a range of oral toxic effects towards different insect hosts or tissues (Waterfield et al. 2007). The *tc* operon[4] has also been found in *X. nematophilus* (Sheets et al. 2011), in other insect-parasitic bacteria, including *P. entomophila*, *Serratia entomophila*, *Paenibacillus* sp. as well as in species not related to insects, i.e. *Pseudomonas syringae* and *P. fluorescens*, *Fibrobacter succinogens*, *Burkholderia* spp., *Yersinia pseudotuberculosis* or even human pathogens (Waterfield et al. 2007).

Xenorhabdus nematophilus also encodes a 42-kDa secreted protein, a toxin capable to kill larvae of *Galleria mellonella* and *Helicoverpa armigera* when injected at doses of 30–40 ng · g^{-1} of larvae (Brown et al. 2004). Further compounds from *Photorhabdus* have an effect on insects, i.e. urea-lipids which inhibit the juvenile hormone epoxide hydrolase, a fundamental enzyme involved in insect development and growth (Nollmann et al. 2015).

[4] *Operon* = a cluster of genes including promoting and regulatory elements with coordinated expression, typical of prokaryotes.

Other nematode-bacteria associations include *Serratia* spp. toxic to larvae of *Galleria melonella* when injected into their hemocel, and isolated from *Caenorhabditis* spp. associated to insect bait traps in soil (Abebe et al. 2011). Entomopathogenic *S. marcescens* also produce cytotoxins (Escobar et al. 2001; Farrar et al. 2001). The bacteria producing toxins include *Chromobacterium subtsugae*, a Gram-negative, violet-pigmented species whose cells showed a toxic effect upon ingestion by larvae of the Colorado potato beetle, *Leptinotarsa decemlineata*, *Diabrotica* spp. or other insects (Martin et al. 2007).

3.3.6 Vibrio spp.

Vibrio is a genus of motile, Gram negative, rod shaped and curved bacteria that includes the human pathogen *V. cholerae*. Other species within the genus, associated to aquatic environments, are pathogens of fishes and shrimps.

Vibrio parahaemolyticus is associated to salt water and seafood, causing a disease in humans after its accidental assumption with contaminated food. Together with further related pathogenic species from the genus *Vibrio*, it is a severe pathogen of shrimps, and the causal agents of shrimp vibriosis. Diseases are a severe limiting factor in shrimp production industries, worldwide. A number of *Vibrio* spp. (*V. penaeicida*, *V. alginolyticus* and *V. nigripulchritudo*) produce factors toxic to the shrimp *Litopenaeus stylirostris*, causing a pathology also amplified by rapid changes in temperatures (from 30 to 20 °C) (Goarant et al. 2000).

Two homologs of the *Photorhabdus* insect-related (Pir) toxin-like gene were found in the 69 kb plasmid pVPA3-1 of a *V. parahaemolyticus* strain, causing acute hepatopancreatic necrosis disease (AHPND) of paeneid shrimps, also known as early mortality syndrome (EMS) (Han et al. 2015). The two genes (*pirA*- and *pirB*-like), coding for two secreted proteins of 13 and 50 kDa, showed a GC content lower than the other plasmid genes, indicative of a recent acquisition.

The luminescent *V. harveyi* is the causal agent of a prawn disease (also called lumineous bacteriosis) invading the host hemolymph and hepatopancreas (Liu et al. 1996). The bacteria isolated from diseased prawns have proteolytic (due to a 30 kDa cysteine protease), phospholipase and hemolytic activities stronger than those of reference strains. The hemolytic exotoxins produced (protease and hemolysin) are thermolabile at 60 °C for 2–10 min. They are produced and released after gene transduction or on reception of quorum sensing signals involving expression of luminescence, in the late bacterial growth phase (Nakayama et al. 2006).

4 Diseases

The bacterial species previously examined originate a wide range of infective pathogenic processes that in many cases have a dramatic impact on their hosts' populations. They may also yield detrimental economic effects on activities like i.e.

fisheries or, on the opposite, may result beneficial in the management of plant pests or other noxious insects. In the latter case, however, not all the pathogenic bacteria are suitable for exploitation and only a few species or strains have been proposed, and are currently applied with success, in biological control strategies. Biological control became by itself an emerging discipline in the last decades, integrating basic microbiology with pest control and invertebrate ecology. This pest management approach has been underpinned by a vast literature reflecting a growing research interest in this field of study, due to the possibility to manage a productive agroecosystem relying only a limited number of biological resources.

Many examples are given by published data on the use of species like Bt or other bacteria for control of plant pests or human disease vectors. All these studies afford the identification of major pathogenicity factors which confer the selected species or strains suitable traits for industrial production and practical applications. These factors include host specificity and virulence, the efficacy in killing the target as well as the bacterium persistance in the environment, coupled to the possibility of an easy culturing. Some of them are herein examined, in reference to some examples of invertebrate bacterial diseases.

4.1 Pathogenicity and Virulence

The terms used in invertebrate pathology have been often discussed and emended, as far as the scientific knowledge in this field of study progressed. Pathogenicity and virulence were reviewed by Shapiro-Ilan et al. (2005). The definition of "*pathogenicity*" has been enriched from the original, simple capability of a microorganism to be pathogenic and induce a disease—provided by Steinhaus and Martignoni (1970) and validated by Shapiro-Ilan et al. (2005)—integrating in the term also a qualitative concept based on the capacity to "*invade and injure the host's tissues*" and applied to groups of species (Tanada and Fuxa 1987). Thomas and Elkinton (2004) defined it as the "*number of dead individuals relative to the number exposed to the pathogen*" in a more quantitative and measurable view. However, pathogenicity is considered as an absolute, present/absent qualitative trait, whereas "*virulence*" may be considered as a variable, quantitative and comparative measure of the disease level (Shapiro-Ilan et al. 2005). In this view, a bacterium may be pathogenic to its host if it induces a disease, but different strains may show varying degrees of virulence, if the measured incidence of the disease (also referred to as *prevalence*) varies among them.

In plant and animal pathogenic bacteria, virulence factors like toxins, invasins, adhesins, type III and IV secretion systems and other proteins, are mostly encoded by genes aggregated in some defined genome regions identified as "pathogenicity islands" (Haker and Kaper 2000). These regions are present only in the genome of pathogenic strains and differ from the conserved, core-genome sequences, being characterized by a different G+C content. The pathogenicity islands are the result of HGT events, conferring the recipient strains new pathogenic capacities to infect and/or adapt to a wider range of hosts, or to colonize new environmental or metabolic niches. They add to other virulence determinants which are encoded by plasmids or chromosome regions called "virulence blocks" or "cassettes" and may be acquired through the activity of transposases, transmitted through a plasmid integration or by phages (Haker and Kaper 2000).

However, being dependent on one or more factors, virulence may often, but not invariably, be associated to genetic determinants. Assays performed by feeding different pathogenic bacteria to larvae of *G. mellonella* showed occurrence of specific virulence factors in various strains of *B. cereus*, Bt and *P. entomophila*. They were considered as indicative of co-evolutive processes linking the pathogens and their host (Fedhilaa et al. 2010). The occurrence of virulence factors may be detected experimentally in relatively simple exposure assays using *G. mellonella* or *C. elegans* as models, and several protocols have been made available in the literature at this regard (Tan 2002; Baldini et al. 2002).

The identification of virulence-conferring genes or other co-occurring factors may require, however, more complex experimental procedures, based on gene or even whole genome sequencing. Further factors may also be present, like the occurrence of other microorganisms that may affect virulence or interfere with one or more related genes. For example, differences observed between virulent and non-virulent strains of the prawn pathogen *V. harvey* were associated to the presence of two bacteriophages, from the family *Myoviridae* and *Siphoviridae*, mediating the acquisition and expression of virulent traits (Flegel et al. 2005).

4.2 Specificity

The definitions of "*host range*" and "*specificity*" can be borrowed by the descriptions provided by Van Klinken (2000) in reference to insect host plants. The *host range* may be defined as the sum of species acting as hosts for a given organism. These may or may not coincide in space and time. The *host specificity* was consequently defined as the result of positioning a pathogenic species along a continuum varying from an extremely specialized (single host) to a generalistic (several hosts) range. The latter should, however, be considered not only in terms of breadth of host species, but also in reference to their quality for the pathogen(s) development and/or reproduction (Van Klinken 2000). There is hence a varying degree of specificity observable, ranging from species with a very narrow host range, even reduced to a few populations or genotypes, til species characterized by a generalistic trophic behaviour, capable to attack hosts from very distant lineages. Many intermediate situations may also occur.

The capacity of bacteria to recognize a specific host, and the underpinning genetic mechanisms involved, are important ecologic drivers. Specificity is a useful trait in biocontrol because a pest-specific pathogen has a very low chance to become a severe limiting factor of other, non-target or even useful species (these may include other predatory invertebrates feeding on the target pest or even some pollinators).

This situation is different from the application of a generalist pathogen, characterized by a broader host range, which may spread over many hosts, including the useful species. Specificity has also implications in the insurgence of host-switch mechanisms, as well as in a pathogen's adaptation to new species, populations or sub-populations (Le Clec'h et al. 2012). It also has a role in the host dispersal dynamics, since more dispersed progenies are likely less exposed to a specific parasite, mainly located at the highest host concentration spots. A generalist pathogen instead may show a less aggregated distribution, given its potential to multiply in many different hosts. Specificity may also affect the community structure composition or act in the competition among species, by inducing selective changes among the species present in a given environment. Finally, it plays a role in the natural regulation of the host population or in population genetics and selection processes (Luijckx et al. 2011).

The mechanisms responsible for adaptation to and recognition of target hosts are fundamental steps in the life-cycle of parasitic bacteria, and show $G_H \times G_P$ selection processes (Little et al. 2007). These are mediated by one or more factors active at the host-parasite interface, in most cases present on the preferential site for host attack, often the gut or the body surface. These factors include residues present on the cuticle, epicuticle or other surface proteins, or other membrane adhesion sites and gut receptors, in case of ingested cells. Selection for specificity also occurs at the subsequent metabolic level, during the early vegetative growth phase of a pathogen, as shown by i.e. the specificity and mechanisms of action of Bt crystals or of the toxins produced by other pathogenic bacteria.

Specificity in Bt has been the focus of an intense debate about either the bacterium pathogenicity and host coevolutionary history, due to its capacity to grow in soil and/or colonize germinating plants. This complex behaviour led to the definition of Bt as an "environmental pathogen". These traits are considered to favor the spreading and persistance of the bacterium, as well as its capacity to colonize new environments and survive outside its preferential hosts, prior to infection (Costa Argôlo-Filho and Lopes Loguercio 2014).

4.3 Examples of Invertebrate Diseases

4.3.1 Crustaceans

Several bacterial diseases are known from crustaceans (Wang 2011). Various chitinoclastic bacteria, including *Vibrio* and *Pseudomonas* spp., cause a syndrome known (with a varying terminology) as "shell disease", "rust disease", "black spot" or "brown spot". The disease is manifested through a number of necrotic lesions with dark spots, found on the exoskeleton of crabs and other crustaceans (Messick and Sindermann 1992). The crustaceans in these conditions cannot be commercialized or may be directly killed by the bacteria or other secondary organisms, eventually developing in the lesions.

A second category of crustacean disease is a systemic bacterial infection, which often shows density-dependent prevalence levels. This situation occurs when the host crabs are kept for some time at high numbers in limited space conditions. Symptoms include lethargy, weakness and color changes or whitening of legs and/or gills. *Post-mortem* visual inspections may show low numbers of hemocytes with reduced aggregation capacity and nodules. *Vibrio parahaemolyticus*, *Pseudomonas*, *Acinetobacter*, *Bacillus*, *Flavobacterium* spp. and other coliform species are considered as possible causal agents of the disease, likely penetrating the crabs through capture lesions or other mechanical injuries (Messick and Sindermann 1992).

The Gram-positive micrococcus *Aerococcus viridans* var. *homari* is responsible of a lethal disease observed in the lobster *Homarus americanus* and named "Gaffkemia". The bacterium is capable of circumventing the host defense responses. After penetrating through lesions and wounds (also due to the cannibalistic behaviour of the lobster), the bacteria spread in the host circulatory system and blood vessels, reaching the heart. A few cells of *Aerococcus* are capable to kill the lobster in 2 weeks at 15 °C, likely through a massive disfunction of the hepatopancreas and of its energetic metabolism. The bacterium experimentally infected several other species of shrimps and crustaceans. It was also found, at low prevalence levels, in crabs preyed by lobsters (Johnson 1983).

The diseases caused by *V. harveyi* and other species are a limiting factor for the world industry of the black tiger shrimp, *Penaeus monodon* and other species. Indirectly, *Vibrio* diseases cause more danger when they induce farmers to abuse in antibiotic treatments. These often are either ineffective or increase the frequencies of virulent pathogens. As for mollusks and other farming industries (see below) any abuse in antibiotic treatments represents a danger for the environment and also for the human health. This is due to the risk of selecting antibiotics resistance traits in human pathogens or in other naturally occurring bacterial populations (Moriarty 1999).

A potential alternative to the use of antibiotics to control vibriosis may be the introduction of useful bacteria, like the marine antagonistic *Bacteriovorax* sp. DA5, applied as a biocontrol agent against the vibriosis of *Litopenaeus vannamei* larvae (Wen et al. 2014).

The Gram positive *Pasteuria ramosa* is a member of a bacterial lineage closely related to the genera *Bacillus* and *Clostridium*, found in the hemolymph of the water flea *Daphnia pulex*, *D. magna* or *D. longispina*. This was the first member of the genus *Pasteuria* described by Metchnikoff in 1888. *Pasteuria ramosa* develops "cauliflower" like vegetative stages after host infection and dramatically reduces its fecundity, ultimately filling the body with durable transmission propagules (Ebert et al. 1996). A further *Daphnia* spp. hemolymph infecting bacterium is *Spirobacillus cienkowskii* which induces a color shift in the host, caused by carotenoids that turn the crustacean body pink to scarlet red (Green 1959). Undescribed coccoid pathogens have been also reported as causal agents of the lethal White Fat Cell Disease of *Daphnia* spp. (Decaestecker et al. 2003).

Other pathogens causing crustacean diseases include *Ca*. 'Rhabdochlamydia porcellionis', an intracellular bacterium found in the hepatopancreas of the terrestrial isopod *Porcellio scaber* (Kostanjšek et al. 2004).

4.3.2 Molluscs

Rickettsia like organisms (RLO) have been reported from epithelial cells of gills and from gastric diverticula of bivalves (Bower et al. 1994). RLO parasitism was not associated, in general, to increased host reactions and its effects appeared limited to the induction of cell hypertrophy (Boehs et al. 2010). However, damage to digestive tubules were found in association to large RLO colonies in the commercial clam *Pitar rostrata* (Cremonte et al. 2005). Heavy branchial inflammatory reactions, possibly lethal, were also reported from the clam *Venerupis rhomboides* (Villalba et al. 1999).

A RLO described as "*Ca.* Xenohaliotis californiensis" is the causal agent of a disease causing the withering syndrome of the black abalone *Haliotis cracherodii* in California and of the yellow abalone *H. corrugata* in Mexico (Cáceres-Martínez et al. 2011). A morphologically distinct RLO variant was also observed on the red abalone *H. rufescens* from the same region. This variant was characterized by pleomorphic bacteria with different reactions to stains (Friedman and Crosson 2012). TEM observations and sequence data (resulting in approx. 98–99 % homology between the two cell lines) showed that the variant was the same RLO infected by a spherical to icosahedral phage hyperparasite, whose infection induced the different RLO morphologies observed (Friedman and Crosson 2012). Various phage hyperparasites have been reported from RLO and *Chlamydia*-like pathogens that infect bivalves (Friedman and Crosson 2012), including *Mercenaria mercenaria* (Harshbarger et al. 1977), *Tellina tenuis* (Buchanan 1978), *Meretrix lusoria* (Wen et al. 1994), *Mytilus galloprovincialis* (Comps and Tigé 1999) and *Crassostrea ariakensis* (Sun and Wu 2004). The presence of phages was often associated to bacterial pleomorphism and possibly to the attenuation of RLO pathogenicity and virulence.

During the late 1990s a severe epidemic occurred in the French regions of Normandy and Brittany, resulting in a mass mortality of the European abalone *Haliotis tuberculata*. The prevalence of the disease was high and, depending on the year and location, varied between 60 and 90 % of abalones. The pathogen was identified as a strain of *Vibrio carchariae*. Its presence appeared difficult to detect in living abalones, apart of some white pustules observable on their foot. The effects of the disease, however, were easily visible as many shells resulted empty, due to the predation or decomposition of the dead molluscs. Genetic studies showed *V. carchariae* was close to a further strain known to induce high mortality rates in the Japanese abalone *Sulculus diversicolor supratexta* (Nicolas et al. 2002). A further bacterial species, *V. fluvialis*, was also found to produce pustules and kill the abalone *H. discus hannai* in China (Li et al. 1998).

Vibrio and *Pseudomonas* spp. represent a serious concern in aquaculture and are severe limiting factors affecting industrial production of mollusks (Romalde and Barja 2010). Abalones are the object of an active aquaculture farming industry, and mass mortalities caused by *V. parahaemolyticus*, *V. anguillarum* and *V. carchariae* present challenges to producers with some eventual important threats for human health. The widespread addition of antibiotics to feed and water, applied by producers and aiming at controlling mainly *Vibrio* spp., unfortunately induced the selection of resistant bacterial lineages (Letchumanan et al. 2015). The transfer of plasmids carrying resistance genes to human pathogens like *V. cholerae* or pathogenic *E. coli* lines has been verified *in vitro*. This potential negative effects derive from the indiscriminate use and dispersal of antibiotics in water and environment, with a possible entry in the human body as antibiotic-treated abalones are consumed as fresh food. As a consequence of this situation, more stringent regulations on the use of antibiotics in aquaculture have been promulgated and applied (Moodley et al. 2014).

Parasitic bacteria of mollusks are present in all marine environments where their hosts live. They have also been reported from deep-sea bathymodiolin mussels living on sulfur- and methane-oxidizing bacteria, associated to hydrothermal vents and cold seeps. The filamentous *Ca.* "Endonucleobacter bathymodioli" kills the host infected cells after multiplying in their nuclei, with the only exception of the gill filament cells hosting the ectosymbiotic bacteria (Zielinski et al. 2009).

4.3.3 Insects

Bacterial diseases of insects are known since many centuries, and their study was endeavoured in the nineteenth century by some microbiology pioneers including L. Pasteur, W. Kirby and A. Bassi (Ibrahim et al. 2010). Most common diseases are caused by pathogenic bacteria acquired through ingestion of infective propagules, although other infection mechanisms may also occur. Members of the genera *Paenibacillus*, *Bacillus*, *Pseudomonas*, *Erwinia*, *Serratia* and several Rickettsiales are among the most frequent causal agents of bacterial diseases. Considering the very large phylogenetic radiations of insects and the enormous number of bacteria yet unknown or undescribed, the pathogens of insects account for a very large number of species. Some of them are herein reviewed, mainly in relation to their impact on insect pests or other economically relevant insects. For the biology and effects of insect endosymbionts see Chap. 3.

Paenibacillus popilliae and *P. lentimorbus*, the causal agents of the milky disease of insects, are so called because of the whitish, milky-like aspect assumed by the hemolymph of the infected scarabaeid larvae at the completion of the bacterium vegetative growth. The *P. popilliae* infection process starts when the insect ingests the infective spores during foraging. Spores germination occurs in the hindgut, producing vegetative cells. These reach the midgut where they penetrate the epithelium, to reach in a few days the hemolymph where their vegetative growth occurs. At the end the infection produces several billion spores (up to $5 \cdot 10^9$) that fill the host body (Priest and Dewar 2000). *Paenibacillus popilliae* was originally found on the Japanese beetle *Popillia japonica*, an invasive species accidentally introduced in North America in 1916. This was the first bacterium to be registered for biological control purposes, 15 years later. A *P. popilliae* isolate from the common cockchafer

(*Melolontha melolontha*) showed a Bt homolog gene producing a 79 kDa coded protein (Cry18Aa1), with significant homology to the Cry2Aa endotoxin of Bt (Zhang et al. 1997).

Paenibacillu larvae may be found in honey. Apart of some few particular cases reported in drug addicts making use of honey for methadone injections (Rieg et al. 2010), the bacterium is mainly a honeybee (*Apis mellifera*) pathogen. *Paenibacillus larvae* is responsible of "foulbrood", a septicemia killing the larvae in the hive. It produces highly resistant spores dispersed in the honey or scattered in the comb or on wax, where they can remain for years. Infection of the young honeybee larvae (up to 1.5 days old) occurs when they feed on contaminated honey. The ingestion is followed by the germination of spores in the midgut. The eventual bacterial vegetative growth in the hemocel kills the larvae in a short time (up to 8–11 days old), and sporulation leaves million spores in the insect carcass (Priest and Dewar 2000). The latter can be visually detected by examining the larvae inside the hive cells. The infected insects show a darker, beige-brown color, lose segmentation and, when probed, release a viscous glue-like substance made of million spores (Rieg et al. 2010). The destruction of the entire hive and contaminated equipment by burning is the only remedy allowing sanitation (Genersch 2010; Schild et al. 2014). A similar disease (European fulbrood) is caused by *Melissococcus plutonius*, invading the peritrophic membrane of the gut wall of 4–5 days old larval stages. Dead larvae may be recognized by their flaccid, brownish and curled bodie. *Streptococcus faecalis* and other facultative or opportunistic parasites may also act as secondary invading bacteria (Schmid-Hempel 1998).

Lysinibacillus sphaericus (former *Bacillus sphaericus*) was originally reported by Kellen et al. (1965) as a pathogen of mosquitoes. The bacterium is a motile, aerobic, Gram-positive and rod-shaped member of the genus, forming round-ellipsoid spores with a swollen sporangium (Ahmed et al. 2007). Five main groups have been identified in *Lysinibacillus*, including a toxin-producing line. Genome analyses of mosquitocidal strains showed that they form a cluster of closely related isolates, likely representing a single species. Data indicated an ancient origin, possibly deriving from a common ancestor (Xu et al. 2015). Given its spore durability and resistance, the bacterium was putatively isolated and cultured by Cano and Borucki from the cadaver of a bee species today extinct, embedded and conserved in a 25–49 Myr old Dominican amber (see Chap. 2).

Bt is one of the first and most popular bacteria applied in pest control, due to its easy mass culturing, the broad range of hosts in which it may induce a septicemia, and its wide genotype diversity. It is also non-pathogenic to higher animals (Roh et al. 2007; Pinto et al. 2012). The irreversible activation of the ingested Bt spores (crystal-forming strains) occurs in the alkaline environment of the insect gut. Their eventual germination starts as a response to a number of germinant factors, including nutrients, L-alanine or ribonucleosides (inosine) encountered in this microenvironment (Wilson and Benoit 1993; Benoit et al. 1995). Activation is mediated by an external coat protoxin. A *ger* operon of the pBtoxis plasmid encoding Cry and Cyt toxins also includes further proteins inducing a higher germination rate, after spore alkaline activation (Abdoarrahem et al. 2009).

The induction of pores in the host membrane by Bt toxins is the primary mode of infection, causing the loss of homeostasis and insect death. During infection, Cry toxins of Bt also interact with a number of Ca-dependent transmembrane glycoproteins (cadherins) located in the insect midgut. The multi-functional cadherins act as Cry toxin receptors in several hosts. These proteins are structurally involved in many morphogenetic processes during the growth and development of the larvae, including aggregation of epithelial cells, cell growth, division and death. In particular, the apoptotic[5] role played by cadherins is required during the larval-pupal body re-structuring, and is exploited by the Cry toxins. Their binding to a specific cadherin receptor domain triggers the host programmed cell death and the eventual destruction of the epithelial matrix, facilitating host invasion and retrival of more nutrients (Ibrahim et al. 2010). When interacting with the cadherin protein BT-R_1, the monomeric three-domain Cry toxin triggers a Mg^{++}-dependent pathway inducing a number of cytological, lethal events that also affect membrane integrity and induce cell lysis (Zhang et al. 2006; George and Crickmore 2012).

The Bt life-cycle is conventionally divided in the following phases: (I) vegetative growth, (II) transition to sporulation, (III) sporulation and (IV) spore maturation and cell lysis. Bt produces also virulence factors, including a phospholipase C, proteases and hemolysins, controlled by a pleiotropic[6] *plcR* gene regulating their expression (George and Crickmore 2012). Given the large amount of biological and genetic informations produced on Bt, further data on this bacterium may be obtained from many comprehensive reviews available in the literature (Bravo et al. 2011; George and Crickmore 2012; Costa Argôlo-Filho and Lopes Loguercio 2014).

Further diseases of economically important insects include *Pseudomonas chlororaphis* subsp. *aurantiaca*, a pathogen of *Bombyx mori* causing high mortality of silkworms in natural conditions (Tao et al. 2011). Bacterial diseases of silkworms are known as "flacherie", a generic term describing the loss of turgor shown by infected and dead host larvae. Common septicemia-causing bacteria of silkworms include *Serratia marcescens*, *Bacillus* spp., *Aeromonas* spp. and *Klebsiella granulomatis* (Tao et al. 2011; Mohanta et al. 2015).

In Nigeria, the Akure-China and Akure-Japan strains of *B. mori* showed a higher susceptibility to flacherie than local hybrid strains. The species causing the disease were identified as *Bacillus badius*, *Citrobacter freundii*, *C. amalomaticus*, *Staphylococcus epidermidis*, *Enterobacter cloacae* and *Serratia marcescens* (Ayoade et al. 2014). The latter two species were also reported from diseased silkworms in Cuba, together with *Acinetobacter baumanii* (Díaz Sánchez et al. 2014). Further pathogens inducing septicemia in *B. mori* include *Providencia rettgeri* (Zhang et al. 2013) and *Bacillus bombysepticus*, the causal agent of silkworm "Black Chest Septicemia". The genome of this species, which is close to Bt and produces spores and toxic parasporal crystals, was recently sequenced (Cheng et al. 2014).

---

[5] *Apoptosis* = programmed cell death.

[6] *Pleiotropy* = property of a gene that affects multiple phenotypic traits.

Entomopathogenic bacteria include *Chromobacterium subtsugae*, lethal when ingested by the larvae of the Colorado potato beetle *Leptinotarsa decemlineata* and other insects (Martin et al. 2007), and *Dickeya dadantii* (syn. *Erwinia chrysanthemi*) a pathogen of plants which is also lethal to the aphid *Acyrthosiphon pisum* through septic injury or oral infection (Grenier et al. 2006).

The relationships linking entomopathogenic nematodes (EPNs) to the bacteria that they carry and introduce in their insect hosts (see Chap. 5) have been the object of many studies, given the potential of EPNs in the management of many insect pests (Burnell and Stock 2000). The bacteria belong to the genera *Xenorhabdus* and to some lineages of the bioluminescent genus *Photorhabdus* (γ-Proteobacteria, Enterobacteriales) (Silva et al. 2002). They are Gram-negative rods characterized by a specific phoretic relationship with nematodes of the genera *Steinernema* and *Heterorhabditis*, respectively (Burnell and Stock 2000). The invading EPN stages (third stage dauer juveniles) enter the host body through the natural apertures or by producing direct abrasions through a special tooth. They subsequently inoculate the bacteria that are stored in and released from a specific vesicle located in their intestine (*Steinernema*) or directly from the anterior part of the intestine (*Heterorhabditis*). The resulting septicemia develops very fast and kills the victim, allowing the nematodes to feed on the bacterial cells that represent their food, required to complete their life-cycle (Burnell and Stock 2000).

Members of the genus *Spiroplasma* (Mollicutes: Spiroplasmataceae) are parasitic in social insects. *Spiroplasma melliferum* is parasitic in honeybee and is acquired *per os*.[7] It produces a septicemia killing the bees in a week and is often acquired in late spring, from flowers and other plant organs on which it is present and in direct contact with the visiting bees. Different strains of *Spiroplasma* have been found in several Hymenoptera, including *Bombus* spp. and other solitary bees (Clark et al. 1985; Schmid-Hempel 1998).

4.3.4 Nematodes

Several nematode parasitic or antagonistic bacteria include members of genera like *Bacillus*, *Pseudomonas* or *Streptomyces*, and the nematode pathogens of the genus *Pasteuria*. It is likely that a large fraction of the several soil bacteria pathogenic to nematodes has yet to be discovered. Some species were observed during studies on nematodes biology or on biological antagonism. *Bacillus nematocida* was discovered for the toxic effects were observed in *in vitro* studies on the free living *Panagrellus redivivus* (Huang et al. 2005).

Other *Bacillus* spp. were active *in vitro* on *Caenorhabditis elegans* or *Pristionchus pacificus* (Rae et al. 2010). Field observations on nematode dynamics may also lead to the identification of new taxa, like *Streptomyces costaricanus* identified from a nematode suppressive soil (Esnard et al. 1995) or endophytic bacteria and

[7] Latin for orally.

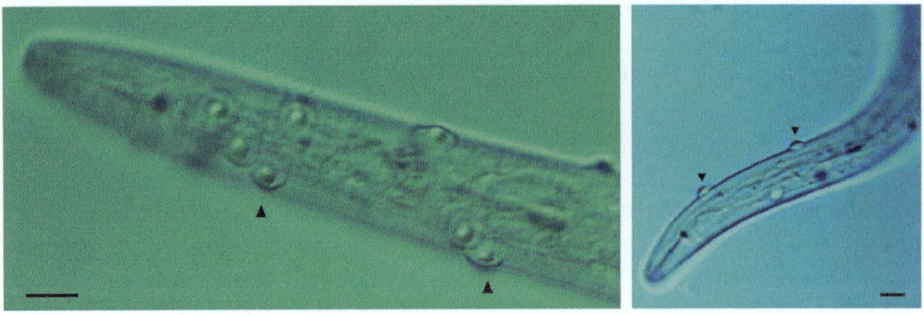

Fig. 4.6 Infective, cup-shaped, resting endospores (*arrowheads*) of the parasitic bacterium *Pasteuria penetrans* adhering to juveniles of the root-knot nematode *Meloidogyne incognita*, extracted from soil before penetrating roots. Scale bars: 5 µm

Pseudomonas spp., including strains of *P. fluorescens*, also active against *Meloidogyne* spp. (Siddiqui and Mahmood 1999; Norabadi et al. 2014).

A large group of nematode parasites is given by the genus *Pasteuria*. Apart of *Pasteuria ramosa*, all the other members of the genus known thus far are specific parasites of nematodes. They are characterized by infective and durable endospores, measuring around 3.5–6 µm in diameter and showing a typical cup-like or falciform shape. The endospores evolved their structure to stick to the host cuticle through adhesive parasporal fibers (Fig. 4.6), and adhere until activation and germination phases are completed (Fig. 4.7). This is a critical phase in the bacterium life-cycle, since the friction forced caused by the hosts' movements among soil particles push to induce their detachment. This peculiar shape also facilitates their detection when examining living or mounted nematodes in light microscopy, at 200–400× magnifications.

Several *Pasteuria* spp. attack a wide range of plant parasitic nematodes. The lineages also includes species parasitic in predatory or free living nematodes. Members of this bacterial group have been found in either cultivated and uncultivated soils, or in aquatic environments on *Plectus* spp. (Sturhan et al. 2005; Franco-Navarro and Godinez-Vidal 2008).

The biodiversity of *Pasteuria* spp. is still underestimated and, given the huge radiation of the phylum Nematoda, it is possible that many hundred or thousand species are present in this lineage. The first nematode-parasitic species described in the genus, *P. penetrans*, attacks root-knot nematodes of the genus *Meloidogyne*. Other species are: *P. thornei*, parasitic in the lesion nematode *Pratylenchus brachyurus*, *P. nishizawae*, attacking cyst nematodes like *Heterodera glycines*, *Ca.* "P. usgae", parasitic in the sting nematode *Belonolaimus longicaudatus*, *Ca.* "P. aldrichii" parasitic in *Bursilla* spp. bacteriovorous nematodes (Giblin-Davis et al. 2011) and *P. hartismeri*, a further species parasitic in root-knot nematodes, in particular *M. ardenensis* (Bishop et al. 2007).

Several species not yet described have been also observed and studied in other nematodes of the genera *Heterodera*, *Globodera* and *Xiphinema*, in the citrus nema-

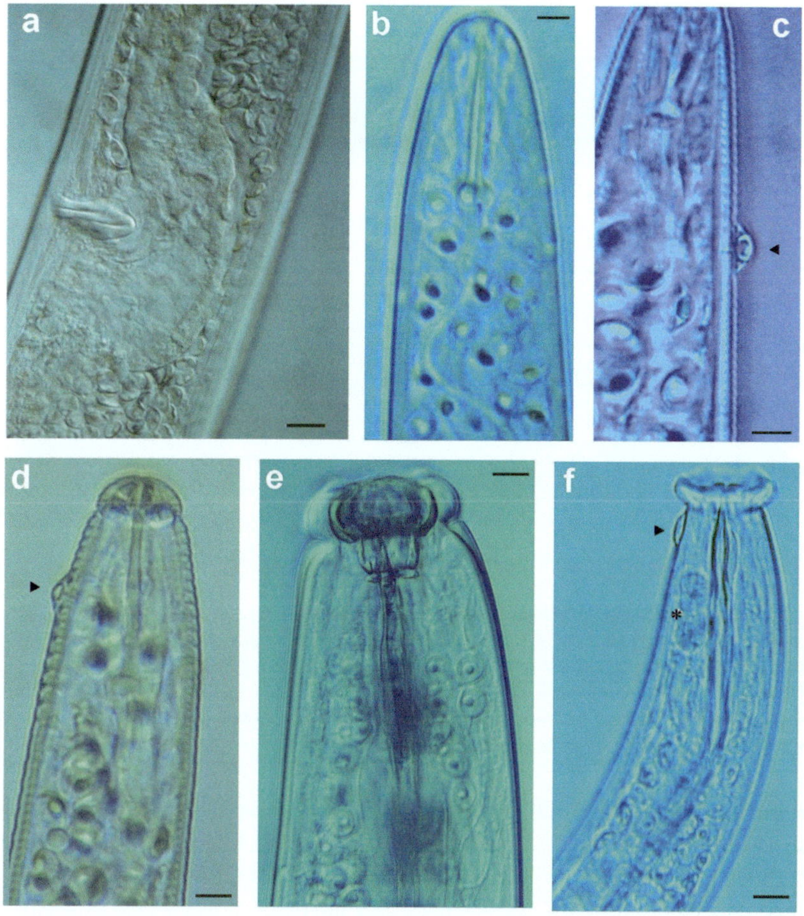

Fig. 4.7 *Pasteuria* spp. endospores from an adult female of the plant parasite *Xiphinema diversicaudatum* (**a**), a juvenile of the citrus nematode *Tylenchulus semipenetrans* (**b**), the plant ectoparasite *Tylenchorhynchus cylindricus* (**c**), a juvenile of the pea cyst nematode *Heterodera goettingiana* (**d**), and from the predatory nematodes *Labronema* (**e**) and *Discolaimus* sp. (**f**). *Arrowheads* show endospores adhering to the cuticle of already parasitized hosts (**c**, **d**, **f**) and germinating propagules from secondary infections (**c**, **d**). *Asterisk* (**f**) shows vegetative thalli. Scale bars: 10 μm (**a**), 3 μm (**b**), 4 μm (**c**–**d**), 5 μm (**e**–**f**)

tode *Tylenchulus semipenetrans* as well as in a number of free-living nematodes, i.e. microbiovoros or predators (Fig. 4.7).

All *Pasteuria* species evolved durable endospores bearing different types of parasporal fibers, that represent either a dormant and an infective propagule (Fig. 4.8). The central core cell is surrounded by a cortex and several wall layers, that confer the endospore its durability and resistance. Experimental tests showed that *P. penetrans* endospores support high temperatures (up to 120 °C for 30 min) without losing infectivity. Similarly they can withstand dehydration, remaining viable and infective for many years (Stirling 2014).

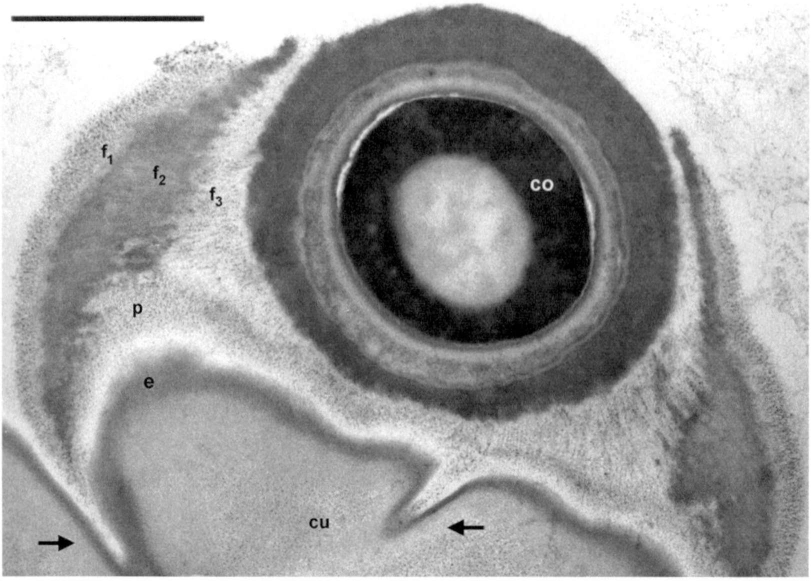

Fig. 4.8 TEM image from a transversal section of a *Pasteuria* sp. infective endospore adhering to the external cuticle (*cu*) surface of a pea cyst nematode *Heterodera goettingiana* male. Adhesion occurs through parasporal fibers (*p*). Further endospore coating fibers are also visible (f_1–f_3). Section shows nematode epicuticle (**e**) and endospore cortex (*co*), encircled by the thick core wall layers. Note the parasporal fibers intrusion into annulation grooves (*arrows*) (Scale bar: 1 μm; image by A. Ciancio, M. Cermola and R. Favre, CNR)

Nematode infection by *P. penetrans* occurs passively through an horizontal transmission when the endospore adhere to the host cuticle by means of the parasporal fibers (Fig. 4.9c, d, f). The endospores usually adhere through their basal side or, in some cases, with their convex part. The host adhesion process is highly specific. Adhesion is mediated by collagen-like residues present on the parasporal fibers that match fibrin-like molecules exposed on the host epicuticle, in a proposed velcro-like mechanism (Davies 2009).

After an activation phase, germination starts with the cortex lysis followed by the extrusion of a germ peg through the basal endospore core plane. It penetrates the nematode cuticle and invades the internal tissues, originating the parasitic phase. The bacterium cells proliferate through fragmentation and produce single units and dicotomic thalli. Sporulation occurs through the formation of tetrads and pairs of cells (Fig. 4.9f–h) originating an asymmetric cell division, in which the distal units complete the endospore formation, with the sporangium and exosporium. The cycle ends with the formation of a number of new endospores. At sporulation completion the host is totally or partially filled by endospores and its reproduction is almost totally inhibited. The propagules are later dispersed in soil at the host's death after cuticle decomposition or rupture, starting new cycles after exosporium decomposition and endospore interception by new susceptible hosts (Fig. 4.9m, a–c).

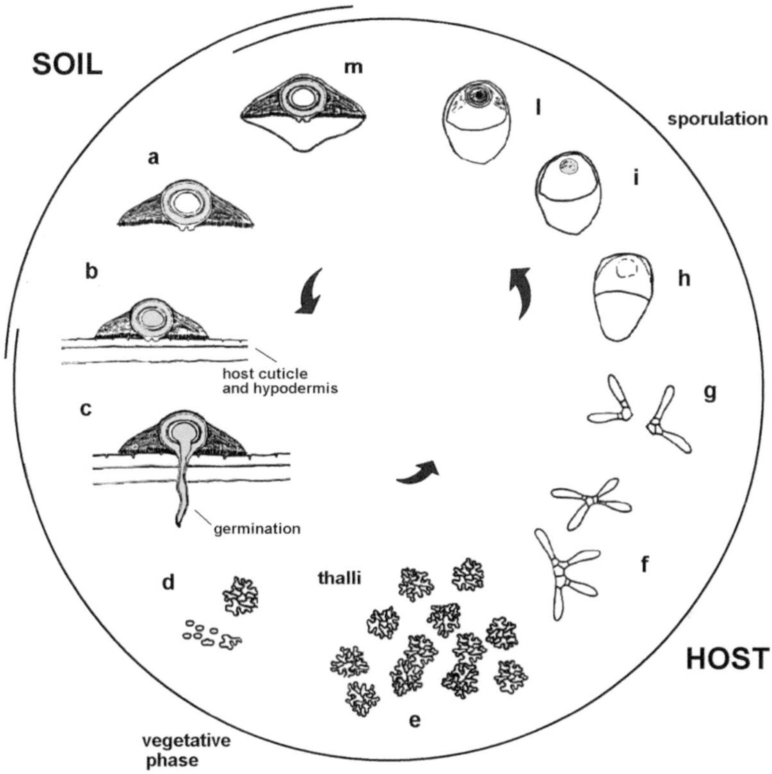

Fig. 4.9 A schematic representation of the lifecycle of *Pasteuria penetrans*, with stages in soil (**a**, **b**, **m**) and within an infected root-knot nematode (**c**–**l**). The endospores in soil are passively intercepted by moving nematodes, a match that is followed by cuticle adhesion (**b**) germ peg penetration (**c**), and by the bacterium development inside the host (**c**–**e**). The cycle is completed by the sporulation phase (**f**–**l**) and subsequent endospores release in soil, at host's death and cuticle rupture (**m**). Stages are shown at different scales

Given the obligate nature of parasitism, *Pasteuria* spp. have been considered as fastidious or unculturable species for many years. A simple method proposed by G. Stirling for production of *P. penetrans* endospores has been adopted for investigation and experimental assays. It consists in the dehydration of roots of tomato or other nematode susceptible plant, artificially infested by adult *Meloidogyne* spp. females that are parasitized and filled with endospores. The plants must be inoculated with juveniles bearing at least three adhering endospores each. After roots liophylization, a powder is produced that may also be pasteurized (60 °C for 30 min) and stored dry for a long time, without losing the propagules viability. The method may be used either as inoculum and inundative soil treataments in biocontrol assays, provided the endospore density per g of product is known. Each *Meloidogyne* female may produce approximately $2 \cdot 10^6$ endospores, 25–28 days after juveniles penetrated roots. Optimal temperature for *P. penetrans* development is around 28 °C (Stirling 2014). Actually, the production of endospores in controlled conditions through an industrial fermentation process has been achieved (Pasteuria Ltd, FL) and some products are available for biological control of nematodes.

References

Abdel-Hameed, A., Lounatmaa, K., Carlberg, G., & El-Tayeb, O. M. (1990). Studies on *Bacillus thuringiensis* H-14 strains isolated in Egypt II. Ultrastructure studies. *World Journal of Microbiology and Biotechnology, 6*, 305–312.

Abdoarrahem, M. M., Gammon, K., Dancer, B. N., & Berry, C. (2009). Genetic basis for alkaline activation of germination in *Bacillus thuringiensis* subsp. *israelensis. Applied and Environmental Microbiology, 1*, 75–19.

Abebe, E., Abebe-Akele, F., Morrison, J., Cooper, V., & Thomas, W. K. (2011). An insect pathogenic symbiosis between a *Caenorhabditis* and *Serratia. Virulence, 2*, 158–161.

Ahmed, H. K., Mitchell, W. J., & Priest, F. G. (1995). Regulation of mosquitocidal toxin synthesis in *Bacillus sphaericus. Applied Environmental Microbiology, 43*, 310–314.

Ahmed, I., Yokota, A., Yamazoe, A., & Fujiwara, T. (2007). Proposal of *Lysinibacillus boronitolerans* gen. nov. sp. nov., and transfer of *Bacillus fusiformis* to *Lysinibacillus fusiformis* comb. nov. and *Bacillus sphaericus* to *Lysinibacillus sphaericus* comb. nov. *International Journal of Systematic and Evolutionary Microbiology, 57*, 1117–1125. doi:10.1099/ijs.0.63867-0.

Altermatt, F., & Ebert, D. (2008). Genetic diversity of *Daphnia magna* populations enhances resistance to parasites. *Ecology Letters, 11*, 918–928.

Aronson, A., Beckman, W., & Dunn, P. (1976). *Bacillus thuringiensis* and related insect pathogens. *Microbiological Reviews, 50*, 1–24.

Auld, S. K. J. R., Hall, S. R., & Duffy, M. A. (2012). Epidemiology of a *Daphnia*-multiparasite system and its implications for the Red Queen. *PLoS ONE, 7*, e39564.

Ayoade, F., Oyejide, N. E., & Fayemi, S. O. (2014). Isolation, identification, antibiogram and characterization of bacterial pathogens of the silkworm, *Bombyx mori*, in South-West Nigeria. *Journal of Biological Sciences, 14*, 425–430.

Baldini, R. L., Lau, G. W., & Rahme, L. G. (2002). Use of plants and insect hosts to model bacterial pathogenesis. In *Bacterial pathogenesis, Part C: Identification, regulation and function of virulence factors* (Methods in enzymology, 358, pp. 3–13). San Diego: Academic Press.

Barboza-Corona, J. E., Lopez-Meza, J. E., & Ibarra, J. E. (1998). Cloning and expression of the crylEa4 gene of *Bacillus thuringiensis* and the comparative toxicity of its gene product. *World Journal of Microbiology and Biotechnology, 14*, 437–441.

Bechtel, D. B., & Bulla, L. A., Jr. (1976). Electron microscope study of sporulation and parasporal crystal formation in *Bacillus thuringiensis. Journal of Bacteriology, 127*, 1472–1481.

Benoit, T. G., Newnam, K. A., & Wilson, G. R. (1995). Correlation between alkaline activation of *Bacillus thuringiensis* var. *kurstaki* spores and crystal production. *Current Microbiology, 31*, 301–303.

Bérénos, C., Wegner, K. M., & Schmid-Hempel, P. (2010). Antagonistic coevolution with parasites maintains host genetic diversity: An experimental test. *Proceedings of the Royal Society B, 278*, 218–224.

Berry, C. (2012). The bacterium, *Lysinibacillus sphaericus*, as an insect pathogen. *Journal of Invertebrate Pathology, 109*, 1–10. doi:10.1016/j.jip.2011.11.008.

Berry, C., et al. (2002). Complete sequence and organization of pBtoxis, the toxin-coding plasmid of *Bacillus thuringiensis* subsp. *israelensis. Applied Environmental Microbiology, 68*, 5082–5095.

Bishop, A. H., Gowen, S. R., Pembroke, B., & Trotter, J. R. (2007). Morphological and molecular characteristics of a new species of *Pasteuria* parasitic on *Meloidogyne ardenensis. Journal of Invertebrate Pathology, 96*, 28–33.

Boehs, G., Villalba, A., Ceuta, L. O., & Luz, J. R. (2010). Parasites of three commercially exploited bivalve mollusc species of the estuarine region of the Cachoeira river (Ilhéus, Bahia, Brazil). *Journal of Invertebrate Pathology, 103*, 43–47.

Bose, J., & Schulte, R. D. (2014). Testing G × G interactions between coinfecting microbial parasite genotypes within hosts. *Frontiers in Genetics, 5*, 124.

Bossé, M., Masson, L., & Brousseau, R. (1990). Nucleotide sequence of a novel crystal protein gene isolated from *Bacillus thuringiensis* subspecies *kenyae. Nucleic Acids Research, 18*, 7443.

Bower, S. M., McGladdery, S. E., & Price, I. M. (1994). Synopsis of infectious diseases and parasites of commercially exploited shellfish. *Annual Review of Fish Diseases, 4*, 1–199.

Bravo, A., & Soberón, M. (2008). How to cope with resistance to Bt toxins? *Trends in Biotechnology, 26*, 573–579.

Bravo, A., Gill, S. S., & Soberón, M. (2007). Mode of action of *Bacillus thuringiensis* Cry and Cyt toxins and their potential for insect control. *Toxicon, 49*, 423–435.

Bravo, A., Likitvivatanavong, S., Gill, S. S., & Soberón, M. (2011). *Bacillus thuringiensis*: A story of a successful bioinsecticide. *Insect Biochemistry and Molecular Biology, 41*, 423–431.

Brey, P. T., et al. (1993). Role of the integument in insect immunity: Epicuticular abrasion and induction of cecropin synthesis in cuticular epithelial cells. *Proceedings of the National Academy of Sciences, USA, 90*, 6275–6279.

Brizzard, B. L., & Whiteley, H. R. (1988). Nucleotide sequence of an additional crystal protein gene cloned from *Bacillus thuringiensis* subsp. *thuringiensis. Nucleic Acids Research, 16*, 2723–2724.

Brockhurst, M. A., Rainey, P. B., & Buckling, A. (2004). The effect of spatial heterogeneity and parasites on the evolution of host diversity. *Proceedings of the Royal Society of London B: Biological Sciences, 271*, 107–111.

Brown, K. L., & Whiteley, H. R. (1992). Molecular characterization of two novel crystal protein genes from *Bacillus thuringiensis* subsp. *thompsoni. Journal of Bacteriology, 174*, 549–557.

Brown, S. E., Cao, A. T., Hines, E. R., Akhurst, R. J., & East, P. D. (2004). A novel secreted protein toxin from the insect pathogenic bacterium *Xenorhabdus nematophila. The Journal of Biological Chemistry, 279*, 14595–14601.

Buchanan, J. S. (1978). Cytological studies on a new species of rickettsia found in association with a phage in the digestive gland of the marine mollusk, *Tellina tenuis. Journal of Fish Diseases, 1*, 27–43.

Buchon, N., Broderick, N. A., & Lemaitre, B. (2013). Gut homeostasis in a microbial world: Insights from *Drosophila melanogaster. Nature Reviews Microbiology, 11*, 615–626.

Buckling, A., & Rainey, P. B. (2002). The role of parasites in sympatric and allopatric host diversification. *Nature, 420*, 496–499.

Burgents, J. E., Burnett, K. G., & Burnett, L. E. (2004). Disease resistance of Pacific white shrimp, *Litopenaeus vannamei*, following the dietary administration of a yeast culture food supplement. *Aquaculture, 231*, 1–8.

Burnell, A. M., & Stock, P. S. (2000). *Heterorhabditis*, *Steinernema* and their bacterial symbionts—Lethal pathogens of insects. *Nematology, 2*, 31–42.

Byers, D. L. (2005). Evolution in heterogeneous environments and the potential of maintenance of genetic variation in traits of adaptive significance. *Genetica, 123*, 107–124.

Cáceres-Martínez, J., Vásquez-Yeomans, R., & Flores-Saaib, R. D. (2011). Intracellular prokaryote *Xenohaliotis californiensis* in abalone *Haliotis* spp. from Baja California, México. *Ciencia Pesquera, 19*, 5–11.

Chak, K. F., & Chen, J. C. (1993). Complete nucleotide sequence and identification of a putative promoter region for the expression in *Escherichia coli* of the *cryIA*(*b*) gene from *Bacillus thuringiensis* var. *aizawai* HD133. *Proceedings of the National Science Council, Republic of China, 17*, 7–14.

Chambers, J. A., et al. (1991). Isolation and characterization of a novel insecticidal crystal protein gene from *Bacillus thuringiensis* subsp. *aizawai. Journal of Bacteriology, 173*, 3966–3976.

Chan, S. W., Thanabalun, T., Wee, B. Y., & Porter, A. G. (1996). Unusual amino acid determinants of host range in the Mtx2 family of mosquitocidal toxins. *Journal of Biological Chemistry, 271*, 14183–14187.

Chang, C. F., et al. (1999). Effect of dietary beta-1,3-glucan on resistance to white spot syndrome virus (WSSV) in postlarval and juvenile *Penaeus monodon. Diseases of Aquatic Organisms, 36*, 163–168.

Cheng, T., et al. (2014). Complete genome sequence of *Bacillus bombysepticus*, a pathogen leading to *Bombyx mori* Black Chest Septicemia. *Genome Announcements, 2*, e00312–e00314.

Ciancio, A. (1995). Phenotypic adaptations in Pasteuria spp. nematode parasites. *Journal of Nematology, 27*, 328–338.

Clark, T. B., et al. (1985). *Spiroplasma melliferum*, a new species from the honeybee (*Apis mellifera*). *International Journal of Systematic Bacteriology, 35*, 296–308.

Comps, M., & Tigé, G. (1999). Procaryotic infections in the mussel *Mytilus galloprovincialis* and its parasite the turbellarian *Urastoma cyprinae. Diseases of Aquatic Organisms, 38*, 211–217.

Costa Argôlo-Filho, R., & Lopes Loguercio, L. (2014). *Bacillus thuringiensis* is an environmental pathogen and host-specificity has developed as an adaptation to human-generated ecological niches. *Insects, 5*, 62–91.

Cremonte, F., Balseiro, P., & Figueras, A. (2005). Occurrence of *Perkinsus olseni* (Protozoa: Apicomplexa) and other parasites in the venerid commercial clam *Pitar rostrata* from Uruguay, southwestern Atlantic coast. *Diseases of Aquatic Organisms, 64*, 85–90.

Crickmore, N., et al. (1998). Revision of the nomenclature for the *Bacillus thuringiensis* pesticidal crystal proteins. *Microbiology and Molecular Biology Reviews, 62*, 807–813.

Crickmore, N., et al. (2016). *Bacillus thuringiensis toxin nomenclature*. http://www.btnomenclature.info/.

Curtis, R. H. C., Jones, J. T., Davies, K. G., Sharon, E., & Spiegel, Y. (2011). Plant nematode surface. In K. G. Davies & Y. Spiegel (Eds.), *Biological control of plant-parasitic nematodes: Building coherence between microbial ecology and molecular mechanisms* (pp. 115–144). Dordrecht: Springer.

Darboux, I., et al. (2002). Loss of the membrane anchor of the target receptor is a mechanism of bioinsecticide resistance. *Proceedings of the National Academy of Science, USA, 99*, 5830–5835.

Darboux, I., Charles, J. F., Pauchet, Y., Warot, S., & Pauron, D. (2007). Transposon-mediated resistance to *Bacillus sphaericus* in a field-evolved population of *Culex pipiens* (Diptera: Culicidae). *Cell Microbiology, 9*, 2022–2029.

Davies, K. G. (2009). Understanding the interaction between an obligate hyperparasitic bacterium, *Pasteuria penetrans* and its obligate plant-parasitic nematode host, *Meloidogyne* spp. *Advances in Parasitology, 68*, 211–245.

De Barjac, H., & Frachon, E. (1990). Classification of *Bacillus thuringiensis* strains. *Entomophaga, 35*, 233–240.

Debro, L., Fitz-James, P. C., & Aronson, A. (1986). Two different parasporal inclusions are produced by *Bacillus thuringiensis* subsp. finitimus. *Journal of Bacteriology, 165*, 258–268.

Decaestecker, E., Vergote, A., Ebert, D., & De Meester, L. (2003). Evidence for strong host clone-parasite species interactions in the *Daphnia* microparasite system. *Evolution, 57*, 784–792.

Decaestecker, E., et al. (2007). Host–parasite 'Red Queen' dynamics archived in pond sediment. *Nature, 450*, 870–873.

Delécluse, A., Rosso, M. L., & Ragni, A. (1995). Cloning and expression of a novel toxin gene from *Bacillus thuringiensis* subsp. *jegathesan* encoding a highly mosquitocidal protein. *Applied Environmental Microbiology, 61*, 4230–4235.

Delécluse, A., Juarez-Perez, V., & Berry, C. (2000). Vector-active toxins: Structure and diversity. In J. F. Charles, A. Delécluse, & C. Nielsen-LaRoux (Eds.), *Entomopathogenic bacteria: From laboratory to field application* (pp. 101–125). Dordrecht: Kluwer.

Díaz Sánchez, A. A., et al. (2014). Bacterias patógenas de larvas de *Bombyx mori* L. en áreas de reproducción en Cuba. *Revista de Protección Vegetal, 29*, 216–219.

Dieppois, G., Opota, O., Lalucat, J., & Lemaitre, B. (2015). *Pseudomonas entomophila*: A versatile bacterium with entomopathogenic properties. In J. L. Ramos et al. (Eds.), *Pseudomonas* (pp. 25–49). Dordrecht: Springer.

Djukic, M., Poehlein, A., Thürmer, A., & Daniel, R. (2011). Genome sequence of *Brevibacillus laterosporus* LMG 15441, a pathogen of invertebrates. *Journal of Bacteriology, 193*, 5535–5536.

Djukic, M., et al. (2014). How to kill the honey bee larva: Genomic potential and virulence mechanisms of *Paenibacillus larvae. PLoS ONE, 9*, e90914. doi:10.1371/journal.pone.0090914.

Ebert, D. (2008). Host–parasite coevolution: Insights from the *Daphnia*–parasite model system. *Current Opinion in Microbiology, 11*, 290–301.

Ebert, D., Rainey, P., Embley, T. M., & Scholz, D. (1996). Development, life cycle, ultrastructure and phylogenetic position of *Pasteuria ramosa* Metchnikoff 1888: Rediscovery of an obligate endoparasite of *Daphnia magna* Straus. *Philosophical Transactions of the Royal Society of London, Series B, 351*, 1689–1701.

Escobar, M. M., Carbonell, G. V., Beriam, L. O., Siqueira, W. J., & Yano, T. (2001). Cytotoxin production in phytopathogenic and entomopathogenic *Serratia marcescens. Revista Latinoamericana de Microbiología, 43*, 165–170.

Esnard, J., Potter, T. L., & Zuckerman, B. M. (1995). Streptomyces costaricanus sp. nov., isolated from nematode-suppressive soil. *International Journal of Systematic Bacteriology, 45*, 775–779.

Estruch, J. J., et al. (1996). Vip3A, a novel *Bacillus thuringiensis* vegetative insecticidal protein with a wide spectrum of activities against lepidopteran insects. *Proceedings of the National Academy of Sciences, USA, 93*, 5389–5394.

Farrar, R. R., Martin, P. A. W., & Ridgway, R. L. (2001). A strain of *Serratia marcescens* (Enterobacteriaceae) with high virulence *per os* to larvae of a laboratory colony of the corn earworm (Lepidoptera: Noctuidae). *Journal of Entomological Sciences, 36*, 380–390.

Federici, B. A., Park, H. W., & Sakano, Y. (2006). Insecticidal protein crystals of *Bacillus thuringiensis*. In J. M. Shively (Ed.), *Inclusions in prokaryotes* (pp. 196–236). Berlin: Springer-Verlag.

Fedhilaa, S., et al. (2010). Comparative analysis of the virulence of invertebrate and mammalian pathogenic bacteria in the oral insect infection model *Galleria mellonella. Journal of Invertebrate Pathology, 103*, 24–29.

Feitelson, J. S. (1993). The *Bacillus thuringiensis* family tree. In L. Kim (Ed.), *Advanced engineered pesticides* (pp. 63–71). New York: Marcel Dekker.

Feng, K. C., Liu, B. L., Chan, H. S., & Tzeng, Y. M. (2001). Morphology of a spectrum of parasporal endotoxin crystals from cultures of *Bacillus thuringiensis* ssp. *kurstaki* isolate A3-4. *World Journal of Microbiology and Biotechnology, 17*, 119–123.

ffrench-Constant, R. H., Eleftherianos, I., & Reynolds, S. E. (2007). A nematode symbiont sheds light on invertebrate immunity. *Trends in Parasitology, 23*, 514–517.

Flegel, T., Pasharawipas, T., Owens, L., & Oakey, H. J. (2005). Evidence for phage-induced virulence in the shrimp pathogen *Vibrio harveyi*. In P. Walker, R. Lester, & M. G. Bondad-Reantaso (Eds.), *Diseases in Asian aquaculture V* (pp. 329–337). Manila: Fish Health Section, Asian Fisheries Society.

Foottit, R. G., & Adler, P. H. (2009). *Insect biodiversity: Science and society* (Eds.). Blackwell, UK, 632 pp.

Franco-Navarro, F., & Godinez-Vidal, D. (2008). Occurrence of *Pasteuria* forms from a biosphere reserve in Mexico. *Nematropica, 38*, 187–194.

Friedman, C. S., & Crosson, L. M. (2012). Putative phage hyperparasite in the rickettsial pathogen of abalone, "*Candidatus* Xenohaliotis californiensis". *Microbial Ecology, 64*, 1064–1072.

Frost, P. C., Ebert, D., & Smith, V. H. (2008). Responses of a bacterial pathogen to phosphorus limitation of its aquatic invertebrate host. *Ecology, 89*, 313–318.

Fünfhaus, A., Poppinga, L., & Genersch, E. (2013). Identification and characterization of two novel toxins expressed by the lethal honey bee pathogen *Paenibacillus larvae*, the causative agent of American foulbrood. *Environmental Microbiology, 15*, 2951–2965.

Genersch, E. (2010). American foulbrood in honeybees and its causative agent, *Paenibacillus larvae. Journal of Invertebrate Pathology, 103*(Supplement 1), 10–19.

George, Z., & Crickmore, N. (2012). *Bacillus thuringiensis* applications in agriculture. In E. Sansinenea (Ed.), *Bacillus thuringiensis biotechnology* (pp. 19–39). New York: Springer Science+Business Media. NL.

Giblin-Davis, R. M., et al. (2011). '*Candidatus* Pasteuria aldrichii', an obligate endoparasite of the bacterivorous nematode *Bursilla. International Journal of Systematic and Evolutionary Microbiology, 61*, 2073–2080.

Goarant, C., et al. (2000). Toxic factors of *Vibrio* strains pathogenic to shrimp. *Diseases of Aquatic Organisms, 40*, 101–107.

Green, J. (1959). Carotenoid pigment in *Spirobacillus cienkowskii* Metchnikoff, a pathogen of Cladocera. *Nature, 183*, 56–57.

Grenier, A. M., Duport, G., Pagès, S., Condemine, G., & Rahbe, Y. (2006). The phytopathogen *Dickeya dadantii* (*Erwinia chrysanthemi*) is a pathogen of the pea aphid. *Applied and Environmental Microbiology, 72*, 1956–1965.

Haider, M. Z., & Ellar, D. J. (1988). Nucleotide sequence of a *Bacillus thuringiensis aizawai* ICI entomocidal crystal protein gene. *Nucleic Acids Research, 16*, 10927.

Haker, J., & Kaper, J. B. (2000). Pathogenicity islands and the evolution of microbes. *Annual Reviews in Microbiology, 54*, 641–679.

Haldane, J. B. S. (1949). Disease and evolution. *La Ricerca Scientifica, 19*, 68–76.

Hamilton, W. D. (1980). Sex versus non-sex versus parasite. *Oikos, 35*, 282–290.

Han, J. E., Tang, K. F. J., Tran, L. H., & Lightner, D. V. (2015). *Photorhabdus* insect-related (Pir) toxin-like genes in a plasmid of *Vibrio parahaemolyticus*, the causative agent of acute hepatopancreatic necrosis disease (AHPND) of shrimp. *Diseases of Aquatic Organisms, 113*, 33–40.

Harshbarger, J. C., Chang, S. C., & Otto, S. V. (1977). Chlamydiae (with phages), mycoplasmas, and rickettsiae in Chesapeake Bay bivalves. *Science, 196*, 666–668.

Hasshoff, M., Bohnisch, C., Tonn, D., Hasert, B., & Schulenburg, H. (2007). The role of *Caenorhabditis elegans* insulin-like signaling in the behavioural avoidance of pathogenic *Bacillus thuringiensis. FASEB Journal, 21*, 1801–1812.

Hefford, M. A., Brousseau, R., Prefontaine, G., Hanna, Z., Condie, J. A., & Lau, P. C. K. (1987). Sequence of a lepidopteran toxin gene of Bacillus thuringiensis subsp kurstaki NRD-12. *Journal of Biotechnology, 6*, 307–322.

Herrnstadt, C., Gilroy, T. E., Sobieski, D. A., Bennett, B. D., & Gaertner, F. H. (1987). Nucleotide sequence and deduced amino acid sequence of a coleopteran-active delta-endotoxin gene from *Bacillus thuringiensis* subsp. *san diego. Gene, 57*, 37–46.

Hinchliffe, S. J., Hares, M. C., Dowling, A. J., & ffrench-Constant, R. H. (2010). Insecticidal toxins from the Photorhabdus and Xenorhabdus bacteria. *The Open Toxinology Journal, 3*, 83–100.

Höfte, H., Soetaert, P., Jansens, S., & Peferoen, M. (1990). Nucleotide sequence and deduced amino acid sequence of a new Lepidoptera-specific crystal protein gene from *Bacillus thuringiensis. Nucleic Acids Research, 18*, 5545.

Howard, R. S., & Lively, C. M. (1998). The maintenance of sex by parasitism and mutation accumulation under epistatic fitness functions. *Evolution, 52*, 604–610.

Huang, X. W., Niu, Q. H., Zhou, W., & Zhang, K. Q. (2005). *Bacillus nematocida* sp. nov., a novel bacterial strain with nematotoxic activity isolated from soil in Yunnan, China. *Systematic and Applied Microbiology, 28*, 323–327.

Ibrahim, M. A., Griko, N., Junker, M., & Bulla, L. A. (2010). *Bacillus thuringiensis*. A genomics and proteomics perspective. *Bioengineered Bugs, 1*, 31–50.

Itami, T., Takahashi, Y., Tsuchihira, E., Igusa, H., & Konda, M. (1994). Enhancement of disease resistance of kuruma prawn *Penaeus japonicus* and increase in phagocytic activity of prawn

hemocytes after oral administration of h-1,3-glucan (Schizophyllan). In L. M. Chou et al. (Eds.), *The third Asian fisheries forum* (pp. 375–378). Manila: Asian Fisheries Society.
Johnson, P. T. (1983). Diseases caused by viruses, rickettsiae, bacteria and fungi. In A. J. Provenzano (Ed.), *The biology of crustacea: Pathobiology* (pp. 2–78). New York: Academic.
Johnson, K. P., Bush, S. E., & Clayton, D. H. (2005). Correlated evolution of host and parasite body size: Tests of Harrison's rule using birds and lice. *Evolution, 59*, 1744–1753.
Johnstone, I. L. (1994). The cuticle of the nematode *Caenorhabditis elegans*: A complex collagen structure. *Bioessays, 16*, 171–178.
Jones, G. W., et al. (2007). A new cry toxin with a unique two-component dependency from *Bacillus sphaericus. FASEB Journal, 21*, 4112–4120.
Kellen, W. R., et al. (1965). *Bacillus sphaericus* Neide as a pathogen of mosquitoes. *Journal of Invertebrate Pathology, 7*, 442–448.
Kim, H. S., et al. (2003). Cloning and characterization of two novel crystal protein genes from a *Bacillus thuringiensis* serovar *dakota* Strain. *Current Microbiology, 46*, 33–38.
Klaphake, E. (2009). Bacterial and parasitic diseases of selected invertebrates. *Veterinary Clinics: Exotic Animal Practice, 12*, 639–648.
Koo, B. T., et al. (1995). Cloning of a novel crystal protein gene cry1K from *Bacillus thuringiensis* subsp *morrisoni. FEMS Microbiology Letters, 134*, 159–164.
Kostanjšek, R., Štrus, J., Drobne, D., & Avguštin, G. (2004). '*Candidatus* Rhabdochlamydia porcellionis', an intracellular bacterium from the hepatopancreas of the terrestrial isopod *Porcellio scaber* (Crustacea: Isopoda). *International Journal of Systematic and Evolutionary Microbiology, 54*, 543–549.
Kronstad, J. W., Schnepf, H. E., & Whiteley, H. R. (1983). Diversity of locations for *Bacillus thuringiensis* crystal protein genes. *Journal of Bacteriology, 154*, 419–428.
Krska, D., Ravulapalli, R., Fieldhouse, R. J., Lugo, M. R., & Merrill, A. R. (2015). C3larvin toxin, an ADP-ribosyltransferase from *Paenibacillus larvae. Journal of Biological Chemistry, 290*, 1639–1653.
Lazzaro, B. P., & Little, T. J. (2009). Immunity in a variable world. *Philosophical Transactions of the Royal Society of London, B Biological Sciences, 364*, 15–26.
Le Clec'h, W., et al. (2012). High virulence of *Wolbachia* after host switching: When autophagy hurts. *PLoS Pathogens, 8*, e1002844.
Lecadet, M. M., et al. (1999). Updating the H-antigen classification of *Bacillus thuringiensis. Journal of Applied Microbiology, 86*, 660–672.
Letchumanan, V., et al. (2015). Occurrence and antibiotic resistance of *Vibrio parahaemolyticus* from shellfish in Selangor, Malaysia. *Frontiers in Microbiology, 6*, 1417.
Li, J., Pandelakis, A. K., & Ellar, D. J. (1996). Structure of the mosquitocidal ä-endotoxin CytB from *Bacillus thuringiensis* sp. *kyushuensis* and implications for membrane pore formation. *Journal of Molecular Biology, 257*, 129–152.
Li, T., Ding, M., Zhang, J., Xiang, J., & Liu, R. (1998). Studies on the pustule disease of abalone (Haliotis discus hannai Ino) on the Dalian coast. *Journal of Shellfish Research, 17*, 707–711.
Li, H. R., et al. (2003). Transgenic plants expressiong *Bacillus thuringiensis* delta-endotoxins. *Entomologia Sinica, 10*, 155–166.
Liddell, H. G., & Scott, R. (1940). *A Greek-English lexicon*. Oxford: Clarendon Press.
Likitvivatanavong, S., Chen, J., Bravo, A., Soberón, M., & Gill, S. S. (2010). Role of cadherin, alkaline phosphatase and aminopeptidase N as receptors of Cry11Ba toxin from Bacillus thuringiensis jegathesan in Aedes aegypti. *Applied and Environmental Microbiology, 77*, 24–31.
Little, T., Birch, J., Vale, P., & Tseng, M. (2007). Parasite transgenerational effects on infection. *Evolutionary Ecology Research, 9*, 459–469.
Liu, P. C., Lee, K. K., & Chen, S. N. (1996). Pathogenicity of different isolates of *Vibrio harveyi* in tiger shrimp, *Penaeus monodon. Letters in Applied Microbiology, 22*, 413–416.

Luijckx, P., Ben-Ami, F., Mouton, L., Du Pasquier, L., & Ebert, D. (2011). Cloning of the unculturable parasite *Pasteuria ramosa* and its *Daphnia* host reveals extreme genotype–genotype interactions. *Ecology Letters, 14*, 125–131.

Luijckx, P., Fienberg, H., Duneau, D., & Ebert, D. (2012). Resistance to a bacterial parasite in the crustacean *Daphnia magna* shows Mendelian segregation with dominance. *Heredity, 108*, 547–551.

Luijckx, P., Fienberg, H., Duneau, D., & Ebert, D. (2013). A matching-allele model explains host resistance to parasites. *Current Biology, 23*, 1085–1088.

Margalith, Y., & Ben-Dov, E. (2000). Biological control by Bacillus thuringiensis subsp. israelensis. In J. E. Rechcigl & N. A. Rechcigl (Eds.), *Insect pest management: Techniques for environmental protection* (pp. 243–301). New York: Lewis Publishers.

Martin, P. A. W., Gundersen-Rindal, D., Blackburn, M., & Buyer, J. (2007). *Chromobacterium subtsugae* sp. nov., a betaproteobacterium toxic to Colorado potato beetle and other insect pests. *International Journal of Systematic and Evolutionary Microbiology, 57*, 993–999. doi:10.1099/ijs.0.64611-0.

Masri, L., et al. (2013). Sex differences in host defence interfere with parasite-mediated selection for outcrossing during host–parasite coevolution. *Ecology Letters, 16*, 461–468.

Masson, L., et al. (1994). Specificity domain localization of *Bacillus thuringiensis* insecticidal toxins is highly dependent on the bioassay system. *Molecular Microbiology, 14*, 851–860.

Mathavan, S., Velpandi, A., & Johnson, J. C. (1987). Sub-toxic effects of *Bacillus sphaericus* 1593 M on feeding growth and reproduction of *Laccotrephes griseus* (Hemiptera: Nepidae). *Experimental Biology, 46*, 149–153.

McPherson, S. A., et al. (1988). Characterization of the coleopteran-specific protein gene of *Bacillus thuringiensis* var. *tenebrionis. Bio/Technology, 6*, 61–66.

Meiyalaghan, S., Jacobs, J. M. E., Butler, R. C., Wratten, S. D., & Conner, A. J. (2006). Transgenic potato lines expressing cry1Ba1 or cry1Ca5 Genes are resistant to potato tuber moth. *Potato Research, 49*, 203–216.

Messick, G. A., & Sindermann, C. J. (1992). Synopsis of principal diseases of the blue crab, *Callinectes sapidus.* NOAA Technical Memorandum NMFS-F/NEC-88. U.S. Department of Commerce, National Marine Fisheries Service, Woods Hole, Massachusetts, 25 pp.

Milutinović, B., Stolpe, C., Peub, R., Armitage, S. A. O., & Kurtz, J. (2013). The red flour beetle as a model for bacterial oral infections. *PLoS ONE, 8*, e64638.

Mitchell, S. E., & Read, A. F. (2005). Poor maternal environment enhances offspring disease resistance in an invertebrate. *Proceedings of the Royal Society B, 272*, 2601–2607.

Mitchell, S. E., Rogers, E. S., Little, T. J., & Read, A. F. (2005). Host-parasite and genotype-by-environment interactions: Temperature modifies potential for selection by a sterilizing pathogen. *Evolution, 59*, 70–80.

Mohanta, M. K., et al. (2015). Characterization of *Klebsiella granulomatis* pathogenic to silkworm, *Bombyx mori* L. *3. Biotech, 5*, 577–583.

Moodley, G., Mashigo, L., Lalloo, R., & Singh, S. (2014). Application of biological agents in abalone aquaculture: A South African perspective. In M. Hernandez-Vergara (Ed.), *Sustainable aquaculture techniques* (pp. 207–237). InTech, Rijeka, Croatia.

Moret, Y., & Moreau, J. (2012). The immune role of the arthropod exoskeleton. *Invertebrate Survival Journal, 9*, 200–206.

Moriarty, D. J. W. (1999). Disease control in shrimp aquaculture with probiotic bacteria. In C. R. Bell, M. Brylinsky, & P. Johnson-Green (Eds.), *Microbial biosystems: New frontiers.* Proceedings of the 8th international symposium on microbial ecology. Atlantic Canada Society for Microbial Ecology, Halifax, Canada.

Morran, L. T., Schmidt, O. G., Gelarden, I. A., Parrish, R. C., & Lively, C. M. (2011). Running with the Red Queen: Host-parasite coevolution selects for biparental sex. *Science, 333*, 216–218.

Morran, L. T., Parrish, R. C., Gelarden, I. A., & Lively, C. M. (2012). Temporal dynamics of outcrossing and host mortality rates in host-pathogen experimental coevolution. *Evolution, 67*, 1860–1868.

Mulla, M. S., Thavara, U., Tawatsin, A., Chomposri, J., & Su, T. (2003). Emergence of resistance and resistance management in field populations of tropical *Culex quinquefasciatus* to the microbial control agent *Bacillus sphaericus. Journal of the American Mosquito Control Association, 19*, 39–46.

Muller, H. J. (1964). The relation of recombination to mutational advance. *Mutation Research, 106*, 2–9.

Murray, E. E., et al. (1991). Analysis of unstable RNA transcripts of insecticidal crystal protein genes of Bacillus thuringiensis in transgenic plants and electroporated protoplasts. *Plant Molecular Biology, 16*, 1035–1050.

Myers, P. S., & Yousten, A. A. (1980). Localization of a mosquito-larval toxin of *Bacillus sphaericus* 1593. *Applied and Environmental Microbiology, 39*, 1205–1211.

Nakayama, T., Nomura, N., & Matsumura, M. (2006). Study on the relationship of protease production and luminescence in *Vibrio harveyi. Journal of Applied Microbiology, 101*, 200–205.

Narva, K. E., et al. (1991). Novel *Bacillus thuringiensis* microbes active against nematodes, and genes encoding novel nematodes-active toxin from *Bacillus thuringiensis* isolates. European Patent Office: EP 0462721.

Nicolas, J. L., Basuyaux, O., Mazurié, J., & Thébault, A. (2002). *Vibrio carchariae*, a pathogen of the abalone *Haliotis tuberculata. Diseases of Aquatic Organisms, 50*, 35–43.

Nielsen-Leroux, C., Charles, J. F., Thiéry, I., & Georghiou, G. P. (1995). Resistance in a laboratory population of *Culex quinquefasciatus* (Diptera: Culicidae) to *Bacillus sphaericus* binary toxin is due to a change in the receptor on midgut brush-border membranes. *European Journal of Biochemistry, 228*, 206–210.

Nishiwaki, H., Nakashima, K., Ishida, C., Kawamura, T., & Matsuda, K. (2007). Cloning, functional characterization, and mode of action of a novel insecticidal pore- forming toxin, Sphaericolysin, produced by *Bacillus sphaericus. Applied and Environmental Microbiology, 73*, 3404–3411.

Noguera, P. A., & Ibarra, J. E. (2010). Detection of new *cry* genes of *Bacillus thuringiensis* by use of a novel PCR primer system. *Applied and Environmental Microbiology, 76*, 6150–6155.

Nollmann, F. I., et al. (2015). A *Photorhabdus* natural product inhibits insect juvenile hormone epoxide hydrolase. *ChemBioChem, 16*, 766–771.

Norabadi, M. T., Sahebani, N., & Etebarian, H. R. (2014). Biological control of root-knot nematode (*Meloidogyne javanica*) disease by *Pseudomonas fluorescens* (Chao). *Archives of Phytopathology and Plant Protection, 47*, 615–621.

Norris, J. R. (1964). The classification of *Bacillus thuringiensis. Journal of Applied Bacteriology, 27*, 439–447.

Oeda, K., et al. (1987). Nucleotide sequence of the insecticidal protein gene of *Bacillus thuringiensis* strain *aizawai* IPL7 and its high-level expression in *Escherichia coli. Gene, 53*, 113–119.

Ohba, M., Mizuki, E., & Uemori, A. (2009). Parasporin, a new anticancer protein group from *Bacillus thuringiensis. Anticancer Research, 29*, 427–433.

Oliveira, C. M., Filho, F. C., Beltran, J. E., Silva-Filha, M. H., & Regis, L. (2003). Biological fitness of a *Culex quinquefasciatus* population and its resistance to *Bacillus sphaericus. Journal of the American Mosquito Control Association, 19*, 125–129.

Oliveira, C. M., Silva-Filha, M. H., Nielsen-Leroux, C., Pei, G., Yuan, Z., & Regis, L. (2004). Inheritance and mechanism of resistance to *Bacillus sphaericus* in *Culex quinquefasciatus* (Diptera: Culicidae) from China and Brazil. *Journal of Medical Entomology, 41*, 58–64.

Opota, O., et al. (2011). Monalysin, a novel ß-pore-forming toxin from the *Drosophila* pathogen *Pseudomonas entomophila*, contributes to host intestinal damage and lethality. *Plos Pathogens, 7*, e1002259.

Orlova, M. V., Smirnova, T. A., Ganushkina, L. A., Yacubovich, V. Y., & Azizbekyan, R. R. (1998). Insecticidal activity of *Bacillus laterosporus*. *Applied and Environmental Microbiology, 64*, 2723–2725.

Page, A. P., & Johnstone, I. L. (2007). The cuticle. In WormBook (Ed.), *The C. elegans research community*, WormBook, doi:10.1895/wormbook.1.138.1, http://www.wormbook.org.

Palma, L., Muñoz, D., Berry, C., Murillo, J., & Caballero, P. (2014). *Bacillus thuringiensis* toxins: An overview of their biocidal activity. *Toxins, 6*, 3296–3325.

Parker, M. W., & Feil, S. C. (2005). Pore-forming protein toxins: From structure to function. *Progress in Biophys & Molecular Biology, 88*, 91–142.

Perlak, F. J. R., Fuchs, R. L., Dean, D. A., McPherson, S. L., & Fischhoff, D. A. (1991). Modification of the coding sequence enhances plant expression of insect control protein genes. *Proceedings of the National Academy of Sciences, USA, 88*, 3324–3328.

Peters, W. (1972). Occurrence of chitin in mollusca. *Comparative Biochemistry and Physiology Part B: Comparative Biochemistry, 41*, 541–544.

Pettersson, B., Rippere, K. E., Yousten, A. A., & Priest, F. G. (1999). Transfer of *Bacillus lentimorbus* and *Bacillus popilliae* to the genus *Paenibacillus* with emended descriptions of *Paenibacillus lentimorbus* comb. nov. and *Paenibacillus popilliae* comb. nov. *International Journal of Systematic Bacteriology, 49*, 531–540.

Pinto, L. M. N., et al. (2012). *Bacillus thuringiensis* monogenic strains: Screening and interactions with insecticides used against rice pests. *Brazilian Journal of Microbiology, 43*, 618–626.

Priest, F. G., & Dewar, S. J. (2000). Bacteria and insects. In F. Priest & M. Goodfellow (Eds.), *Applied microbial systematics* (pp. 165–202). Dordrecht: Springer-Science + Business Media.

Rae, R., Iatsenko, I., Witte, H., & Sommer, R. J. (2010). A subset of naturally isolated *Bacillus* strains show extreme virulence to the free-living nematodes *Caenorhabditis elegans* and *Pristionchus pacificus*. *Environmental Microbiology, 12*, 3007–3021.

Rieg, S., et al. (2010). *Paenibacillus larvae* bacteremia in injection drug users. *Emerging Infectious Diseases, 16*, 487–489.

Rodcharoen, J., & Mulla, M. S. (1997). Biological fitness of *Culex quinquefasciatus* (Diptera:Culicidae) susceptible and resistant to *Bacillus sphaericus*. *Journal of Medical Entomology, 34*, 5–10.

Roh, J. Y., Choi, J. Y., Li, M. S., Jin, B. R., & Je, Y. H. (2007). *Bacillus thuringiensis* as a specific, safe, and effective tool for insect pest control. *Journal of Microbiology and Biotechnology, 17*, 547–559.

Romalde, J. L., & Barja, J. L. (2010). Bacteria in molluscs: Good and bad guys. In Formatex (Ed.), *Current research, technology and education topics in applied microbiology and microbial biotechnology* (Vol. 1, pp. 136–147). Badajoz: Formatex Research Center.

Rosso, M. L., & Delécluse, A. (1997). Contribution of the 65-kilodalton protein encoded by the cloned gene cry19A to the mosquitocidal activity of *Bacillus thuringiensis* subsp. *jegathesan*. *Applied Environmental Microbiology, 63*, 4449–4455.

Ruiu, L. (2013). *Brevibacillus laterosporus*, a pathogen of invertebrates and a broad-spectrum antimicrobial species. *Insects, 4*, 476–492.

Ruiu, L., Satta, A., & Floris, I. (2012). Observations on house fly larvae midgut ultrastructure after *Brevibacillus laterosporus* ingestion. *Journal of Invertebrate Pathology, 111*, 211–216.

Rungrod, A., Tjahaja, N. K., Soonsanga, S., Audtho, M., & Promdonkoy, B. (2009). *Bacillus sphaericus* Mtx1 and Mtx2 toxins co-expressed in *Escherichia coli* are synergistic against *Aedes aegypti* larvae. *Biotechnology Letters, 31*, 551–555.

Sanchis, V., & Bourguet, D. (2008). *Bacillus thuringiensis*: Applications in agriculture and insect resistance management. A review. *Agronomy for Sustainable Development, 28*, 11–20.

Sanchis, V., et al. (1989). Nucleotide sequence and analysis of the N-terminal coding region of the Spodoptera-active d-endotoxin gene of *Bacillus thuringiensis aizawai* 7.29. *Molecular Microbiology, 3*, 229–238.

Sato, R., et al. (1994). Cloning, heterologous expression, and localization of a novel crystal protein gene from *Bacillus thuringiensis* serovar *japonensis* strain Buibui toxic to scarabaeid insects. *Current Microbiology, 28*, 15–19.

Schild, H. A., Fuchs, S. W., Bode, H. B., & Grünewald, B. (2014). Low-molecular-weight metabolites secreted by Paenibacillus larvae as potential virulence factors of American foulbrood. *Applied and Environmental Microbiology, 80*, 2484–2492.

Schirmer, J., Wieden, H.-J., Rodnina, M. V., & Aktories, K. (2002). Inactivation of the elongation factor Tu by mosquitocidal toxin-catalysed mono-ADP-ribosylation. *Applied and Environmental Microbiology, 68*, 4894–4899.

Schmid-Hempel, P. (1998). *Parasites in social insects*. Princeton, New Jersey: Princeton University Press, 413 pp.

Schnepf, H. E., Wong, H. C., & Whiteley, H. R. (1985). The amino acid sequence of a crystal protein from *Bacillus thuringiensis* deduced from the DNA base sequence. *Journal of Biological Chemistry, 260*, 6264–6272.

Schulenburg, H., & Ewbank, J. J. (2007). The genetics of pathogen avoidance in *Caenorhabditis elegans. Molecular Microbiology, 66*, 563–570.

Schulte, R. D., Makus, C., Hasert, B., Michiels, N. K., & Schulenburg, H. (2010). Multiple reciprocal adaptations and rapid genetic change upon experimental coevolution of an animal host and its microbial parasite. *Proceedings of the National Academy of Sciences, USA, 107*, 7359–7364.

Schulte, R. D., Makus, C., Hasert, B., Michiels, N. K., & Schulenburg, H. (2011). Host-parasite local adaptation after experimental coevolution of *Caenorhabditis elegans* and its microparasite *Bacillus thuringiensis. Proceedings of the Royal Society of London B, 278*, 2832–2839.

Schulte, R. D., Hasert, B., Makus, C., Michiels, N. K., & Schulenburg, H. (2012). Increased responsiveness in feeding behaviour of *Caenorhabditis elegans* after experimental coevolution with its microparasite *Bacillus thuringiensis. Biology Letters, 8*, 234–236.

Shapiro-Ilan, D. I., Fuxa, J. R., Lacey, L. A., Onstad, D. W., & Kaya, H. K. (2005). Definitions of pathogenicity and virulence in invertebrate pathology. *Journal of Invertebrate Pathology, 88*, 1–7.

Sharma, V., Singh, P. K., Midha, S., Ranjan, M., Korpole, S., & Patil, P. B. (2012). Genome sequence of *Brevibacillus laterosporus* strain GI-9. *Journal of Bacteriology, 194*, 1279.

Sheets, J. J., et al. (2011). Insecticidal toxin complex proteins from *Xenorhabdus nematophilus*. Structure and pore formation. *The Journal of Biological Chemistry, 286*, 22742–22749.

Shevelev, A. B., et al. (1993). Primary structure of cryX, the novel delta-endotoxin-related gene from *Bacillus thuringiensis* spp. *galleriae. FEBS Letters, 336*, 79–82.

Shida, O., Takagi, H., Kadowaki, K., & Komagata, K. (1996). Proposal for two new genera. Brevibacillus gen. nov. and Aneurinibacillus gen. nov. *International Journal of Systematic Bacteriology, 46*, 939–946.

Shimizu, M., et al. (1988). Cloning and expression in *Escherichia coli* of the 135-kDa insecticidal protein gene from *Bacillus thuringiensis* subsp. *aizawai* IPL7. *Agricultural and Biological Chemistry, 52*, 1565–1573.

Shin, B. S., et al. (1995). Distribution of cryV-type insecticidal protein genes in *Bacillus thuringiensis* and coning of cryV-type genes from *Bacillus thuringiensis* subsp. *kurstaki* and *Bacillus thuringiensis* subsp. *entomocidus. Applied and Environmental Microbiology, 61*, 2402–2407.

Sick, A., Gaertner, F., & Wong, A. (1990). Nucleotide sequence of a coleopteran-active toxin gene from a new isolate of *Bacillus thuringiensis* subsp. *tolworthi. Nucleic Acids Research, 18*, 1305.

Siddiqui, Z. A., & Mahmood, I. (1999). Role of bacteria in the management of plant parasitic nematodes. A review. *Bioresource Technology, 69*, 167–179.

Silva, C. P., et al. (2002). Bacterial infection of a model insect: *Photorhabdus luminescens* and *Manduca sexta. Cellular Microbiology, 4*, 329–339.

Silva-Werneck, J. O., & Ellar, D. J. (2008). Characterization of a novel Cry9Bb δ-endotoxin from *Bacillus thuringiensis. Journal of Invertebrate Pathology, 98*, 320–328.

Silva-Werneck, J. O., De-Souza, M. T., de S. Dias, J. M. C., & Ribeiro, B. M. (1999). Characterization of *Bacillus thuringiensis* subsp. *kurstaki* strain S93 effective against the fall armyworm (*Spodoptera frugiperda*). *Canadian Journal of Microbiology, 45*, 464–471.

Singer, S. (1987). Current status of the microbial larvicide *Bacillus sphericus*. In K. Karamorosch (Ed.), *Biotechnology in invertebrate pathology and cell culture* (pp. 133–156). San Diego: Academic Press.

Singer, S. (1996). The utility of strains of morphological group II *Bacillus*. In S. L. Neidleman & A. I. Laskin (Eds.), *Advances in applied microbiology* (Vol. 42, pp. 219–261). San Diego: Academic Press.

Smulevitch, S. V., et al. (1991). Nucleotide sequence of a novel delta-endotoxin gene cryIG of *Bacillus thuringiensis* ssp. *galleriae*. *FEBS Letters, 293*, 25–28.

Soberón, M., Gill, S. S., & Bravo, A. (2009). Signaling versus punching hole: How do *Bacillus thuringiensis* toxins kill insect midgut cells? *Cell and Molecular Life Sciences, 66*, 1337–1349.

Steinhaus, E. A., & Martignoni, M. E. (1970). *An abridged glossary of terms used in invertebrate pathology*, 2nd edn. USDA Forest Service, PNW Forest and Range Experiment Station.

Stirling, G. (2014). *Biological control of plant-parasitic nematodes* (Soil ecosystem management in sustainable agriculture 2nd ed.). Oxfordshire: CABI.

Sturhan, D., Shutova, T. S., Akimov, V. N., & Subbotin, S. A. (2005). Occurrence, hosts, morphology, and molecular characterisation of *Pasteuria* bacteria parasitic in nematodes of the family Plectidae. *Journal of Invertebrate Pathology, 88*, 17–26.

Sun, J., & Wu, X. (2004). Histology, ultrastructure, and morphogenesis of a rickettsial-like organism causing disease in the oyster *Crassostrea ariakensis* Gould. *Journal of Invertebrate Pathology, 86*, 77–86.

Tabashnik, B., Brévault, T., & Carrière, Y. (2013). Insect resistance to Bt crops: Lessons from the first billion acres. *Nature Biotechnology, 31*, 510–521. doi:10.1038/nbt.2597.

Tan, M. W. (2002). Identificatin of host and pathogen factors involved in virulence using *Caenorhabditis elegans*. In *Bacterial pathogenesis, part C: Identification, regulation and function of virulence factors* (Methods in enzymology, 358, pp. 13–29). San Diego: Academic Press.

Tanada, Y., & Fuxa, J. R. (1987). The host population. In J. R. Fuxa & Y. Tanada (Eds.), *Epizootiology of insect diseases* (pp. 113–157). New York: Wiley.

Tao, H. P., et al. (2011). Isolation and identification of a pathogen of silkworm *Bombyx mori*. *Current Microbiology, 62*, 876–883.

Thomas, M. B., & Blanford, S. (2003). Thermal biology in insect-parasite interactions. *Trends in Ecology and Evolution, 18*, 344–350.

Thomas, S. R., & Elkinton, J. S. (2004). Pathogenicity and virulence. *Journal of Invertebrate Pathology, 85*, 146–151.

Thorne, L., et al. (1986). Structural similarity between the Lepidoptera- and Diptera-specific insecticidal endotoxin genes of *Bacillus thuringiensis* subsp. "*kurstaki*" and "*israelensis*". *Journal of Bacteriology, 166*, 801–811.

Treiber, N., Reinert, D. J., Carpusca, I., Aktories, K., & Schulz, G. E. (2008). Structure and mode of action of a mosquitocidal holotoxin. *Journal of Molecular Biology, 381*, 150–159.

Tyrell, D. J., Bulla, L. A., Andrews, R. E., Kramer, K. J., Davjdson, L. I., & Nordin, P. (1981). Comparative biochemistry of entomocidal parasporal crystal of selected *Bacillus thuringiensis* strains. *Journal of Bacteriology, 145*, 105–1062.

Van Frankenhuyzen, K. (2009). Insecticidal activity of *Bacillus thuringiensis* crystal proteins. *Journal of Invertebrate Pathology, 101*, 1–16.

Van Klinken, R. D. (2000). Host specificity testing: Why do we do it and how we can do it better. In R. Van Driesche, T. A. Heard, A. S. McClay, & R. Reardon (Eds.), *Proceedings of session: Host specificity testing of exotic arthropod biological control agents – the biological basis for improvement in safety* (pp. 54–68). USDA Forest Service, Publication #FHTET-99-1, Morgantown: West Virginia, USA.

Van Valen, L. (1973). A new evolutionary law. *Evolutionary Theory, 1*, 1–30.
Villalba, A., Carballal, M. J., López, C., Cabada, A., Corral, L., & Azevedo, C. (1999). Branchial rickettsia-like infection associated with clam Venerupis rhomboides mortality. *Diseases of Aquatic Organisms, 36*, 53–60.
Visser, B., Munsterman, E., Stoker, A., & Dirkse, W. G. (1990). A novel *Bacillus thuringiensis* gene encoding a *Spodoptera exigua*-specific crystal protein. *Journal of Bacteriology, 172*, 6783–6788.
Vodovar, N., et al. (2006). Complete genome sequence of the entomopathogenic and metabolically versatile soil bacterium *Pseudomonas entomophila. Nature Biotechnology, 24*, 673–679.
Wang, W. (2011). Bacterial diseases of crabs: A review. *Journal of Invertebrate Pathology, 106*, 18–26. doi:10.1016/j.jip.2010.09.018.
Waterfield, N., Hares, M., Hinchliffe, S., Wren, B., & ffrench-Constant, R. (2007). The insect toxin complex of *Yersinia*. In R. D. Perry & J. D. Fetherston (Eds.), *The Genus Yersinia. From genomics to function* (Advances in experimental medicine and biology, 603, pp. 247–257). New York: Springer.
Wen, C. M., Kou, G. H., & Chen, S. N. (1994). Rickettsiaceae-like microorganisms in the gill and digestive gland of the hard clam, *Meretrix lusoria* Röding. *Journal of Invertebrate Pathology, 64*, 138–142.
Wen, C., Xue, M., Liang, H., & Zhou, S. (2014). Evaluating the potential of marine *Bacteriovorax* sp. DA5 as a biocontrol agent against vibriosis in *Litopenaeus vannamei* larvae. *Veterinary Microbiology, 173*, 84–91.
Wilson, G. R., & Benoit, T. G. (1993). Alkaline pH activates *Bacillus thuringiensis* spores. *Journal of Invertebrate Pathology, 62*, 87–89.
Wilson, K., Cotter, S. C., Reeson, A. F., & Pell, J. K. (2001). Melanism and disease resistance in insects. *Ecology Letters, 4*, 637–649.
Wojciechowska, J. A., Lewitin, E., Revina, L. P., Zalunin, I. A., & Chestukhina, G. G. (1999). Two novel delta-endotoxin gene families cry26 and cry28 from *Bacillus thuringiensis* ssp. *finitimus. FEBS Letters, 453*(1–2), 46–48.
Wolinska, J., & King, K. C. (2009). Environment can alter selection in host–parasite interactions. *Trends in Parasitology, 25*, 236–244.
Wolinska, J., & Spaak, P. (2009). The cost of being common: Evidence from natural *Daphnia* populations. *Evolution, 63*, 1893–1901.
Xu, K., Yuan, Z., Rayner, S., & Hu, X. (2015). Genome comparison provides molecular insights into the phylogeny of the reassigned new genus *Lysinibacillus. BMC Genomics, 16*, 140. doi:10.1186/s12864-015-1359-x.
Ye, W., et al. (2012). Mining new crystal protein genes from *Bacillus thuringiensis* on the basis of mixed plasmid-enriched genome sequencing and a computational pipeline. *Applied and Environmental Microbiology, 78*, 4795–4801.
Young, J. A., Yourth, C. P., & Agrawal, A. F. (2009). The effect of pathogens on selection gainst deleterious mutations in *Drosophila melanogaster. Journal of Evolutionary Biology, 22*, 2125–2129.
Yu, X., et al. (2012). Co-expression and synergism analysis of Vip3Aa29 and Cyt2Aa3 insecticidal proteins from *Bacillus thuringiensis. Current Microbiology, 64*, 326–331.
Zeigler, D. (1999). *Bacillus thuringiensis and Bacillus cereus. Bacillus* Genetic Stock Center Catalog of Strains. 7th edn, Part 2. The Ohio State University, USA, 56 pp.
Zeigler, D. (2013). *The Family Paenibacillaceae. Bacillus* Genetic Stock Center Catalog of Strains. Part 5. The Ohio State University, USA, 32 pp.
Zhang, J., Hodgman, T. C., Krieger, L., Schnetter, W., & Schairer, H. U. (1997). Cloning and analysis of the first *cry* gene from *Bacillus popilliae. Journal of Bacteriology, 179*, 4336–4341.
Zhang, X. B., Candas, M., Griko, N. B., Taussig, R., & Bulla, L. A. (2006). A mechanism of cell death involving an adenylyl cyclase/PKA signaling pathway is induced by the Cry1Ab toxin of

Bacillus thuringiensis. Proceedings of the National Academy of Sciences, USA, 103, 9897–9902.
Zhang, J., Shen, Z., Tang, X., Xu, L., & Zhu, F. (2013). Isolation and identification of a pathogen, *Providencia rettgeri*, in *Bombyx mori. Journal of Bacteriology Research, 5*, 22–28.
Zhong, C., et al. (2000). Characterization of a *Bacillus thuringiensis* δ-endotoxin which is toxic to insects in three orders. *Journal of Invertebrate Pathology, 76*, 131–139.
Zhong, W. F., Wu, J., Cai, P. Z., & Yan, W. Z. (2004). Cloning and sequencing of Cry1Aa13 gene from *Bacillus thuringiensis* subsp *sotto. Journal of the Sichuan University (Natural Science Edition), 41*, 1050–1053.
Zhu, Y., et al. (2011). Complete genome sequence of *Bacillus thuringiensis* serovar *finitimus* strain YBT-020. *Journal of Bacteriology, 193*, 2379–2380.
Zielinski, F. U., et al. (2009). Widespread occurrence of an intranuclear bacterial parasite in vent and seep bathymodiolin mussels. *Environmental Microbiology, 11*, 1150–1167.
Zubasheva, M. V., Ganushkina, L. A., Smirnova, T. A., & Azizbekyan, R. R. (2010). Larvicidal activity of crystal-forming strains of *Brevibacillus laterosporus. Applied Biochemistry and Microbiology, 46*, 755–762.

Chapter 5
Travelling Bacteria: Vectors

Abstract Bacterial diseases transmitted by blood-sucking arthropods are reviewed, including the aetiology and management of arthropod-transmitted diseases of man like bubonic plague, lyme disease and relapsing fevers. Transmission of other *Borrelia*-associated diseases is also reviewed, with data on rickettsioses, epidemic or endemic typhus and spotted fever vectors, tularemia, bartonellosis and others. Arthropod-transmitted diseases of animal domestic species and wildlife are also discussed. Aspects concerning vectored plant pathogens causing phloematic, xylematic and other diseases are also reviewed.

Keywords Bedbugs • *Borrelia* • Fever • Human disease • Lice • Pathogens • Phytoplasma • Relapsing fevers • *Rickettsia* • Typhus • Ticks • Transmission • Tularemia • Vectors • *Yersinia*

1 Introduction

Transmitted diseases are a global threat for the direct losses that they claim and the devastating effects that they have on the human society, either in urban environments and marginal settlements or villages. Infections also represent a threat for livestock and plants, on which we all depend for production of food and other commodities. The majority of transmitted bacterial infections are caused by arthropods, among which many insects and mites are vectors of a wide range of pathogens.

Bacteria also show a further, different type of association with invertebrates, mainly confined to transport and called "phoresy".[1] This is a close, not-harmful commensal relationship of one organism (phoront) with another, which behaves as an active transporter. The phoronts have a number of functional adaptations with their carriers, including specific links with some organs (i.e. legs, gut) which they depend on for transport. They phoronts usually do not receive energy from the host, do not multiply in its body nor provide it a direct benefit. The transport or transmission is targeted towards a third organism or a specific environment, and leads to the completion of the phoront life-cycle and its eventual multiplication.

[1] *Phoresy* = movement, transport.

A. Ciancio, *Invertebrate Bacteriology*, DOI 10.1007/978-94-024-0884-3_5

The diversity of relationships installed among pathogens, vectors and hosts of infectious diseases is the result of evolutive adaptations. A similar consideration also holds for the phoretic relationships. Although vectoring and phoresy may appear as very close terms, a phoretic relationship may not invariably induce an infectious disease, leading in general to improved phoront dispersal or to the colonization of new food sources. Transmission by a vector implies instead a direct release in the host tissues of the infectious agent(s), through different mechanisms usually in place during the vector feeding activity. In many cases the pathogens also multiply inside their vectors, although not at harmful levels interfering with their transmission functionality.

The two types of associations may also overlap. This occurs, for example, in entomopathogenic nematodes (EPNs), which transport and transmit to their hosts specific insect-killing bacteria. In this case the relationship is considered at a more evolved stage than phoresy and is defined "necromenic" (Dillman et al. 2012). The complex links between EPNs and the associated bacteria involve the release of toxins and antibiotics into the target insect body, mirrored by a corresponding immunity observable for the carrying nematodes. Also in this case the magnitude of the underlying evolutive radiations implies that many details and characters of these relationships cannot be considered as completely known, and still require further investigations. In this chapter we will review some processes characterizing major diseases transmitted by invertebrates, which have an important role in bacteria spreading, epidemiology as well as in the hosts and invertebrate ecology. See next Chapter for phoresy.

2 Disease Transmission

2.1 Human Diseases

Many bacterial diseases are transmitted by blood-sucking arthropods feeding on human beings. A large number of vectors can be found among Diptera, ticks or mites. The transmission processes are often zoonotic, being associated to other animal reservoirs present in nature, which allow survival of the pathogens in the environments in which they are encountered (Caimano 2006; Merritt et al. 2010; Dantas-Torres et al. 2012; Minard et al. 2013).

Arthropod-transmitted diseases of humans occur since long evolutionary times. They are more widespread in the Tropics, where the lack of management options, poverty and climatic and other environmental conditions favor their persistence. Due to the actual rates of progressive climate warming and global mobilization they also represent a marching threat, spreading outside their natural geographic boundaries, progressively moving towards higher latitudes (Gubler 2009).

Several studies in anthropology focused on arthropod-transmitted diseases because they are highly informative about the human history, as well as about the changes and evolution of our societies (Inhorn and Brown 1990). Infectious diseases and related epidemics – among which plague is the most popular example – have been the first cause of human mortality, claiming the highest death tolls for more than 5000 years, since first urban and agrarian settlements.

Epidemics may be also considered as a source of evolutive pressure through natural selection, and the cause of many cultural changes, whose adoption gave shape to the global society that we know today. The anthropology of infectious diseases is indeed a holistic scientific discipline investigating the effects that diseases had, and still have, on many aspects of the human societies and of the interactions among their members (Inhorn and Brown 1990).

The many infectious diseases that still afflict a large part of the world population are indeed caused by a highly diverse number of agents, of which vector-transmitted bacteria represent an important fraction. Apart of bacterial diseases, arthropod-transmitted pathologies also include many viruses, amoebae, plasmodia and invertebrate parasites like nematodes and other helminths. For an updated list of human diseases see the WHO[2] *ecological distructive of Diseases* at http://www.who.int/classifications/icd/revision/en/.

It is worth to note that many arthropod-transmitted diseases are largely affected by the natural, environmental and socio-political conditions in which the affected people live. The most efficient and wise approaches for their control hence also depend on how (and if) these factors are considered, all along the space in which the transmission events occur.

As indicated more than two decades ago by Inhorn and Brown (1990), management plans cannot avoid to consider the environment, and the way humans behave in it, including many ecological distructive forces, having often anthropic origins. These include not only forest destruction for agriculture, settlement or mining, but also the construction of new roads, travel infrastructures or dams, artificial lakes or irrigation facilities, that may play a role in spreading many vectors, eventually responsible for disease transmission. All these changes concern either the natural reservoirs of many arthropod-transmitted bacteria and the probabilities of people to be exposed to disease aetiologic agents, in the newly colonized environments (Merritt et al. 2010).

The environment plays a role also in the transmission of diseases caused by human parasitic invertebrates. The example of schistosomiasis (a disease, not caused by a bacterium but by an eukaryote, the trematode *Schistosoma* spp.) is pivotal at this regard. The parasite affects more than 250 million people (as reported by WHO) and is transmitted by the snail *Biomphalaria glabrata*, which is the intermediate host for the infective larvae release by the slugs in ponds or small lakes. The larvae eventually get in contact and penetrate the skin of people bathing or rinsing in these contaminated waters, developing inside their organs and reaching the adult stages to release eggs. These will be ingested by the slugs after being eliminated from the host body, closing the cycle. Irrigation canals and artificial lakes, together with lack of hygiene and poor or absent sewage management, largely contribute to

[2] World Health Organization, http://www.who.int/en/.

the disease spreading (Inhorn and Brown 1990). These mechanisms are active at various spatial scales, and may be responsible for unexpected disease outbreaks, in areas previously exempt of vectors or subject to significant local changes.

2.1.1 Plague

When considering arthropod-transmitted bacterial diseases of humans, the most popular events are the three pandemics of the tremendous **bubonic plague** (also called the Black Death) that occurred in Europe, largely known because of several historical or literature accounts. The pandemics caused million deaths in the Middle Age, with first outbreak recorded early during the ruling of the Constantinople emperor Justinian I (532 AD). Plague is also considered as the most plausible reason for the diversion of Attila's route during the Huns invasion of Northern Italy, at the collapse of the Roman Empire. Further later pandemic outbreaks started in the Middle East in 1346, spreading and killing up to 30 % of European population by the year 1369. A third pandemic, known as the Great plague of London, spread in Europe during 1665–1666, with million death tolls again. Plague had a profound impact on the affected societies as it induced many cultural and social changes concerning their demography and economy, as well as the regimes of land ownership and labour provision (Cartwright and Biddiss 2014). Sparse epidemics still occurred in the last decades.

The aetiological agent of plague is the pleomorphic bacterium *Yersinia pestis* (γ-Proteobacteria, Enterobacteriaceae). The genus *Yersinia* includes 11 species, of which only a few induce a disease. *Yersinia pestis* evolved around 10,000–20,000 year ago from the closely related *Y. pseudotuberculosis*, a benign enteric species transmitted through food or water (Achtman et al. 1999; Hinnebusch and Erickson 2008). Further pathogenic species are *Y. enterocolitica* and *Y. ruckeri*. The former induces a benign enteric disease, similar to *Y. pseudotuberculosis*, whereas the latter is responsible of the red mouth disease of rainbow trouts (Hinnebusch and Erickson 2008). In spite of its phylogenetic proximity to the other species, *Y. pestis* is unique in the genus because of its higher virulence and pathogenicity, as well as because of its arthropod transmission route. It is the only causal agent of bubonic plague, lethal not only to humans but also to other mammals (Eisen and Gage 2009).

Bubonic plague is transmitted by blood feeding fleas that previously fed on *Y. pestis* infected rats, and may produce three human disease variants: the first is characterized by the occurrence of bubos, corresponding to swellings of lymphnodes; the second variant is a pneumonic infection, lethal in a few days; a third variant is characterized by a diffused and lethal septicaemia (Cartwright and Biddiss 2014). The pneumonic disease can be transmitted directly from human to human, without intervention of a flea vector, thus causing rapid outbreaks. In many cases the environment in which the epidemics occur is important, being the plague associated to the scarcity or absence of public and personal hygiene, favouring rats multiplication and exposure to fleas. Feigin et al. (2009) report other transmission routes, including direct contacts with the bacterium during evisceration of dead animals or, in case of pneumonic plague, inhalation or assumption of mucosal droplets reaching the eyes or pharynx.

The primary vector of *Y. pestis* is the oriental rat flea *Xenopsylla cheopis* (Pulicidae). However, many other flea species can act as vectors. Transmission occurs when the flea regurgitates the infectious bacterial cells during biting. *Yersinia pestis* can form a biofilm within the gut environment of its vector, a specific capacity favoring the intradermal transmission of the bacteria to the host during biting. A low number of cells is required for a lethal infection, a trait that confers *Y. pestis* a high virulence and spreading capacity, making it one of the most frightening pathogens. This capacity is not found in *Y. pseudotuberculosis*, and plays a fundamental role in epidemics (Hinnebusch and Erickson 2008; Konnov et al. 2010).

Transmission may be simply mechanical, when it proceeds from a dirty flea mouth bite, or biological, in which the bacterium multiplies in large aggregates in the mouth, midgut and preventricolous valve of the vector. In particular, the bacterial biofilms affect or block the valve functionning during blood assumption, inducing its content to be regurgitated on the site, thus originating an infection (Hinnebusch and Erickson 2008). Two genes, present on two distinct plasmids, that are unique for *Y. pestis*, are important for its arthropod transmission. They are the murine toxin Ymt, a phospholipase D allowing the bacterium survival in the vector midgut, and the plasminogen activator Pla, required for systemic dissemination in the recipient host after intradermal transmission. LPS and other factors are also involved, including the formation of an extracellular matrix that avoids the ingestion of the bacteria by the fleas and their eventual elimination during defecation. This occurs thanks to the biofilms which adhere to the proventricolous spines (Hinnebusch and Erickson 2008). The production of biofilms occurs frequently among arthropod vectored bacteria and appears as a common trait, functional to this mode of transmission.

Yersinia pestis can survive in soil for some time, but its persistence in nature occurs through sparse epidemic foci present within populations of several species of wild rodents and associated fleas. Natural reservoirs of sub-populations of rats which are tolerant or resistant to the disease sustain the bacterium persistence. Susceptible rats allow transmission of the bacterium to humans by the fleas leaving their cadavers. Epidemics may also be possibly associated to other natural causes producing a high rats mortality, i.e. floods or other events leading to the destruction of their living habitat.

Treatments with the antibiotics tetracycline and streptomycin are indicated by WHO for patients. Insurgence of antibiotic resistance among *Y. pestis* isolates has also been reported.

2.1.2 Lyme Disease and Relapsing Fevers

All *Borrelia* spp. are transmitted to their vertebrate hosts by hematophagous arthropod vectors, belonging to three groups: human body louse, argasid soft ticks and ixodid hard ticks (Caimano 2006). Ticks (Arthropoda: Arachnida, subclass Acari) have been recognized since ancient times as important human parasites, but their role as disease vectors was identified only by the end of the nineteenth century, when transmission of the protozoan *Babesia bigemina* (the aetiologic agent of Texas cattle fever) was demonstrated for the tick *Boophilus annulatus* (Parola and Raoult 2001).

Many tick species are vectors of Borreliae in wild animal populations (Trout Fryxell et al. 2012). Bites of the blacklegged tick *Ixodes scapularis* (Fig. 5.1) may transmit **Lyme disease**, a severe human pathology whose aetiologic agents are the spirochetes *Borrelia burgdorferi*,[3] *B. afzelii*, *B. mayonii* and *B. garinii* (Gray et al. 1994; Liebisch and Olbrich 1991; Rand et al. 1998; Caimano 2006; Dolan et al.

Fig. 5.1 Stages of *Ixodes scapularis* in ambush (**a**), on a parasitized dog (**b**) and engorged females, after a blood meal (**c**) (Adapted from Spencer et al. (2003))

[3] In honor of W. Burgdorfer, medical entomologist.

2016; Pritt et al. 2016). The epidemiology of Lyme disease matches the geographical distribution of its arthropod vectors (Caimano 2006). *Borrelia burgdorferi* is the causal agent of Lyme disease in North America, whereas *B. afzelii* and *B. garinii* are the causal agents of the disease in Europe and Asia, respectively. *Ixodes scapularis* is also vector of other *Borrelia* spp. like *B. miyamotoi*, the causal agent of tick-borne relapsing fever. Lyme disease is the most common vector-transmitted disease in the northern hemisphere and is most often acquired through the bite of infected nymphal ticks (Stafford et al. 1998). It has been also reported from China, Japan, Australia and Africa (Caimano 2006).

Ixodes scapularis attacks and feeds also on blood of wild animals, including dogs, deer and rodents (Fig. 5.1). In particular, the white-footed mouse *Peromyscus leucopus* is the natural reservoir of *B. burgdorferi* in Northeastern and Upper Midwestern states in the USA (Magnarelli et al. 1988). In the Western USA the vector is the western blacklegged tick *I. pacificus*. In this area the dusky-footed wood rat *Neotoma fuscipes* and the California kangaroo rat *Dipodomys californicus* are the natural reservoirs (Brown and Lane 1992; Wang et al. 2014).

Lyme disease is progressively spreading in the USA due to the increased densities of either the rodent and vector populations. It recently reached states like Ohio, previously considered free from the tick (Wang et al. 2014). In Southern Québec, due to milder winters induced by progressive global warming, the white-footed mouse is expanding its distribution area and is progressively moving poleward at an average speed of 10 km $year^{-1}$. Model predictions arose concerns about the effect of winter length and temperature maxima and minima on the correlated expansion of *B. burgdorferi* (Roy-Dufresne et al. 2013). Lyme disease severely affects humans and laboratory mice, but it has a minimal impact on its natural reservoir hosts (Schwanz et al. 1989). The disease induces fatigue, headache, fever and characteristic skin rashes called erythema migrans. In a more advanced stage it affects joints, heart and the nervous system.

The ticks ecology is well adapted to their feeding behaviour and the biotopes in which they live. Usually, each stage (larval, nymph and adult) has only one host, from which they detach once repleted. This occurs at the end of their single blood meal, for subsequent moulting or eggs release, after a rapid engorgement phase, when the ticks may increase their body size up to >100 times (Fig. 5.1). Feeding activities may last for several days but in total cover less than 10 % of the ticks life, which is mostly exophilic.

In Europe, ticks like *I. ricinus* transmit Lyme disease, *Anaplasma phagocytophilum*, the aetiologic agent of **human granulocytic ehrlichiosis**, and the Tick Borne Encephalitis virus (Rizzoli et al. 2004).

Ticks may live 2–3 years or more, dying after eggs deposition. Their life-cycle includes long host-seeking periods. In this process the ticks have various strategies, including remaining positioned in ambush with the front legs extended, after climbing on grass or other specific, locally available supports (wood, bushes, trees). They have sophisticated host recognition capacities, responding rapidly to steps and vibrations, to a sharp increase of temperature or humidity, or to the emission CO_2

and/or NH_3. All these environmental signals are indicative of the presence of an host in proximity. Depending on the tick strategy, once the host is detected they fall on its body from the ambush position. In alternative they may directly run to attack the host from their resting location or wait for attacking hosts into their nests (Parola and Raoult 2001). In *I. scapularis*, motile *B. burgdorferi* migrate from the midgut to the salivary glands within 48 h after host attachment (Embers and Lopez 2012).

Ixoidid hard ticks produce salivary secretions which include anti-inflammatory, cytolytic, anticoagulant, vasodilators and anesthetic compounds, with enzymes and other substances functional to the anchoring of the dermal penetrating hypostome, a kind of biological needle used for feeding. They may also induce a progressive paralysis through the release, during a blood meal, of neurotoxic substances. The paralysis may even result lethal, although only in a low number of cases. A gentle removal of the tick, i.e. by using fine tweezers to allow a quick detachment of the parasite taking care to pick it up with its feeding apparatus, usually shows recovery in 1–2 days (Felz et al. 2000; Parola and Raoult 2001). Sterilizing agents (i.e. alchool or iodine-based skin desinfectants) should then be applied to the scar, also clearing hands and fingers, having care not to crush the tick that may be sealed in a container with alchool and then disposed of. For more informations on Lyme disease and other arthropod-transmitted diseases, including prevention and treatments, see the CDC[4] related web pages http://www.cdc.gov/lyme/ and http://www.cdc.gov/ticks/.

Other *Borrelia*-associated diseases include **relapsing fevers** (RF), caused by *B. recurrentis* transmitted through the human body louse *Pediculus humanus humanus* and *P. capitus* agents of the epidemic RF. Endemic RF is instead transmitted through bites of soft-body ticks like *Ornithodoros moubata* or hard ticks like *O. tholozani* (Embers and Lopez 2012). In the former tick the transmission also occurs transstadially[5] (Dautel and Kahl 1999; Parola and Raoult 2001). In transstadial transmission the bacterial infection is carried from the larvae to the adults and then transmitted to hosts during blood feeding. In this infective process the vector also acts as natural reservoir for the transmitted pathogen (Toutoungi and Gern 1993; Sonenshine 1994; Gubler 2009).

Louse-borne relapsing fever (LBRF) is caused by *B. recurrentis*, vectored by the human body louse *P. humanus*. The spirochetes multiply in the lice hemolymph, penetrating through the gut epithelium. The bacterium is acquired when the lice feed on humans, which act as natural reservoir. The disease may be transmitted by direct contact of the bacteria with the skin, once released by crushing or scratching the lice (Caimano 2006). Differing from other infective processes, transmission in this case does not require a wound for infection. LBRF recurrent epidemics are often associated to poverty and other humanitarian catastrophic events or wars. They may have a lethal outcome, especially for children, pregnant women or weak

---

[4] Center for Disease Control and Prevention, Atlanta, GA, USA.

[5] In transstadial transmission (also called trans-ovarial transmission) the bacteria vertically infect the developing eggs inside the adult female, eventually passing to the embryo and larval stages. This process has the potential to confer an efficient vector capacity to all larvae descending from a single female.

and debilitated people, and may cause million death tolls (Raoult and Roux 1999; Embers and Lopez 2012).

The **tick-borne relapsing fever** (TBRF) is caused by several *Borrelia* spp., like *B. duttonii*, the causal agent of TBRF in the Old World. Each *Borrelia* species is specifically associated to a vector within the genus *Ornithodoros* and *Argas*, and to their specific natural host reservoirs (Schwan 1996; Mitani et al. 2004). The causal agents of TBRF remain motile in their tick hosts and colonize their salivary glands, a process fundamental for subsequent transmission. They have defined geographic ranges and vector specificity. *Borrelia persica* is associated to the hard tick *O. tholozani* and is common in a region spanning from Middle East to India. The ticks can survive for many years between blood meals, and are often encountered in arid environments or mud houses. Other spirochetes endemic in the region are *B. caucasica* and *B. latyschewii*. Species like *B. hermsii* and *B. turicatae* are encountered in America, associated to the tick *O. parkeri*. A further specific association is found between *B. mazzottii* and *O. talaje* in Mexico (Embers and Lopez 2012).

In either LBRF and TBRF, periods with high fever are alternated to relapsing phases. This alternated fevers are due to the spirochetes capacity to evade the reaction of the host immune system, by modifying the external surface coat components. In *B. hermsii* these have been found to arise by the expression of different members of a gene family coding for variants of variable small proteins (Vsps) and variable large proteins (Vlps). Their recombination is achieved by a process of gene conversion, in which transcriptionally silent gene variants replace those already expressed in active loci. Further mechanisms include partial conversion and recombination of *vsps* and *vlps* genes (Embers and Lopez 2012). An additional strategy to evade the host immune system relies on a process known as erythrocyte rosetting. In this case the spirochetes are covered and hidden by red blood cells aggregates, binding to neolacto-glycans carrying glycoconjugates present on the blood cells surface (Guo et al. 2009).

RF can produce severe bacteremia, associated to liver and speen damage, vomiting, diarrhea and neurologic disorders. Complications may occur in pregnancy with fetal death due to placental transmission. Other patient damage include neuropathy and central nervous invasion, myocarditis or even coma. A symptom observed in LBRF is conjunctival hemorrhaging (Embers and Lopez 2012). If left untreated, the mortality rates due to LBRF and TBRF can reach 70 % and 10 % of patients, respectively. RF can be treated with antibiotics, and those capable to pass the blood-brain barrier are needed in pathologies affecting the central nervous system. For further informations on treatments and associated complications see Dworkin et al. (2008), Embers and Lopez (2012) and CDC webpages.

2.1.3 Rickettsioses

The genus *Rickettsia* (α-Proteobacteria) includes bacteria characterized by an intracellular and obligate endosymbiont life-style (Weiss and Dasch 1991). A multi-gene analysis of its phylogenetic radiation showed the occurrence of various groups, including the typhus and the spotted fever groups, together with other closely related lineages.

The typhus group includes *R. prowazekii*,[6] causal agent of the insect born epidemic typhus and *R. typhi*, the agent of endemic typhus (Robinson et al. 2003). The spotted fever group includes the causal agents of several tick-borne diseases, divided in the monophyletic clusters of *R. rickettsii*, *R. massiliae* and *R. montana*, with a second lineage with *R. australis*, *R. akari* and *R. felis*. Further groups include the monophyletic *R. africae* and *R. sibirica* clusters, the *R. slovaca* and the phylogenetically divergent clusters of *R. japonica*, *R. canadensis* and *R. bellii* (Labruna et al. 2007; Vitorino et al. 2007; Fournier and Raoult 2009). The genus *Rickettsia* includes more than 25 species, and has a close relationship to other disease-associated bacteria of genera *Orientia*, *Ehrlichia*, *Neoerlichia*, *Coxiella*, *Neorickettsia* and *Anaplasma* (Lukin et al. 2001).

By the end of the 1990s the diseases caused by *Rickettsia* spp. and associated taxa (called Rickettsioses) have been recognized as emerging threats (Raoult and Roux 1997). Vectors include several blood feeding tick and insect species, which acquire and transmit the bacteria during their blood meals. In some cases also horizontal transmissions have been reported among vectors, including transmission through ticks co-feeding on the same host, or sexual transmission from an infected male to a recipient female, during mating (Raoult and Roux 1997). Rickettsioses have close relationships with their vectors and hosts, the biotopes in which epidemics occur and the events favoring their spread, including natural or social disasters. Their dispersal is more dependent on hosts local displacements, migrations or aggregation, rather than on the limited movements of the vectors.

Human lice (*Pediculus humanus humanus* and *P. humanus capiti*) may transmit **epidemic typhus**. The disease, whose aetiologic agent is *R. prowazekii*, is present worldwide, and outbreaks are often associated to crowded people conditions related to wars, famine and other catastophic events, with limited medical and humanitarian assistance (Svraka et al. 2006). The natural reservoir of *R. prowazekii* are humans, which become infected when attacked by bacterium-infected lice. Infected lice have a shorter life-span, and die 24 h after their last blood meal, but bacteria may remain viable for some weeks in their body or faeces. Transmission occurs by direct contact of *R. prowazekii* cells present in the louse faeces or proceeding from crushed insects, which enter in direct contact with the host blood in open bite scars. Infection may also occur by inhalation of aerosols containing *R. prowazekii* cells or by their contact with the mouth and eye mucosa.

After an incubation period of 1–2 weeks the disease starts with sudden fever, chills, prostration and myalgia. Clinical signs of the diseases are usually nonspecific, and symptoms include rashes appearing in around 50 % of cases, initially in the upper trunk and ascillary regions, and then visible on the whole body except face, palms and sole. Eventually they may become darker, hemorrhagic or petechial, associated to vomiting, nausea and hypotension. Varying levels of virulence may be encountered among the different hosts as well as among the *Rickettsia* species (Fig. 5.2).

[6]The genus and species names honor H. T. Rickett and S. J. M. von Prowazek, who died during their pioneering studies on rickettsiae.

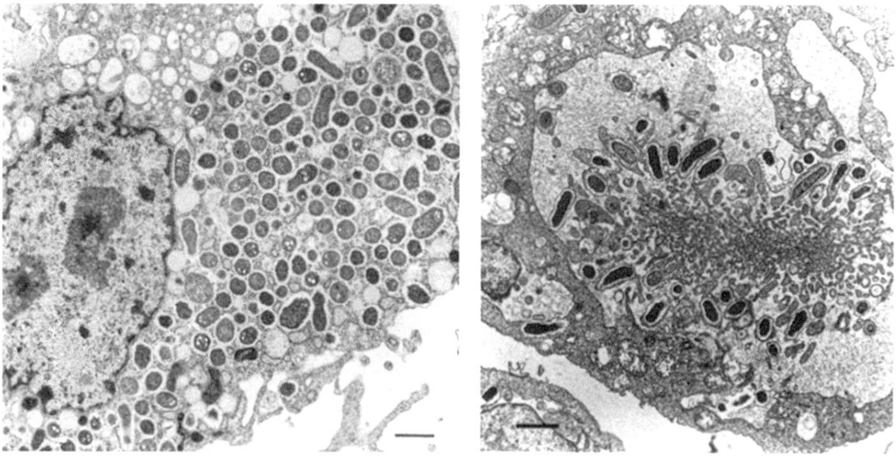

Fig. 5.2 TEM images showing higher abundance and limited cell damage induced by Rickettsia prowazekii in chicken embryo cells (*left*), compared to more severe cell damages observed at lower density for R. rickettsii (*right*) (Adapted from Silverman (1991). Scale bars = 1 μm)

For precise diagnosis advanced biosafety level-3 laboratories with trained personnel are required. However, simpler and efficient paper-blot serologic procedures have also been developed, as well as specific PCR-based methods from clinical blood samples or lice body (Fenollar and Raoult 1999; Svraka et al. 2006). The disease may become latent and reappear again after some years, in a recrudescent, milder form. Personal and clothing hygiene are recommended for prevention and care, flanked by insecticidal treatments to kill the vectors. Usually, the antibiotic treatments are effective against epidemic typhus, whose induced mortality is correlated to increasing age and may reach up to 30–40 % of untreated people. Deaths reached the order of two to three million people in most catastrophic epidemics like those associated to World War I (Raoult and Roux 1999). Further informations on the disease may be found on the CDC and related websites.

Endemic typhus (also called murine or flea-borne typhus) is caused by *R. typhi* (initially denominated *R. mooseri*) and *R. felis*, the aetiologic agent of cat flea typhus. The two species induce similar symptoms characterized by periods with alternate fever, acquired through the bite of infected vector fleas. Transmission occurs when the scars induced by blood feeding fleas become contaminated by the bacteria expelled through the flea faeces in the feeding site, or resulting by scratching their bodies. Other ways of transmission include contact of dry flea faeces with eyes or contamination of respiratory tracts by inhalation.

The disease symptoms are not specific and include fever, headache and rashes. Rats are the *R. typhi* natural reservoirs, most commonly *Rattus rattus* or the Norway rat (*Rattus norvegicus*), together with house mice and other rodents. Domestic cats, dogs or opossums are natural reservoirs of *R. felis*. The rats are involved in the so-called urban cycles in which endemic typhus is transmitted through the rat flea *Xenopsylla cheopis* vectoring *R. typhi* and, with varying prevalence levels, also *R.*

felis. Other flea species hosting *R. typhi* are *Leptosylla segnis* and *Ctenocephalides felis*. The cats and opossums are often associated to endemic typhus cycles in peri-urban environments. Here *R. felis,* the aetiologic agent of the disease, is most commonly transmitted by the cat flea *C. felis* (Azad et al. 1997; Eremeeva et al. 2008). A further *R. felis* and vector association includes the flea *Anomiopsyllus nudata* feeding on the white-throated woodrat *Neotoma albigula* (Stevenson et al. 2005). Endemic typhus can be controlled through antibiotic treatments.

Members of the hard tick family Ixodidae transmit the **spotted fever rickettsioses** (**SFR**), whose aetiologic agents form a group of about 15 species, among which *R. rickettsii*. In North America the bite of *Dermacentor andersoni* and other Ixodidae ticks like *D. reticulatus* (Fig. 5.3) can transmit the **Rocky Mountain Spotted Fever**, induced by *R. rickettsii*. The bacterium shows a low prevalence among ticks, but is common in the human population in the area because vectors feed readily on humans. Transmission occurs via regurgitation and the acquisition occurs transstadially (Parola and Raoult 2001). Further SFR aetiologic agent in the Southern USA are *R. parkeri*, transmitted by the Gulf Coast tick *Amblyomma maculatum* (Fig. 5.3), *R. akari*, the mite-borne agent of **Rickettsialpox** and *R. felis*, the agent of cat-flea SFR (Paddock et al. 2004).

SFR is transmitted in South America by the tick *Amblyomma cajenennse* (the Cayenne tick), and is present in many countries (Schoeler et al. 2005). Rickettsiae multiply in almost any organ and fluid of their vector ticks, including the salivary glands and ovaries, and may be acquired directly during a blood meal or vertically, through a transovarial route (Parola and Raoult 2001). SFR induces fever and malaise with chills, weakness and prostration, breath shortness and cough, arthralgia, abdominal pain and nausea.

Rickettsialpox (also known as vesicular rickettsiosis) is an occasional disease with mild fever, induced by *R. akari* transmitted by the house mouse mite

Fig. 5.3 Adult females of *Dermacentor reticulatus* (*left*) and *Amblyomma maculatum* in ambush on grass (Adapted from Mehlhorn (2008) (*left*) and Parola et al. (2009))

Liponyssoides sanguineus (Schoeler et al. 2005). *Mus musculus* and other rodents act as natural reservoirs. The mites can pass the bacterium vertically to larvae by trans-ovarial transmission. The disease has a worldwide distribution and starts as a cuteneous lesion induced by the mite bite, followed by varicella-like eruptions and later on by fever, myalgia, chills, sweatings, cephalalgia. Maculopapular and maculovesicular eruptions, with vomiting or abdominal pain, may also be present. It may be treated with antibiotics (Acha and Szyfres 2001).

Trans-ovarial transmission of rickettsiae has been reported in ticks of the genus *Amblyomma*, vectors of the **African tick bite fever** caused by *R. africae* (Socolovschi et al. 2009). In sub-Saharan Africa the ticks feed readly on humans present in their biotopes, with high prevalence levels reaching 80 % of sieropositives (Parola and Raoult 2001).

In Australia, SFR-group diseases include the **Queensland tick typhus**, caused by *R. australis*. They are transmitted by the ticks *I. tasmani*, *I. holocyclus* and *I. cornuatus*. **Flinders Island Spotted Fever**, caused by *R. honei*, is transmitted by *Bothriocroton hydrosauri*. The *R. honei* strain "*marmionii*" is associated to the tick *Haemaphysalis novaeguineae* (Stenos et al. 2003; Unsworth et al. 2007; Parola et al. 2012).

The **Mediterranean spotted fever** (boutonneuse fever) is caused by *R. conorii* and is transmitted in Europe and Africa by the brown dog tick *Rhipicephalus sanguineus*, with dogs as natural reservoirs. In spite of its urban occurrence, the vector specificity toward dogs contributes to a low disease prevalence in the human populations (Raoult and Roux 1997; Gubler 2009).

2.1.4 Diseases Transmitted by Gnats and Flies

Eye gnats from the genera *Hippelates* and *Liohippelates* (Diptera: Chloropidae) are non-biting insects feeding on organic fluids from nose, eyes, wounds and natural orifices of mammals, assuming their bacterial contents. They can mechanically transmit the spirochaete pathogen causing **yaws** (*Treponema pertenue*), a disease resulting in skin ulcers, also affecting cartilage and bones. A further disease associated to eye gnats is congiuntivitis (also known as **pink eye**), caused by *Haemophilus aegyptius* (Tondella et al. 1994).

Gnats of the deer fly genus *Chrysops* and horse flies of the genus *Tabanus* (Diptera: Tabanidae) transmit **tularemia**, whose aetiological agent is the Gram negative bacterium *Francisella tularensis* (originally described as *Pasteurella tularensis*). Transmission may be mechanical, from open sores, because of the frequent and interrupted meals taken by the flies on different animals. Also the tick *Dermacentor* sp. acts as aetiologic agent of tularemia. The disease is transmitted transstadially by the tick bite. Horse and deer flies can also transmit anthrax, caused by *Bacillus anthracis* (Braderic and Punda-Polic 1992; Parola and Raoult 2001).

The sand flies of the genus *Phlebotomus* are vectors of the gram negative *Bartonella bacilliformis*, causal agent of **bartonellosis** in South America. In Peru,

Ecuador and Colombia the sandflies of the genus *Phlebotomus* transmit the disease, which is also known as Oroya fever or Carrion's disease.

House flies (Diptera: Muscidae), blow flies (Diptera: Calliphoridae), and flesh flies (Diptera: Sarcophagidae) that often live among filth and garbage can carry on their feet and mouthparts (and mechanically transmit) human pathogens causing **dysentery** (*Shigella dysenteriae*), **typhoid fever** (*Eberthella typhosa*) and **cholera** (*Vibrio cholerae*).

Eye-seeking flies may be important vectors of **trachome**, a severe disease of the eye causing roughening and lesions of the eyelid, with nasal or ocular discharges. Trachome may induce eyelids to turn inward, leading to blindness if untreated. The disease is caused by *Chlamydia trachomatis* and affects more than 80 million people in Africa, Middle East, Asia and South America. It is acquired by direct contact with infected people or contaminated materials (tissues, towels). Flies caught from eyes of children showed that *Musca sorbens* was the main insect vector, as the insect feet and proboscis were contaminated by the bacterium. The insect possibly represents a common transmission route and fly control measures may contribute significantly to reduce the disease prevalence, in particular among children (Emerson and Bailey 1999; Emerson et al. 2000). Trachome may be treated with the antibiotics tetracycline or azithromycin. Due to the transmission modes and prevalence among infected people, treating whole communities at once may contribute to reduce the local incidence of the disease (Evans and Solomon 2011).

A skin ulcerating disease known as **buruli** is emerging as a widespread threat in tropical regions of Africa, Asia, Central and South America. The disease aetiologic agent, *Mycobacterium ulcerans*, is found in ponds and other marshy areas. The disease may have severe devastating effects including the induction of leg and arm deformities, skin lesions, with final dramatic outcomes including amputations. It is probably acquired through the exposure of wounds to contaminated resting waters. Buruli transmission is suspected to depend on *Acanthamoeba* as natural host reservoir, and to occur through the deposition of bacterial biofilms on wounds by insects, including mosquitoes and bugs associated to aquatic and swampy environments (Portaels et al. 1999; Marsollier et al. 2002; Merritt et al. 2010; Wilson et al. 2011).

2.1.5 Other Human Diseases

The ticks *Dermacentor variabilis* and *Amblyomma americanum* transmit **ehrlichiosis** and **Sennetsu fever** (see Sect. 2.2.1) caused by the intracellular bacteria *Neorickettsia* (*Ehrlichia*) *sennetsu*, *E. canis*, *E. chaffeensis*, *E. equi* and *E. phagocytophila* (Anderson et al. 1993; Steiert and Gilfoy 2004; Trout Fryxell and DeBruyn 2016). Erlichiosis is mainly a canine disease, but may also attack humans. *Amblyomma americanum* known as Lone Star tick is also the vector of *Anaplasma* spp., the aetiologic agents of **anaplasmosis** in humans and ruminants (see next section), and of *Francisella tularensis*, the aetiologic agent of tularemia in humans. The tick may harbor many bacterial endosymbionts, including further human pathogenic specied of the genera *Borrelia*, *Coxiella*, *Rickettsia* or *Legionella* (Trout Fryxell and DeBruyn 2016).

Red mites (*Leptotrombidium* spp., Trombiculidae) transmit **scrub typhus** (**Tsutsugamushi** disease) a severe life-threatening pathology. It is characterized by an acute fever, localized cutaneous necrosis, lymphadenopathy, headache, myalgia, cough, gastrointestinal complications, transient hearing loss and rash. The disease aetiologic agent is the intracellular pathogen *Orientia* (former *Rickettsia*) *tsutsugamushi*.

In Asia and Africa scrub typhus is often underdiagnosed, attacking around one million people every year, with a 10 % mortality rate of untreated cases (Paris et al. 2013). The mites are the natural reservoir of the bacterium. The nymphal and adult mites are predators of insect eggs in soil, but larval stages feed only once on vertebrate hosts that include rodents and humans. The transmission of the disease occurs by the feeding larval stages, which are vertically (transstadially) and horizontally infected by co-feeding mites (Frances et al. 2001; Paris et al. 2013).

Adding to previously cited diseases, the body louse *Pediculus humanus* var. *corporis* transmits *Bartonella quintana* (previously knowns as *Rickettsia* or *Rochalimaea quintana*), the aetiologic agent of **trench fever**. The disease was observed during World War I and is today associated to people living with poor hygiene and in marginal social conditions. Apart of acute fever and other non-specific symptoms like prostration, headache or conjunctivitis, it may induce blanching erythematous macular rashes, chronic lymphadenopathy, bacillary angiomatosis or endocarditis. Erythromycin and/or doxycycline are usually applied for treatments (Ohl and Spach 2000). **Bacillary angiomatosis** or **peliosis** (BAP) is a vascular proliferative disease caused by the same aetiologic agent of trench fever (*B. quintana*) in immunocompromised patients (Koehler and Tappero 1993; Mohle-Boetani et al. 1996). BAP disease may be resolved by treatments with antibiotics like erythromycin or tetracycline.

Cat scratch disease, also known as benign inoculation lymphoreticulosis, is associated to cat scratches and is caused by *Bartonella henselae* and the causal agent of trench fever, *B. quintana*, vectored by the cat flea *Ctenocephalidae felis*. The flea can acquire *B. quintana* by feeding and release viable organisms into the faeces from which it can be transmitted to susceptible hosts (Kernif et al. 2014). In Uganda and other regions in Tropical Africa the flea is also a competent vector of plague bacterium *Y. pestis*. However, as a secondary vector, it has an efficiency lower than that of other fleas. It can occasionally infest other potential rodent reservoirs such as the roof rat *Rattus rattus* or the Nile rat *Arvicanthis niloticus* (Eisen et al. 2008).

Q fever is caused by infection with the obligate, intracellular gram negative bacterium *Coxiella burnetii*, commonly found in cattle, sheep and goats in which the disease is called **coxiellosis**. In humans it causes fever, headache and myalgia. It is usually acquired by inhalation of farm dust containing durable spore-like bacterial forms proceeding from excreta or placental remains. The bacteria may be dispersed by wind, reaching potential hosts at locations over a mile far. Transmission to humans by ticks has been reported but is rare (Tissot-Dupont et al. 1999; Herrin et al. 2011). Treatments with antibiotics like doxycycline have high therapeutic efficacy (Anderson et al. 2013).

2.2 *Animal Diseases*

Animals, either domestic or wild, are natural reservoirs of many arthropod transmitted diseases. The specific pathogens they harbor may have a significant impact on cattle or other domestic species of economic or affective value. Zoonoses[7] like plague or Lyme disease have a strong impact on humans, but the risk exists that many unknown diseases present in animal niches in the wild might still be "waiting" for the right conditions to emerge and spread. As a consequence, animal diseases deserve our attention and must be studied and monitored with economic and research investments comparable to those dedicated to other human diseases.

2.2.1 Domestic Species

The pathologies caused by transmitted bacteria represent a fraction of a global toll of around 75 most common aetiologic agents of domestic animal diseases, which also include several protozoa, viruses with further agents like nematodes and other helminths. A large fraction of the bacterial aetiologic agents previously cited for human beings, like *Y. pestis* and rickettsiae, also occur in a broad range of animal hosts which act as reservoirs in nature. They have been treated in the previous section and are not examined here, unless some additional details may elucidate further particular aspects of their biology.

Among the bacterial diseases of animals, *Anaplasma* spp. represent an emerging group of obligate, intracelluar hemoparasites. The pathogens are the aetiologic agents of **anaplasmosis**, an important group of diseases occurring in sheep, cattle, dogs and other animals. In cattle, *A. marginale* and *A. centrale* cause a blood disease of ruminants with destruction of erythrocytes, anemia, fever, lethargy, leukopenia. Anaplasmoses reduce milk production and may induce abortion. Bovine anaplasmosis caused by *A. marginale* is endemic in tropical and subtropical regions, and is actually reaching higher latitudes, worldwide. The disease has been detected in calves, water buffalos, bisons, African antelopes and mule deer. It may infect up to 70 % of erythrocytes and may lead aged animals to death (Rymaszewska and Grenda 2008).

In North America the ticks *Dermacentor andersoni* and *D. variabilis* are the vectors of *A. marginale*. The animals surviving infections are the natural carriers of the disease. The bacteria rely on an antigenic modulation to escape the host immune response, thus causing repeated infective parasitemic cycles. Cattle anaplasmosis is controlled by monitoring through a bovine serology survey and slaughtering of infected animals. This strategy allowed Canada to regain its anaplasmosis-free status (Howden et al. 2010).

Anaplasma centrale is mainly found in cattle, in which it induces a milder disease than that caused by *A. marginale*. It attacks the erythrocytes in which the bacteria form cellular aggregates in the central host cell region. The bacterium has a

[7] Diseases passed from animals to humans.

worldwide distribution similar to that of *A. marginale*, towards which it may occasionally confer resistance. It is transmitted by ticks of the genera *Ixodes* and *Haemaphysalis* (Rymaszewska and Grenda 2008).

Anaplasma bovis, detected in cattle and other small ruminants like sheep or goats, attacks monocytes, causing a disease called **monocytic anaplasmosis**. It is present in Brazil, North America, Africa and Japan. The bacterium is vectored by *Dermacentor* spp. ticks, causing the aspecific symptoms common to other anaplasmoses, together with more specific symptoms like swellings of prescapular lymphnodes, mucus secretions and mucosal paling (Santos and Carvalho 2006).

Anaplasma ovis attacks erythrocytes of small ruminants, and bacterial inclusions may be found either in the central or marginal regions of parasitized cells. The disease may be lethal and can produce severe economic losses to farmers, in particular in tropical regions in which sheep and goats represent the main, if not unique, income source. *Anaplasma ovis* has been found also in USA, Italy and Hungary, and conversion to more traditional pasturing is considered as a factor facilitating transmission by *Dermacentor* spp. vectors (Rymaszewska and Grenda 2008).

In dogs, *I. scapularis* and *I. pacificus* transmit *A. phagocytophilum* (formerly known as *Erlichia phagocytophila*) which infects white blood cells. The tick *Rhipicephalus sanguineus* may transmit *A. platys*, infecting the blood platelets and causing a disease called **canine cyclic thrombocytopenia**. *Anaplasma phagocytophilum* specifically colonizes neutrophils and forms small cellular aggregates within a protective vacuole. It has a positive tropism[8] towards neutrophils and is capable to evade their immunitary reaction avoiding oxidative killing by stimulating neutrophils NADPH oxidase to rapidly detoxify O^{2-} reactive species (Carlyon et al. 2004). *Anaplasma phagocytophilum* may be transmitted also to horses, small ruminants and humans through the bite of *Dermacentor* spp., *I. scapularis* or *I. pacificus*. In humans the bacterium induces the disease known as **human granulocytic anaplasmosis**, with fatal outcomes in weaker patients, in case further complications are also present (Rymaszewska and Grenda 2008; Kocan et al. 2012). The disease is sensitive to antibiotics (doxycycline), but specialized laboratories are needed for its diagnosis.

Further bacterial tick-borne diseases of cattle include **heartwater**, a disease that was transported with African cattle to the Carribean regions, caused by *Rickettsia ruminantium*. The bacterium is transmitted to cattle by the ticks *Amblyomma variegatum* and other *Amblyomma* spp. (Norval et al. 1992).

Avian tick-borne zoonoses include *Borrelia anserina*, the aetiologic agent of **avian spirochetosis** that affects chickens. The bacterium is transmitted by infected ticks *Argas miniatus* or *A. sanchezi* (Ixodoidea: Argasidae). The disease also affects turkeys, geese, ducks, canaries and pheasants (DaMassa and Adler 1979; Lisbôa et al. 2009). Acute, fatal spirochetosis was also observed in Northern spotted owls (*Strix occidentalis caurina*) in the USA (Thomas et al. 2002).

---

[8] *Tropism* = movement. Positive tropism indicates displacement or orienting towards a given factor or space; negative tropism indicate movements in the opposite direction.

Cat flea typhus, caused by *R. felis*, is an emerging global threat also for humans. Prevalence of the disease, vectored by *C. felis*, is probably underestimated, due to the similarity of symptoms to those of murine typhus or dengue. The bacterium was described in 1990 and has some peculiar traits like its thermal requirements, with an optimal temperature range of 28–32 °C and the presence of a plasmid (Adams et al. 1990; Pérez-Osorio et al. 2008). *Rickettsia felis* has been also found in vector *C. felis* collected from monkeys, hedgehogs, opossums and rodents, in different world regions (Pérez-Osorio et al. 2008).

In India, stray and pet dogs show high prevalence for *C. felis orientis* and *C. felis*, which were found positive for the *ompB* gene used to detect the presence of *Rickettsia* spp. in their bodies. Both ticks add to *C. canis*, positive for spotted fever group rickettsiae, confirming that the more than 35 million dogs present in the country represent a vast reservoir for a number of zoonotic rickettsiae, including, among others, also *R. rickettsii*, *R. conorii* and *R. felis* (Hii et al. 2015). Other rickettsiae occurring in dogs are *R. canada* and *R. prowazekii* (Breitschwerdt et al. 1995).

Canine ehrlichiosis, caused by *Ehrlichia canis*, *E. chaffeensis*, *E. ewingii*, and Panola Mountain *Ehrlichia* sp., is prevalent in the Southeastern USA, where it is vectored, among other bacteria, by the lone star tick *A. americanum*. Metagenomic studies on the ticks microbiome with deep Illumina-based Next Generation Sequencing showed that the associated bacteria species composition was affected by previous vector life-history factors like environment, life-stage, population structure and host choice (Trout-Fryxell and DeBruyn 2016).

Apart of arthropod vectors, other associations have been found in farming in relation to the transmission of a bacterial disease. Ferguson et al. (2010) identified the shallow seawater and small jellyfish *Phialella quadrata* as vector of a bacterial disease of farmed salmon (*Salmo salar*) occurring in Scotland. The disease followed the gill damage induced by the jellyfish nematocysts toxin as a secondary infection. This was caused by the filamentous bacterium *Tenacibaculum maritimum*, present on the jellyfish mouth parts. The bacterium is the aetiologic agent of ulcerative diseases in many species of marine fishes (Avedaño-Herrera et al. 2006). The reduced dimension of the jellyfish allowed its passing through the meshes of the sea cages and its ingestion by the salmons, which acquired the bacterium, caused the eventual gill damage and related mortality (Ferguson et al. 2010).

A further vector-pathogen association has been studied in trematodes. The digenean trematode *Nanophyetus salmincola* requires three hosts for its life-cycle, including snails, fishes and dogs as final hosts. The trematode transmits the pathogenic bacterium *Neorickettsia helminthoeca* (Headley et al. 2011). Its cycle includes the river snail *Oxytrema silicula* (hosting the trematode stages known as rediae and cercariae), fishes (mainly salmonids, that are parasitized by the free-swimming trematode cercariae that rapidly penetrate their abdominal skin), and mammals or birds, eating the fishes, as final hosts. The trematode cysts are mainly located in the fish kidney, heart, liver, intestine and tail flesh. Salmons retain the trematode metacercariae for years during their migrations. The metacercariae complete their life-cycle in the mammals or birds that feed on the salmons or other fishes. Infected hosts eventually release the trematode eggs, that infect the snails, completing the cycle. *Neorickettsia helminthoeca* is maintained in the trematode during all stages. It is released in the cytoplasm of intestinal fish histiocytes and then transferred through the circulatory and lymphatic systems to visceral and somatic lymphnodes. It is an obligate intracytoplasmic pathogen of macrophage vacuoles, causing the so-called **salmon poisoning disease** (SPD), characterized by a severe lymphadenopathy and edema in final hosts. SPD has an acute, febrile and fatal outcome only for the final hosts (dogs, coyotes or foxes) that acquire the disease by feeding on infected, uncooked or even badly smoked salmons.

SPD has been reported in British Columbia, Canada, from Northwest Pacific USA and Southern Brazil. It was observed to cause swelling and edematous reactions in dogs lymphnodes, with hypertrophy, proliferation of lymphoid tissue within the intestine, causing an enteritis with intestinal hemorrage. Death of the animals usually occurs around 18 days after the consumption of the infected fish. Since Neorickettsiae appear resistant to many antibiotics, oxytetracycline and more effectively deoxycycline have been used to control the infection (7 mg/kg at 8 h intervals for 3–5 days). Praziquantel is highly effective against the trematode as a single 10–30 mg/kg oral or subcutaneous dose (Headley et al. 2011). A gastrointestinal disease (**nanophyetiasis**) has been reported for humans ingesting uncooked or smoked salmons or trouts eggs. Recent data suggest that more zoonoses are expected to emerge, including novel Anaplasmataceae, as a consequence of the increase and spread of crude fish consumption (Eastburn et al. 1987; Headley et al. 2011).

A similar disease is caused in humans by *Neorickettsia sennetsu* the aetiologic agent of **Sennetsu fever** in Japan and Malaysia. Fluke metacercariae parasitize the grey mullet fish *Mugil cephalus*, but the bacterium was not isolated from the trematode (Rikihisa et al. 2005). *Neorickettsia risticii* is the causative agent of **Potomac horse fever** (PHF), also called equine monocytic ehrlichiosis. PHF is caused by Letichodendoriidae trematodes infecting Pleuroceridae snails. Insects are secondary hosts and insectivorous birds and/or bats are the possible definitive hosts. The disease is probably acquired by horses when feeding on contaminated pastures (Rikihisa et al. 2005; Headley et al. 2011).

2.2.2 Wildlife

Wild animals provide many examples of diseases for which they act as hosts and/or natural reservoirs of pathogens that can spread among domesticated species, including pets, cattle or even humans.

Tularemia (see Sect. 2.1.4) has been reported from 250 wild animal species, including Scandinavian beavers (*Castor fiber*), foxes and hares. The latter are severely affected, showing damages to spleen, liver, bone marrow and other internal organs, together with hemorrhagic enteritis (Frölich et al. 2002).

Infectious **keratoconjunctivitis** caused by *Mycoplasma conjunctivae* is a severe eye infection affecting chamois and ibex in the Alps. The disease indirectly affects their survival in a difficult environment in which partial or total blindness may result fatal, inducing high mortalities. The disease is associated to flies considered to transmit the bacterium from contiguous sheep flocks (Frölich et al. 2002).

The causal agent of Lyme disease, *B. burgdorferi* affects passerine birds in the USA and was found in larvae of associated *I. scapularis* (Anderson et al. 1990). During spring migrations, the birds migrating to Scandinavia and Denmark showed high prevalence of Lyme disease agents like *B. garinii*, found in infesting *I. ricinus* and other ticks. The birds were considered involved in the dispersion and hetrogeneous distribution of the bacterium strains, observed in Northern Europe (Olsén et al. 1993, 1995). Similarly, seabirds from Artic Norway were found to be affected by Lyme disease, whose aetiologic agent was *B. garinii*. The vector *Ixodes uriae* is a common tick of seagulls and other colonial seabirds like the guillemot *Uria aalge*. The bacterium was similar to other strains obtained from birds or clinical specimens proceeding from Northern Europe, confirming the role of birds in its spatial dispersal over long distances (Caimano 2006; Larsson et al. 2007).

Other diseases including epidemics of sylvatic plague caused by *Y. pestis*, have been reported in black-tailed prairy dog populations in USA, in which it may induce high mortalities causing relevant conservation concerns (Hanson et al. 2007). Other reports in the USA include typhus caused by *R. prowazekii*, endemic in flying squirrels (*Glaucomys volans*). The disease is transmitted mainly during wintering in nests by the squirrel lice *Neohaematopinus scuiropteri*. The lice do not attack humans, but susceptible fleas like *Orchopeas howardi* may transmit the disease (Bozeman et al. 1975). Rodents like *Mus musculus* and *Meriones crassus* have been reported as reservoirs for spotted fever group rickettsiae in the Middle East (Mumcuoglu et al. 1993).

2.3 Plant Pathogens

Associations of pathogenic bacteria with arthropods can be found also in a number of severe plant diseases, depending on transmission by many insects as vectors. The most studied diseases concern the phloematic and xylematic plant systems, due to the importance and losses induced in many economically important crops. The host-vector relationships can be grouped following a number of factors valid either for plant pathogenic viruses and bacteria, resumed in the scheme presented in Table 5.1 (Kwon et al. 1999; Blanc 2004; James and Perry 2004; Nadarasah and Stavrinides 2011).

The relationships among plants, pathogens and vectors may be complex. For example, the Begomovirus coat protein interacts with proteins of its whitefly vector *Bemisia tabaci* at the gut and salivary gland membranes and with one protein encoded by its primary endosymbiont, to facilitate transmission. The primary endosymbiont encodes a 60S heat shock protein (HSP60) that interacts with the nucleocapsid of virions circulating in the haemolymph toward the salivary glands. The protein binds to the virus and promotes its stability by masking its surface, thus limiting the whitefly innate immune response (Brown 2007).

Table 5.1 Categorization of biological relationships of plant pathogenic bacteria with insect vectors

	Movement	
Replication	Circulative	Non circulative
Propagative	Ingested bacteria replicate in vectors and cross their membranes; enter the haemolymph and salivary glands; transmitted at feeding	Physical association with stylet or legs; transmitted at feeding
Non propagative	Bacteria ingested by feeding vectors migrate into midgut or hindgut epithelium; released into the haemolymph they enter salivary glands; transmitted at feeding	Physical association with stylet or legs; transmitted at feeding

2.3.1 Phloematic Diseases

Huanglongbing (HLB), formerly known as "greening disease", is a pathology of citrus trees caused by members of the genus *Ca.* "Liberibacter", a group of fastidious and phloem-inhabiting α-Proteobacteria. Members of this lineage, including *Ca.* "L. asiaticus", *Ca.* "L. americanus" and *Ca.* "L. africanus" are actually causing severe losses in citrus crops in several producing regions in Asia, Africa, and Central or North America. The disease is transmitted by the Asian citrus psyllid *Diaphorina citri* (Hemiptera: Psyllidae), which is vector of *Ca.* L. asiaticus and *Ca.* L. americanus in Asia and America, respectively. In Africa, *Ca.* L. africanus is vectored by the African citrus psyllid *Trioza erytreae* (Hemiptera: Triozidae) (Jagoueix et al. 1994; Teixeira et al. 2005; Bové 2006; Gottwald 2010; Grafton-Cardwell et al. 2013).

HLB is a global threat to the world citrus industry and is present on the EPPO[9] A1 and other alert quarantine lists (Bové 2006). The disease is graft-transmissible and was initially reported by Lin Kung Hsiang in Southern China as early as 1919. It was later reported in other citrus growing areas in the Philippines (1921), South Africa (1928), India (1967), Brazil (2004) and Florida (2005) (Bové 2006). Symptoms can be observed during the whole year on infected trees whose leaves show blotchy mottle, with tpical asymmetric and random discolorations between their left and right sides. Other leaf symptoms include vein discoloration and yellowing, with corking and occurrence of green islands. Fruit symptoms include changes of taste (bitter or salt), coloration starting at the peduncular area (rather than at the opposite side, as normal), appearence of darkening and staining marks in the area under the calix button, reduced size and altered, asymmetric shapes (Bové 2006). The disease induces yellow shootings on stunted trees, which decline in a few years. Being the symptoms non-specific, PCR is required to confirm diagnosis (Subandiyah et al. 2000).

A further fastidious species, *Ca.* L. solanacearum, has been sequenced from potato and other solanaceous hosts. In the Americas and New Zealand it is the causal agent of the potato zebra chip disease, and is transmitted by its vector *Bactericera cockerelli* (Hemiptera: Psyllidae) (Lin et al. 2011).

[9] European Plant Protection Organization.

Phytoplasmas are pleomorphic bacteria deprived of a cell wall, classified in the phylum Tenericutes (Class: Mollicutes). Their classification is based on the 16S rRNA ribosomal gene sequence and on the informations obtained by applying Restriction Fragment Length Polymorphism (RFLP) analysis. They are classified in 33 groups, with more than 105 subgroups actually recognized (Wei et al. 2007; Zhao et al. 2010; Gurr et al. 2015). Phytoplasmas are all '*Candidatus*' species acting as specific or polyphagous aetiologic agents of plant diseases (Lee et al. 2000; Duduk and Bertaccini 2011). They are transmitted by phloem sap-sucking insects of the families Cicadellidae, Cixiidae, Psyllidae, Delphacidae and Derbidae (Weintraub and Beanland 2006).

After insect acquisition through feeding, the phytoplasmas undergo a period of latency in their vectors, which may last up to 80 days, during which the bacteria multiply inside the insect. They reach the hemocoel passing through the midgut epitelial cells or between them, and circulate in the hemolymph reaching other organs like the Malpighian tubules, fat bodies, brain or the reproductive apparatus. Colonization of salivary glands, and in particular of posterior acinar cells, is required for transmission to plants. This organ, and related tissues, are considered as a substantial barriers in part responsible for vectors specificity (Weintraub and Beanland 2006). Vertical, transovarial transmission to eggs has also been reported (Alma et al. 1997). The effects of infection on vectors include the possibility of a longer lifespan and increase suitability of the host plant by i.e. reducing its chemical defense arsenal (Weintraub and Beanland 2006).

The phytoplasma-induced plant pathologies are characterized by massive changes of the phloem physiology altering the translocation of nutrients. This is due to an influx of Ca^{++} into the sieve tubes, inducing the occlusion of the sieve-plates because of callose deposition or formation of protein plugs (Musetti et al. 2013). Visible symptoms of infection often include yellowings and stunting, reduced productivity and yields. Typical symptoms include phylloidy, the production of leaf-like structures instead of flowers, virescence showing green petals with flower sterility, and witches' broom, a stress related symptom due to the loss of apical dominance with proliferation of ascillary and elongated small branches, observed in woody plants.

The phloematic infections caused by phytoplasmas severely compromize plants productivity and yields, and the range of host plants and new taxa reported as affected by phytoplasmas is still expanding (Agrios 2008; Lee et al. 2011). A list showing some of the most important and widespread diseases caused by phytoplasma and related vectors is given in Table 5.2. One of best known and studied example is the grapevine Flavescence dorée, a quarantine disease present in the EU, caused by *Ca.* "Phytoplasma vitis" vectored by cicadellids like *Scaphoideus titanus* and *Euscelidius variegatus* (Lessio and Alma 2006; Morone et al. 2007; Musetti et al. 2013). For more comprehensive reviews of these plant pathogens see Lee et al. (2000), Bertaccini et al. (2014) and Gurr et al. (2015).

Spiroplasmas are Mollicutes characterized by helical, motile cells. They are insect pathogens, but three species are also known as plant pathogens. These are *Spiroplasma citri*, pathogenic in *Citrus* spp., in which it is responsible of the disease

Table 5.2 Examples of phytoplasma diseases, host plants and vectors

Diseases	Host plants	Vector	Reference
American elm yellows	*Ulmus* spp.	*Scaphoideus luteolus* (Cicadellidae)	Lee et al. (2000)
Apple proliferation	Apple	*Cacopsylla melanoneura*, *C. picta* (Psyllidae),	Tedeschi and Alma (2004, 2006)
		Fieberiella florii (Auchenorrhyncha)	
Aster yellows complex, broccoli phyllody, cabbage witches' broom, poplar witches broom, turnip virescence	Several hosts, i.e. aster, cabbage, carrot, celery, onion, strawberry tomato	Several, i.e. *Macrosteles fascifrons*, *M. quadrilineatus* (Cicadellidae)	Lee et al. (2000)
Flavescence doreé	*Vitis* spp., Faba bean	*Scaphoideus titanus*, *Euscelidius variegatus* (Cicadellidae)	Morone et al. (2007)
Purple-top wilt	Potato	*Scleroracus flavopictus* (Cicadellidae)	Shiumi and Sugiura (1984)
Stolbur	Potato	*Hyalesthes obsoletus*, *Reptalus panzeri*, *R. quinquecostatus* (Cixiidae)	Mitrović et al. (2015)
		Euscelis incisus, *Psammotettix alienus* (Cicadellidae)	
Sugarcane white leaf	Sugarcane	*Matsumuratettix hiroglyphicus* (Cicadellidae)	Hanboonsong et al. (2002)

known as Citrus stubborn, *S. kunkelii*, the agent of corn stunt, and *S. phoeniceum*, found in periwinkle (*Catharanthus roseus*) plants in the Middle East (Saillard et al. 1987; Bové 1997; Bové et al. 2003).

Similarly to phytoplasma, also the plant pathogenic spiroplasmas are transmitted by leafhoppers (Cicadellidae), in which they also multiply. Vectors of *S. citri* include *Circulifer tenellus*, *Scaphytopius nitridus* and *S. acutus delongi* in USA, and *Neoaliturus haematoceps* and *C. tenellus* in the Mediterranean region. *Spiroplasma kunkelii* is transmitted by *Dalbulus maidis* (on which it may result pathogenic, shortening its lifespan) and other *Dalbulus* spp., by *Graminella nigrifons*, *Stirellus bicolor*, *Exitianus exitiosus* and *E. obscurinervis*. *Spiroplasma phoeniceum* is vectored by *Macrosfeles fascifrons* (Saillard et al. 1987; Oldfield 1988; Bové 1986; Klein et al. 1988; Golino and Oldfield 1990; Carloni et al. 2011; Moya-Raygoza et al. 2007).

A further phloematic pathology transmitted by insects is the cucurbit yellow vine disease, caused by *Serratia marcescens*, which injure leaves and may induce fruit collapse. The bacterium is transmitted by the squash bug *Anasa tristis* through its piercing-sucking activity, reaching the intracellular content. The insect piercing activity may also reach the xylem. The bacterium transmission appears to be of a non circulative type (Nadarasah and Stavrinides 2011).

2.3.2 Xylematic Diseases

Xylella fastidiosa is one of the most important and severe plant pathogenic bacteria transmitted by insects. The bacterium is non-circulative, and is retained in vectors' foregut (in which it may also multiply) from where it may be readly transmitted. The pathogen attacks more than 300 plant species on which it induces symptoms similar to a severe water losses, leaf scorch and drying leading to defoliation, stunting, almost complete yield losses and eventual plant death.

The pathogen has been reported from many world regions, including Southern, Central and North America, Asia and recently Europe. The most studied pathology caused by *X. fastidiosa*, Pierce's disease of grapevine, is known since the end of the nineteenth century, severely affecting grapevine vineyards in North America. Another important disease caused by *X. fastidiosa* is the citrus variegated chlorosis, of which a recent epidemic was reported in Brazil (Redak et al. 2004). The losses are enormous, and eradication is very difficult even on small initial foci, and almost impossible once established. For more informations on the pathology see the reviews by Purcell (1997), Redak et al. (2004), Janse and Obradovic (2010), and Martelli et al. (2015).

The phylogeny of *X. fastidiosa* is complex, with at least six subspecies that have been recognized, differing for host ranges and pathogenicity. They are *X. f.* subsp. *fastidiosa*, *X. f.* subsp. *multiplex*, *X. f.* subsp. *pauca*, *X. f.* subsp. *sandy*, *X. f.* subsp. *tashke* and *X. f.* subsp. *morus* (Schaad et al. 2004; Janse and Obradovic 2010). Multilocus sequence typing showed that homologous recombinations and genetic transfer occur among subspecies, favoring the bacterium diversification. They appear useful when reconstructing the phylogenetic relationships among isolates and their origins as well (Yuan et al. 2010; Loconsole et al. 2016).

Several xylophagous sharpshooters and spittlebugs of the family Cicadellidae are vectors of *X. fastidiosa*. In North America they include *Xyphon fulgida*, *Draeculacephala minerva*, *Graphocephala atropunctata*, *Homalodisca coagulata* and *Oncometopia* spp. In Brazil *Dilobopterus costalimai*, *Oncometopia facialis* and *Acrogonia citrina* are common under citrus tree canopies, and *Bucephalogonia xanthophis* is often found in nurseries (Redak et al. 2004). Polyphagous cicadellids feeding on weeds are abundant in South America citrus groves, including *Hortensia similis*, *Ciminius albolineatus*, *Plesiommata* spp., *Ferrariana trivittata*, *Scopogonalia subolivacea* and *Sonesimia grossa*. It is estimated that thousand insect species may act as vectors of *X. fastidiosa* (Purcell 2008). For a more detailed list of vectors and data on their geographic occurrence see reviews by Redak et al. (2004), and Janse and Obradovic (2010).

In the insect body, *X. fastidiosa* is located in the foregut where it multiplies. There are no reports of transovarial transmission of *X. fastidiosa* and the bacterium is lost at insect moult, becoming permanently established in the adults. The pathogen is acquired during feeding, its cells adhering to the vector foregut external cuticle. In *G. atropunctata* the bacterium was observed in the cibarium and pre-cibarium, a few days after feeding on infected plants, developing extensive colonies within 2 weeks. The pathogen is then transmitted from the external foregut surface, when the insect feeds on the xylem vessels of susceptible plants (Purcell et al. 1979). The vessels are eventually colonized, leading to their progressive occlusions by bacterial cells and biofilms, with a complete loss of their vascular function (Redak et al. 2004).

A severe outbreak of *X. fastidiosa* subsp. *pauca* was recently reported in Apulia (Southern Italy), on host plants including olive and many other plants including oleander, cherry, myrtleleaf and rosemary (Loconsole et al. 2014; Martelli et al. 2015). The isolates obtained from infected plants showed that the outbreak was due to the accidental introduction in the region of a single subspecies, *X. f.* subsp. *pauca*, possibly through imported ornamental coffee plants (Cariddi et al. 2014; Loconsole et al. 2016) likely proceeding from Central America on which the disease was asymptomatic. The disease remained undetected until the bacterium was transmitted in the field to highly susceptible olive trees and/to other perennial hosts. *Xylella fastidiosa* was transmitted by a local endemic hemipteran, the splittlebug *Philaenus spumarius* (Fig. 5.4) a polyphagous species previously considered of secondary importance as a plant herbivore, that became an important pest due to the newly acquired status of *X. f.* subsp. *pauca* vector (Cariddi et al. 2014; Saponari et al. 2014; Loconsole et al. 2016). A further *X. fastidiosa* subsp. *multiplex* has been recently detected in Southern France and Corsica. Also in this case the epidemic was due to imported, asymptomatic ornamental plants (Legendre et al. 2014).

The olive pathology observed in Apulia has been identified as a new disease complex called CODIRO (Olive quick decline syndrome or *Complesso del Disseccamento Rapido dell'Olivo* [italian]) (Martelli et al. 2015). Actually, due to the high number of susceptible plants in the region and to the local farmers' resistance to uprooting (the only management strategy reducing the spread of the epidemic foci), the efforts deployed to reduce the epidemic progression towards pathogen-free areas could not be implemented efficiently. This occurred in spite of the early detection of the pathogen, the media attention and many institutional actions, including the possibility of a disease spatial tracking trough a GPS referenced database (see http://webapps.sit.puglia.it/freewebapps/MonitoraggioXFSintesi/). It is expected that in future the epidemic spread will slow down, due to the reduction in the frequency of newly infected plants in the areas already colonized by the pathogen. The most demanding task will be reducing the speed of the epidemic northbound boundaries spreading from the initial infection foci. Actually *X. fastidiosa* is decimating thousand olive trees in the Southern Apulia provinces of Lecce and Brindisi (Fig. 5.5). In some cases the trees are more than a century old and central to a traditional agricultural system, whose destruction is causing economic losses and significant social turmoils.

The outbreak of *X. f.* subsp. *pauca* and the associated disease complex had a severe impact on agriculture in Apulia. It highlighted the difficulties inherent the implementation of efficient quarantine inspections at the EU level, especially when

Fig. 5.4 Typical foams (**a**, **b**) produced on grasses by the nymphal stages (**c**, **d**) of *Philaenus spumarius*, the vector of *Xylella fastidiosa* subsp. *pauca* in Apulia (Italy). The spittlebug transmits the bacterium when penetrating host plants with its foregut spear visible in ventral view (**d**, *arrow*) (Photos by the author)

Fig. 5.5 Early symptoms of CODIRO observed on olive trees positive for *Xylella fastidiosa* subsp. pauca at Torchiarolo (Italy). The disease appears with apical dryings on leaves (**a**), followed by a progressive drying visible on branches (**b**) and whole trees (**c**, **d**). *Arrowhead* in (**a**) shows a further bacterial disease caused by Pseudomonas savastanoi

plants are asymptomatic, and the importance of controlling imported plants and goods (Janse and Obradovic 2010; Loconsole et al. 2016).

Although no curative remedy is today available for the disease, research efforts are considered for the selection of tolerant host plants, use of vector control strategies, early detection of infective foci and monitoring, with uprooting of infected plants and surrounding susceptible ones. Phages have been reported in *X. fastidiosa*, including four phages depending on type IV pilus for infection of *X. fastidiosa* and *Xanthomonas* spp. (Summer et al. 2010; Ahern et al. 2014). However, although no data are yet available on field applications capable to slow down the epidemics, promising results were obtained with a mixture of four different phages applied to control of *X. f.* subsp. *fastidiosa* on infected grapevine plants, suggesting a possible route for field exploitation of phages in applied plant therapy and *Xylella* biocontrol (Das et al. 2015).

All events leading to an epidemic have a certain likelihood for happening, and accurate inspections may indeed contribute to reduce, or even eliminate, the risk of their occurrence. The statistical probability of an epidemic like that of *X. fastidiosa* in Apulia may be considered as the product of a series of conditions that have to occur in order to induce an invasive outbreak. These include the likelihood of: (1) exporting the pathogen in plants which are infected (initial infection prevalence); (2) the pathogen reproduction and/or survival in the travel conditions (i.e. in response to changes in temperature or other environmental parameters during shipment), and during the time period spent until the final destination is reached; (3) its introduction in a new previously exempt region, characterized by climatic conditions that are favourable for the pathogen; (4) encountering, in the region of arrival, one or more suitable vectors; (5) being then acquired by these new vector(s) when the shipped plants reach the field; (6) encountering in the field other susceptible host plants (risk of infection) and finally (7) the pathogen likelihood of being efficiently transmitted, multiply and be spatially dispersed at rates high enough to allow the infection of new hosts and the eventual outbreak. If one of these events has a zero probability (for example due to the absence of a vector or an unfavorable climate) the event series reaches a dead end. When all the step conditions are matched (with different likelihoods) the epidemic instead takes place, with a given probability of occurrence. The final likelihood for an epidemic event to start (P_E) is then given by the product of all steps probabilities:

$$P_E = P_1 \cdot P_2 \cdot P_3 \cdot P_4 \cdot P_5 \cdot P_6 \cdot P_7$$

The series events are not exaustive, and may be integrated by further ones accounting for environmental factors affecting the recipient host plants (i.e. their density, susceptibility or tolerance), the pathogen (virulence, pathogenicity) and the vector(s) (transmission efficiency, survival). Some steps may have a high probability to occur for *X. fastidiosa*, due to the high number of vectors or plant hosts. Inferring a mean estimated 0.5 likelihood (a coin toss) for each one of the seven steps previously cited (these are just estimated values adopted in this example for clarity, the considered probabilities in the real world may be higher or lower than 0.5, and need to be experimentally or statistically determined), the final epidemic likelihood P_E is 0.5^7, that is 0.78 % (equivalent to around 1/128 cases). This low probability might be considered as overestimated, since only a few steps may have a >0.5 likelihood, and other factors may also be active. In any case, given the large number of plants globally commercialized today and the number of intercontinental shipments, these probabilities can be considered as real, as demonstrated by the Apulian epidemic. They suggest that, *at the hypothetical conditions given* (including shipping of infected plant material with an unprobable 50 % disease prevalence, that is without any previous inspection), the transcontinental movement of the pathogen, followed by its introduction and spreading in fields in a previously exempt area, has a chance to occur only in one out of around 128 shipments. This estimate suggests that any programme for controlling and shipping healthy, certified plants (step 1) may largely reduce the final overall risk of an epidemic.

Much more improbable appears, furthermore, the independent replication of the same event in the same area, for a second or even third *Xylella* isolate or subspecies.

In this case the probability that two or three distinct and separated isolates have to co-occur in the same area (and at the same time) should be equal to the product of each individual P_E, corresponding (always considering "coin toss" likelihoods) to 0.006 % (1/16,400 shipments) or 0.000048 % (1/2,100,000 shipments), for two and three isolates or subspecies, respectively.

2.3.3 Other Plant Diseases

Disease transmission routes include further examples in which the pathogenic bacteria multiply into the vectors or are hosted in external body parts. These ways of transmission involve the passive transport of bacterial cells, which may be moved from one parasitized plant to a susceptible one by adhering externally to some insect body parts like the foregut, legs or ovipositor, or be internally retained in the intestine, after acquisition through feeding. In some cases the bacteria may also invade and colonize internal body parts and be recognized by the host defense system as in the case of *Pectobacterium carotovorum* in *Drosophila* (Acosta Muniz et al. 2007). Many species involved are *Enterobacteriaceae*, and include a number of economically important plant pathogens. The mechanisms of transmission and further biological host/vector relationships have been reviewed in detail by Agrios (2008) and Nadarasah and Stavrinides (2011).

Bacterial soft rots of many important crops like potato, fleshy fruits or other vegetables are passively transmitted. Soft rot disease is characterized by the pectolytic activity of exoenzymes produced by the invading bacteria, causing the rupture of the cell walls and the rotting of the infected tissues. The aetiologic agents include bacteria like *Pectobacterium carotovorum* (formerly known as *Erwinia carotovora* pv. *carotovora*) and other occasional pathogens like *Pseudomonas fluorescens*, *P. chrysanthemi* and *Bacillus* or *Clostridium* spp. The seedcorn maggot *Delia platura* (Diptera: Anthomyiidae) is involved in the dissemination and transmission of the bacteria on field or stored potatoes, together with other dipterans like *D. florilega* or *Drosophila busckii*. In *Drosophila*, *P. carotovorum* is capable to use the host as a vector by colonizing the insect body, actively penetrating the gut epithelium and invading the body cavity. The *P. carotovorum evf* gene product is active in this mechanism, allowing the colonization and eventual bacterial cells accretion within the host midgut, by targeting and disrupting the gut peritrophic membrane (Acosta Muniz et al. 2007).

In soft rot disease the maggot larvae get contaminated when they feed on already infected potato seeds, which represent the initial field inoculum. The maggots retain the bacterial cells in their body and transmit the disease to healthy plants through the wounds they produce during feeding. The bacteria overwinter in the insects and by their release through the faeces in the wounds when insects feed, they may overcome the potato corky layers defense barrier, eventually spreading to the soft tissues. Similar relationships have also been observed for other Diptera, like the cabbage maggot *Delia radicum* and soft rot of Brassicaceae, *D. antiqua*, the onion black fly *Tritoxa flexa* and the onion bulb fly *Eumerus strigatus* associated to the onion soft rot, and the iris borer, *Macronoctua onusta* (Lepidoptera) that transmits the iris soft rot (Agrios 2008; Nadarasah and Stavrinides 2011).

The **bacterial wilt of cucurbits** is caused by *Erwinia tracheiphila* which overwinters in the intestine of its vectors, the striped cucumber beetle *Acalymma vittatum* and the spotted cucumber beetle *Diabrotica undecimpunctata* (Coleoptera: Chrysomelidae). *Erwinia tracheiphila* is acquired during feeding, and migrates to the insect gut epithelium. The beetles deposit the bacterial cells in the wounds through their faeces when feeding is resumed on leaves during spring. The cells of *E. tracheiphila* rapidly colonize xylem vessels, eventually infecting the rest of the plant, clogging the vessels due to the accumulation of products arising from the degradation of cell walls, and thus causing the plant wilting. The beetles become contaminated when feeding on infected plants and can also transport the bacteria on the stylet, rapidly transmitting the disease when attacking healthy plants (Agrios 2008; Nadarasah and Stavrinides 2011).

Similar mechanisms also occur in other plant-bacteria associations, like the **bacterial wilt of corn**, causing wilt and leaf blight, due to infection by *Pantoea stewartii*. The bacterium cells colonize the xylem producing stewartan, an exo-capsular polysaccharide reducing the water flow and causing wilting. The host infection involves a type III secretion system (SS), a 201 kDa protein similar to other bacterial virulence factors like the DspE of *E. amylovora*, and a quorum sensing system reacting to the density of cells and controlling the expression of other pathogenicity factors (Nadarasah and Stavrinides 2011). *Pantoea stewartii* overwinters in the vector intestine and is then transmitted when the insects feed and release its cells through their faeces in the feeding wounds. Vectors are the corn flea beetle *Chaetocnema pulicaria*, the toothed flea beetle *C. denticulata*, the spotted cucumber beetle *D. undecimpunctata*, the larvae of the seed corn maggot *D. platura*, the wheat wireworm *Agriotes mancus* and *Phyllophaga* spp. beetles (Agrios 2008; Nadarasah and Stavrinides 2011).

Strain DC283 of *P. stewartii* is also pathogenic for the pea aphid *Acyrthosiphon pisum*, acting through the expression of an adhesion protein coded by the gene *ucp1*, and used by the bacterium to bind to the host cells (Stavrinides et al. 2009, 2010). Other plant pathogens also lethal for the pea aphid include *Dickeya dadantii*, causing a soft rot disease in potato, maize and other crops, provided with a number of genes coding for insecticidal toxins, and *E. aphidicola*, the causal agent of leaf spots on pea and bean. Strains of *Pseudomonas syringae*, the causal agent of many plant diseases including kiwi and tomato bacterioses, were also found to be entomopathogenic, using the pea aphid as primary host, due to strain-specific virulence factors. The bacteria could be acquired from artificially inoculated plants by feeding aphids when probing the plant surface with their stylet. The bacterial cells colonized the aphid digestive tracts and were eventually excreted through their honeydew, to be deposited again on the plant surface (Stavrinides et al. 2009). These and other pathogenic links of enterobacteria with plant feeders are indicative of long term evolutionary relationships. They probably involve some original components of ancestral entomopathogens, that were retained while new plant pathogenic bacteria evolved at the issue of the hosts herbivorous lifestyle (Nadarasah and Stavrinides 2011).

Fire blight of pears, apples and other rosaceous is caused by *E. amylovora* which affects flowers, young shoots and twigs producing cankers in which the cells overwinter. The blossoms and stigma are the primary route of plant invasion. The proliferating cells proceeding from the stigma and nectaria destroy the bark and then enter the phloem and xylem vessels (Spinelli et al. 2005). Disease-induced polysaccharides like amylovoran and levan are responsible for plugging of the vascular tissue and shoot wilting, together with other type III SS factors (Oh and Beer 2005). Although many mechanisms of infection occur, starting from overwintering cells oozing out from twig and branch cankers (including rain splash), more than 200 insect species have been reported as passive carrier of *E. amylovora* (Agrios 2008). Pollinators, including honey bees and other insects are aspecifically associated to transmission transferring during their visits the bacterial cells, that adhere to their bodies, to blossoms and shoots on which the disease is mechanically transmitted.

Ralstonia solanacearum, the aetiologic agent of the **bacterial wilt of solanaceous** and other crops, is often associated in the tropics to hymenopterans like bees (*Trigona corvine*), wasps (*Polybia* spp.), and dipterans (*Drosophila* spp.) visiting infected plant sites and probably responsible for spreading the bacterium to other plants.

Further **bacterioses** include a severe disease affecting banana and enset in tropical Africa, caused by *Xanthomonas vasicola* pv. *musacearum*, transmitted to healthy flowering plants by males of the stingless bee *Plebeina denoiti* and honey bee *Apis mellifera*, when entering flowers, and by other fruit flies too (Shimelash et al. 2008; Tinzaara et al. 2006). Papaya Bunchy Top (PBT) is widespread in the American tropics and affects the xylem, reaching the phloem and cortex in severely diseased papaya plants. PBT is caused by a Rickettsia-like pathogen, and is associated to the leafhoppers *Empoasca papayae* and *E. stevensi* (Davis et al. 1996; Pantoja et al. 2015).

References

Acha, P. N., & Szyfres, B. (2001). *Zoonoses and communicable diseases common to man and animals* (Vol. 2). Washington, DC: Pan American Health Organization, 387 pp.

Achtman, M., Zurth, K., Morelli, G., Torrea, G., Guiyoule, A., & Carniel, E. (1999). *Yersinia pestis*, the cause of plague, is a recently emerged clone of *Yersinia pseudotuberculosis*. *Proceedings of the National Academy of Science USA, 96*, 14043–14048.

Acosta Muniz, C., Jaillard, D., Lemaitre, B., & Boccard, F. (2007). *Erwinia carotovora* Evf antagonizes the elimination of bacteria in the gut of *Drosophila* larvae. *Cell Microbiology, 9*, 106–119.

Adams, J. R., Schmidtmann, E. T., & Azad, A. F. (1990). Infection of colonized cat fleas, *Ctenocephalides felis* (Bouche), with a rickettsia-like microorganism. *American Journal of Tropical Medicine and Hygiene, 43*, 400–409.

Agrios, G. N. (2008). Transmission of plant diseases by insects. In *Encyclopedia of entomology* (pp. 3853–3885). Dordrecht: Springer.

Ahern, S. J., Das, M., Bhowmick, T. S., Young, R., & Gonzalez, C. F. (2014). Characterization of novel virulent broad-host-range phages of *Xylella fastidiosa* and *Xanthomonas*. *Journal of Bacteriology, 196*, 459–471.

Alma, A., et al. (1997). Identification of phytoplasmas in eggs, nymphs and adults of *Scaphoideus titanus* ball reared on healthy plants. *Insect Molecular Biology, 6*, 115–121.

Anderson, J. F., Magnarelli, L. A., & Stafford, K. C. (1990). Bird-feeding ticks transstadially transmit *Borrelia burgdorferi* that infect Syrian hamsters. *Journal of Wildlife Disease, 26*, 1–10.

Anderson, B. E., et al. (1993). *Amblyomma americanum*: A potential vector of human ehrlichiosis. *American Journal of Tropical Medicine and Hygiene, 49*, 239–244.

Anderson, A., et al. (2013). Diagnosis and management of Q fever. Recommendations from the CDC and the Q fever working group. *Morbidity and Mortality Weekly Report, 62*(RR03), 1–23.

Avedaño-Herrera, R., Toranzo, A. E., & Magariños, B. (2006). Tenacibaculosis infection in marine fish caused by *Tenacibaculum maritimum*: A review. *Disease of Aquatic Organisms, 71*, 255–266.

Azad, A. F., Radulovic, S., Higgins, J. A., Noden, B. H., & Troyer, J. M. (1997). Flea-borne rickettsioses: Ecologic considerations. *Emerging Infectious Diseases, 3*, 319–327.

Bertaccini, A., Duduk, B., Paltrinieri, S., & Contaldo, N. (2014). Phytoplasmas and phytoplasma diseases: A severe threat to agriculture. *American Journal of Plant Sciences, 5*, 1763–1788.

Blanc, S. (2004). *Insect transmission of viruses* (Microbe–vector interactions in vector-borne diseases, Vol. 63, pp. 43–62). Cambridge: Cambridge University Press.

Bové, J. M. (1986). Stubborn and its natural transmission in the Mediterranean area and in the near east. *FAO Plant Protection Bulletin, 34*, 15–23.

Bové, J. M. (1997). Spiroplasmas: Infectious agents of plants, arthropods and vertebrates. *Wiener Klinische Wochenschrift, 109*, 604–612.

Bové, J. M. (2006). Huanglongbing. *Journal of Plant Pathology, 88*, 7–37.

Bové, J. M., Renaudin, J., Saillard, C., Foissac, X., & Garnier, M. (2003). *Spiroplasma citri*, a plant pathogenic mollicute: Relationships with its two hosts, the plant and the leafhopper vector. *Annual Review of Phytopathology, 41*, 483–500.

Bozeman, F. M., Masiello, S. A., Williams, M. S., & Elisberg, B. L. (1975). Epidemic typhus rickettsiae isolated from flying squirrels. *Nature, 255*, 545–547.

Braderic, N., & Punda-Polic, V. (1992). Cutaneous anthrax due to penicillin-resistant *Bacillus anthracis* transmitted by an insect bite. *Lancet, 340*, 306–307.

Breitschwerdt, E. B., Hegarty, B. C., Davidson, M. G., & Szabados, N. S. A. (1995). Evaluation of the pathogenic potential of *Rickettsia canada* and *Rickettsia prowazekii* organisms in dogs. *Journal of the American Veterinary Medical Association, 207*, 58–63.

Brown, J. K. (2007). The *Bemisia tabaci* complex: Genetic and phenotypic variation and relevance to TYLCV–vector interactions. In H. Czosnek (Ed.), *Tomato yellow leaf curl virus disease* (pp. 25–56). NL: Springer.

Brown, R. N., & Lane, R. S. (1992). Lyme disease in California: A novel enzootic transmission cycle of *Borrelia burgdorferi*. *Science, 256*, 1439–1442.

Caimano, M. (2006). The genus *Borrelia*. *Prokaryotes, 7*, 235–293.

Cariddi, C., et al. (2014). Isolation of a *Xylella fastidiosa* strain infecting olive and oleander in Apulia, Italy. *Journal of Plant Pathology, 96*, 1–5.

Carloni, E., et al. (2011). *Exitianus obscurinervis* (Hemiptera: Cicadellidae), a new experimental vector of *Spiroplasma kunkelii*. *Journal of Economic Entomology, 104*, 1793–1799.

Carlyon, J. A., Latif, D. A., Pypaert, M., Lacy, P., & Fikrig, E. (2004). *Anaplasma phagocytophilum* utilizes multiple host evasion mechanisms to thwart NADPH oxidase-mediated killing during neutrophil infection. *Infection and Immunity, 72*, 4772–4783.

Cartwright, F. F., & Biddiss, M. (2014). *Disease and history* (3rd ed.). London: Thistle Publishing Ltd., 252 pp.

DaMassa, A. J., & Adler, H. E. (1979). Avian spirochetosis: Natural transmission by *Argas* (*Persicargas*) *sanchezi* (Ixodoidea: Argasidae) and existence of different serologic and immunologic types of *Borrelia anserina* in the United States. *American Journal of Veterinary Research, 40*, 154–157.

Dantas-Torres, F., Chomel, B. B., & Otranto, D. (2012). Ticks and tick-borne diseases: A one health perspective. *Trends in Parasitology, 28*, 437–446.

Das, M., Bhowmick, T. S., Ahern, S. J., Young, R., & Gonzalez, C. F. (2015). Control of Pierce's disease by phage. *PLoS ONE, 10*, e0128902.

Dautel, H., & Kahl, O. (1999). Ticks (Acari: Ixodoidea) and their medical importance in the urban environment. In *Proceedings of the third international conference on urban pests*, pp .73–82.

Davis, M. J., Kramer, J. B., Fewerda, F. H., & Brunner, B. R. (1996). Association of a bacterium and not a phytoplasma with papaya bunchy top disease. *Phytopathology, 86*, 102–109.

Dillman, A. R., et al. (2012). An entomopathogenic nematode by any other name. *PLoS Pathogens, 8*, e1002527. doi:10.1371/journal.ppat.1002527.

Dolan, M. C., et al. (2016). Vector competence of the blacklegged tick, *Ixodes scapularis*, for the recently recognized Lyme borreliosis spirochete *Candidatus* Borrelia mayonii. *Ticks and Tick-Borne Diseases, 7*. http://dx.doi.org/10.1016/j.ttbdis.2016.02.012.

Duduk, B., & Bertaccini, A. (2011). Phytoplasma classification: Taxonomy based on 16S ribosomal gene, is it enough? *Phytopathogenic Mollicutes, 1*, 3–13.

Dworkin, M. S., Schwan, T. G., Anderson, D. E., Jr., & Borchardt, S. M. (2008). Tick-borne relapsing fever. *Infectious Disease Clininc of North America, 22*, 449–468.

Eastburn, R. L., Fritsche, T. F., & Terhune, C. A. (1987). Human intestinal infection with *Nanophyetus salmincola* from salmonid fishes. *American Journal of Tropical Medicine and Hygiene, 36*, 586–591.

Eisen, R. J., & Gage, K. L. (2009). Adaptive strategies of *Yersinia pestis* to persist during inter-epizootic and epizootic periods. *Veterinary Research, 40*, 01.

Eisen, R. J., et al. (2008). Early-phase transmission of *Yersinia pestis* by cat fleas (*Ctenocephalides felis*) and their potential role as vectors in a plague-endemic region of Uganda. *The American Journal of Tropical Medicine and Hygiene, 78*, 949–956.

Embers, M. E., & Lopez, J. E. (2012). Immune resistance by relapsing fever spirochetes. In M. E. Embers (Ed.), *The pathogenic spirochetes: Strategies for evasion of host immunity and persistence* (pp. 173–190). New York: Springer.

Emerson, P. M., & Bailey, R. L. (1999). Trachoma and fly control. *Community Eye Health, 12*, 57.

Emerson, P. M., Bailey, R. L., Mahadi, O. S., Walraven, G. E., & Lindsay, S. W. (2000). Transmission ecology of the fly *Musca sorbens*, a putative vector of trachoma. *Transactions of the Royal Society of Tropical Medicine and Hygiene, 94*, 28–32.

Eremeeva, M. E., et al. (2008). *Rickettsia typhi* and *R. felis* in rat fleas (*Xenopsylla cheopis*), Oahu, Hawaii. *Emerging Infectious Diseases, 14*, 1613–1615.

Evans, J. R., & Solomon, A. W. (2011). Antibiotics for trachoma. *The Cochrane Database of Systematic Reviews, 16*, CD001860.

Feigin, R., Cherry, J., Demmler-Harrison, G., & Kaplan, S. (2009). *Feigin & Cherry's textbook of pediatric infectious diseases* (6th ed.). Philadelphia: Saunders, Elsevier.

Felz, M. W., Smith, C. D., & Swift, T. R. (2000). A six-year-old girl with tick paralysis. *New England Journal of Medicine, 342*, 90–94.

Fenollar, F., & Raoult, D. (1999). Diagnosis of rickettsial diseases using samples dried on blotting paper. *Clinical Diagnostic and Laboratory Immunology, 6*, 483–488.

Ferguson, H. W., et al. (2010). Jellyfish as vectors of bacterial disease for farmed salmon (*Salmo salar*). *Journal of Veterinary Diagnostic Investigation, 22*, 376–382.

Fournier, P. E., & Raoult, D. (2009). Current knowledge on phylogeny and taxonomy of *Rickettsia* spp. *Annals of the New York Academy of Science, 1166*, 1–11.

Frances, S. P., Watcharapichat, P., & Phulsuksombati, D. (2001). Vertical transmission of *Orientia tsutsugamushi* in two lines of naturally infected *Leptotrombidium deliense* (Acari: Trombiculidae). *Journal of Medical Entomology, 38*, 17–21.

Frölich, K., Thiede, S., Kozikowski, T., & Jakob, W. (2002). A review of mutual transmission of important infectious diseases between livestock and wildlife in Europe. *Annals of the New York Academy of Sciences, 969*, 4–13.

Golino, D. A., & Oldfield, G. N. (1990). Plant pathogenic spiroplasmas and their leafhopper vectors. *Advances in Disease Vector Research, 6*, 267–299.

Gottwald, T. R. (2010). Current epidemiological understanding of citrus Huanglongbing. *Annual Review of Phytopathology, 48*, 119–139.

Grafton-Cardwell, E. E., Stelinski, L. L., & Stansly, P. A. (2013). Biology and management of Asian citrus psyllid, vector of the huanglongbing pathogens. *Annual Reviews of Entomology, 58*, 413–432.

Gray, J. S., Kahl, O., Janetzki-Mittmann, C., Stein, J., & Guy, E. (1994). Acquisition of *Borrelia burgdorferi* by *Ixodes ricinus* ticks fed on the European hedgehog, *Erinaceus europaeus* L. *Experimental and Applied Acarology, 18*, 485–491.

Gubler, D. J. (2009). Vector-borne diseases. *Revue Scientifique et Technique (International Office of Epizootics), 28*, 583–588.

Guo, B. P., et al. (2009). Relapsing fever *Borrelia* binds to neolacto glycans and mediates rosetting of human erythrocytes. *Proceedings of the National Academy of Science USA, 106*, 19280–19285.

Gurr, G. M., et al. (2015). Phytoplasmas and thier insect vectors: Implications for date palm. In W. Wakil, J. Romeno Falero, & T. A. Miller (Eds.), *Sustainable pest managment in date palm: Current status and emerging challenges* (pp. 287–314). Cham: Springer.

Hanboonsong, Y., Choosai, C., Panyim, S., & Damark, S. (2002). Transovarial transmission of sugarcane white leaf phytoplasma in the insect vector *Matsumuratettrix hiroglyphicus* (Matsumura). *Insect Molecular Biology, 11*, 97–103.

Hanson, D. A., Britten, H. B., Restani, M., & Washburn, L. R. (2007). High prevalence of *Yersinia pestis* in black-tailed prairie dog colonies during an apparent enzootic phase of sylvatic plague. *Conservation Genetics, 8*, 789–795.

Headley, S. A., Scorpio, D. G., Vidotto, O., & Dumler, J. S. (2011). *Neorickettsia helminthoeca* and salmon poisoning disease: A review. *The Veterinary Journal, 187*, 165–173.

Herrin, B., Mahapatra, S., Blouin, E., & Shaw, E. (2011). Growth of *Coxiella burnetii* in the *Ixodes scapularis*-derived IDE8 tick cell line. *Vector Borne Zoonotic Diseases, 11*, 917–922.

Hii, S. F., Lawrence, A. L., Cuttell, L., Tynas, R., Megat Abd Rani, P. A., Šlapeta, J., & Traub, R. J. (2015). Evidence for a specific host-endosymbiont relationship between '*Rickettsia* sp. genotype RF2125' and *Ctenocephalides felis orientis* infesting dogs in India. *Parasites and Vectors, 8*, 169.

Hinnebusch, B. J., & Erickson, D. L. (2008). *Yersinia pestis* biofilm in the flea vector and its role in the transmission of plague. *Current Topics in Microbiology and Immunology, 322*, 229–248.

Howden, K. J., Geale, D. W., Paré, J., Golsteyn-Thomas, E. J., & Gajadhar, A. A. (2010). An update on bovine anaplasmosis (*Anaplasma marginale*) in Canada. *Canadian Veterinary Journal, 51*, 837–840.

Inhorn, M. C., & Brown, P. J. (1990). The anthropology of infectious disease. *Annual Review of Anthropology, 19*, 89–117.

Jagoueix, S., Bové, J. M., & Garnier, M. (1994). The phloem-limited bacterium of greening disease of citrus is a member of the alpha subdivision of the Proteobacteria. *International Journal of Systematic Bacteriology, 44*, 379–386.

James, C. K. N. G., & Perry, K. L. (2004). Transmission of plant viruses by aphid vectors. *Molecular Plant Pathology, 5*, 505–511.

Janse, J. D., & Obradovic, A. (2010). *Xylella fastidiosa*: Its biology, diagnosis, control and risks. *Journal of Plant Pathology, 92*, S1.35–S1.48.

Kernif, T., et al. (2014). Acquisition and excretion of *Bartonella quintana* by the cat flea, *Ctenocephalides felis felis*. *Molecular Ecology, 23*, 1204–1212.

Klein, M., Rasooly, P., & Raccah, B. (1988). New findings on the transmission of *Spiroplasma citri*, the citrus stubborn disease agent in Israel, by a beet leafhopper from the Jordan valley. *Hassadeh, 68*, 1736–1737. (In Jewish).

Kocan, K. M., et al. (2012). Sheep experimentally infected with a human isolate of *Anaplasma phagocytophilum* serve as a host for infection of *Ixodes scapularis* ticks. *Ticks and Tick-Borne Diseases, 3*, 147–153.

Koehler, J. E., & Tappero, J. W. (1993). Bacillary angiomatosis and bacillary peliosis in patients infected with human immunodeficiency virus. *Clinical Infectious Diseases, 17*, 612–624.

Konnov, N. P., Popov, N. V., Velichko, L. N., & Knyazeva, T. V. (2010). The phenomenon of *Yersinia pestis* biofilm formation in the organism of fleas. *Entomological Review, 90*, 638–642.

Kwon, M. O., Wayadande, A. C., & Fletcher, J. (1999). *Spiroplasma citri* movement into the intestines and salivary glands of its leafhopper vector, *Circulifer tenellus. Bacteriology, 89*, 1144–1151.

Labruna, M. B., et al. (2007). Infection by *Rickettsia bellii* and *Candidatus* "Rickettsia amblyommii" in *Amblyomma neumanni* ticks from Argentina. *Microbial Ecology, 54*, 126–133.

Larsson, C., Comstedt, P., Olsen, B., & Bergstrom, S. (2007). First record of Lyme disease *Borrelia* in the Arctic. *Vector-Borne and Zoonotic Diseases, 7*, 453–456.

Lee, I. M., Davis, R. E., & Gundersen-Rinda, D. E. (2000). Phytoplasma: Phytopathogenic Mollicutes. *Annual Review of Microbiology, 54*, 221–255.

Lee, I. M., Bottner-Parker, K. D., Zhao, Y., Villalobos, W., & Moreira, L. (2011). '*Candidatus* Phytoplasma costaricanum' a novel phytoplasma associated with an emerging disease in soybean (*Glycine max*). *International Journal of Systematic and Evolutonary Microbiology, 61*, 2822–2826.

Legendre, B., et al. (2014). Identification and characterisation of *Xylella fastidiosa* isolated from coffee plants in France. *Journal of Plant Pathology, 96*, S4.100.

Lessio, F., & Alma, A. (2006). Spatial distribution of nymphs of *Scaphoideus titanus* (Homoptera: Cicadellidae) in grapes, and evaluation of sequential sampling plans. *Journal of Economic Entomology, 99*, 578–582.

Liebisch, A., & Olbrich, S. (1991). The hedgehog tick, *Ixodes hexagonus* Leach, 1815, as a vector of *Borrelia burgdorferi* in Europe. In F. Dusbabek & V. Bukva (Eds.), *Modern acarology* (pp. 67–71). Prague: Academia.

Lin, H., et al. (2011). The complete genome sequence of '*Candidatus* Liberibacter solanacearum', the bacterium associated with potato zebra chip disease. *PLoS ONE, 6*, e19135.

Lisbôa, R. S., et al. (2009). Avian spirochetosis in chickens following experimental transmission of *Borrelia anserina* by *Argas* (*Persicargas*) *miniatus. Avian Diseases, 53*, 166–168.

Loconsole, G., et al. (2014). A *Xylella fastidiosa* strain with unique biology and phylogeny is associated with a severe disease of olive in Southern Apulia. *Journal of Plant Pathology, 96*, S4.38.

Loconsole, G., et al. (2016). Intercepted isolates of *Xylella fastidiosa* in Europe reveal novel genetic diversity. *European Journal of Plant Pathology, 144.* In print. doi 10.1007/s10658-016-0894-x.

Lukin, E. P., Vorobev, A. A., & Bykov, A. S. (2001). Taxonomy and classification of Rickettsiae. *Zhurnal Mikrobiologii, Epidemiologii, i Immunobiologii, 2*, 105–110. [In Russian].

Magnarelli, L. A., Anderson, J. F., Hyland, K. E., Fish, D., & Mcaninch, J. B. (1988). Serologic analyses of *Peromyscus leucopus*, a rodent reservoir for *Borrelia burgdorferi*, in northeastern United States. *Journal of Clinical Microbiology, 26*, 1138–1141.

Marsollier, L., et al. (2002). Aquatic insects as a vector for *Mycobacterium ulcerans. Applied and Environmental Microbiology, 68*, 4623–4628.

Martelli, G. P., Boscia, D., Porcelli, F., & Saponari, M. (2015). The olive quick decline syndrome in south-east Italy: A threatening phytosanitary emergency. *European Journal of Plant Pathology, 144*, 235–243.

Mehlhorn, H. (2008). Dermacentor reticulatus. In H. Mehlhorn (Ed.), *Encyclopedia of parasitology* (pp. 324–325). Berlin: Springer.
Merritt, R. W., et al. (2010). Ecology and transmission of buruli ulcer disease: A systematic review. *PLoS Neglected Tropical Diseases, 4*, e911.
Minard, G., Mavingui, P., & Valiente Moro, C. (2013). Diversity and function of bacterial microbiota in the mosquito holobiont. *Parasites and Vectors, 6*, 146.
Mitani, H., Talbert, A., & Fukunaga, M. (2004). New world relapsing fever *Borrelia* found in *Ornithodoros porcinus* ticks in Central Tanzania. *Microbiology and Immunology, 48*, 501–505.
Mitrović, M., et al. (2015). Potential hemipteran vectors of stolbur phytoplasma in potato fields in Serbia. *Phytopathogenic Mollicutes, 5*, S49–S50.
Mohle-Boetani, J. C., et al. (1996). Bacillary angiomatosis and bacillary peliosis in patients infected with human immunodeficiency virus: Clinical characteristics in a case-control study. *Clinical Infectious Diseases, 22*, 794–800.
Morone, C., et al. (2007). Epidemiology of flavescence dorée in vineyards in northwestern Italy. *Phytopathology, 97*, 1422–1427.
Moya-Raygoza, G., Palomera-Avalos, V., & Galaviz-Mejia, C. (2007). Field overwintering biology of *Spiroplasma kunkelii* (Mycoplasmatales: Spiroplasmataceae) and its vector *Dalbulus maidis* (Hemiptera: Cicadellidae). *Annals of Applied Biology, 151*, 373–379.
Mumcuoglu, K. Y., et al. (1993). Ecological studies on the Brown Dog Tick *Rhipicephalus sanguineus* (Acari: Ixodidae) in southern Israel and its relationship to spotted fever group rickettsiae. *Journal of Medical Entomology, 30*, 114–121.
Musetti, R., et al. (2013). Phytoplasma-triggered Ca^{2+} influx is involved in sieve-tube blockage. *Molecular Plant-Microbes Interactions, 26*, 379–386.
Nadarasah, G., & Stavrinides, J. (2011). Insects as alternative hosts for phytopathogenic bacteria. *FEMS Microbiology Reviews, 35*, 555–575.
Norval, R. A. I., Andrew, H. R., Yunker, C. E., & Burridge, M. J. (1992). Biological processes in the epidemiology of heartwater. In B. Fivaz, T. Petney, & I. Horak (Eds.), *Tick vector biology* (pp. 71–86). Berlin: Springer Verlag.
Oh, C. S., & Beer, S. (2005). Molecular genetis of *Erwinia amylovora* involved in the development of fire blight. *FEMS Microbiology Letters, 253*, 185–192.
Ohl, M. E., & Spach, D. H. (2000). *Bartonella quintana* and urban trench fever. *Clinical Infectious Diseases, 31*, 131–135.
Oldfield, G. N. (1988). Ecological associations of *Spiroplasma citri* with insects, plants and other plant mycoplasmas in the western United States. In K. Maramorosch & S. P. Raychaudhuri (Eds.), *Mycoplasma diseases of crops. Basic and applied aspects* (pp. 175–191). New York: Springer-Verlag.
Olsén, B., Jaenson, T. G. T., Noppa, L., Bunikis, J., & Bergström, S. (1993). A Lyme borreliosis cycle in seabirds and *Ixodes uriae* ticks. *Nature, 362*, 340–342.
Olsén, B., Jaenson, T. G. T., & Bergström, S. (1995). Prevalence of *Borrelia burgdorferi* sensu lato-infected ticks on migrating birds. *Applied and Environmental Microbiology, 61*, 3082–3087.
Paddock, C. D., et al. (2004). *Rickettsia parkeri*: A newly recognized cause of spotted fever rickettsiosis in the United States. *Clinical Infectious Diseases, 38*, 805–811.
Pantoja, M. L., et al. (2015). Rickettsia-related bacteria associated with papaya plants showing bunchy top disease in Cuba. *Journal of General Plant Pathology, 81*, 166–168.
Paris, D. H., Shelite, T. R., Day, N. P., & Walker, D. H. (2013). Unresolved problems related to scrub typhus: A seriously neglected life-threatening disease. *American Journal of Tropical Medicine and Hygiene, 89*, 301–307.
Parola, P., & Raoult, D. (2001). Ticks and tickborne bacterial diseases in humans: An emerging infectious threat. *Clinical Infectious Diseases, 32*, 897–928.
Parola, P., Labruna, M. B., & Raoult, D. (2009). Tick-borne Rickettsioses in America: Unanswered questions and emerging diseases. *Current Infectious Disease Reports, 11*, 40–50.

Parola, P., et al. (2012). Update on tick-borne rickettsioses around the world: A geographic approach. *Clinical Microbiology Reviews, 26*, 657–702.

Pérez-Osorio, C. E., Zavala-Velázquez, J. E., Arias León, J. J., & Zavala-Castro, J. E. (2008). *Rickettsia felis* as emergent global threat for humans. *Emerging Infectious Diseases, 14*, 1019–1023.

Portaels, F., Elsen, P., Guimaraes-Peres, A., Fonteyne, P. A., & Meyers, W. M. (1999). Insects in the transmission of *Mycobacterium ulcerans* infection (Buruli ulcer). *Lancet, 353*, 986.

Pritt, B. S., et al. (2016). Identification of a novel pathogenic *Borrelia* species causing Lyme borreliosis with unusually high spirochaetaemia: A descriptive study. *The Lancet Infectious Diseases, 16*, 556–564.

Purcell, A. H. (1997). *Xylella fastidiosa*, a regional problem or global threat? *Journal of Plant Pathology, 79*, 99–105.

Purcell, A. H. (2008). Transmission of *Xylella fastidiosa* bacteria by xylem-feeding insects. In *Encyclopedia of entomology* (pp. 3885–3886). Dordrecht: Springer.

Purcell, A. H., Finlay, A. H., & McClean, D. L. (1979). Pierce's disease bacterium: Mechanism of transmission by leafhopper vectors. *Science, 206*, 839–841.

Rand, P. W., Lacombe, E. H., Smith, R. P., & Ficker, J. (1998). Participation of birds (*Aves*) in the emergence of Lyme disease in southern Maine. *Journal of Medical Entomology, 35*, 270–276.

Raoult, D., & Roux, V. (1997). Rickettsioses as paradigms of new or emerging infectious diseases. *Clinical Microbiology Review, 10*, 694–719.

Raoult, D., & Roux, V. (1999). The body louse as a vector of reemerging human diseases. *Clinical Infectious Diseases, 29*, 888–911.

Redak, R. A., et al. (2004). The biology of xylem fluid-feeding insect vectors of *Xylella fastidiosa* and their relation to disease epidemiology. *Annual Review on Entomology, 49*, 243–270.

Rikihisa, Y., Dumler, J. S., & Dasch, G. A. (2005). Neorickettsia. In G. M. Garrity (Ed.), *Bergey's manual of systemic bacteriology* (2nd ed., Vol. 2, pp. 132–137). New York: Springer.

Rizzoli, A., et al. (2004). *Ixodes ricinus*, malattie trasmesse e reservoirs. *Parassitologia, 46*, 119–122.

Robinson, D., Leo, N., Prociv, P., & Barker, S. C. (2003). Potential role of head lice, *Pediculus humanus capitis*, as vectors of *Rickettsia prowazekii*. *Parasitology Research, 90*, 209–211.

Roy-Dufresne, E., Logan, T., Simon, J. A., Chmura, G. L., & Millien, V. (2013). Poleward expansion of the white-footed mouse (*Peromyscus leucopus*) under climate change: Implications for the spread of Lyme disease. *PLoS ONE, 8*, e80724. doi:10.1371/journal.pone.0080724.

Rymaszewska, A., & Grenda, S. (2008). Bacteria of the genus *Anaplasma* – characteristics of *Anaplasma* and their vectors: A review. *Veterinární Medicína, 53*, 573–584.

Saillard, C., et al. (1987). *Spiroplasma phoeniceum* sp. nov. a new plant-pathogenic species from Syria. *International Journal of Systematic Bacteriology, 37*, 106–115.

Santos, C. F., & Carvalho, C. B. (2006). First report of *Anaplasma bovis* (Donatien and Lestoquard, 1936) Dumler et al. (2001) at micro region of Campos dos Goytacazes, State of Rio de Janeiro, Brazil. *Revista Brasileira de Parasitologia Veterinária, 15*, 126–127.

Saponari, M., et al. (2014). Infectivity and transmission of *Xylella fastidiosa* Salento strain by *Philaenus spumarius* L. (Hemiptera: Aphrophoridae) in Apulia, Italy. *Journal of Economic Entomology, 107*, 1316–1319.

Schaad, N. W., Postnikova, E., Lacy, G., Fatmi, M., & Chang, C. J. (2004). *Xylella fastidiosa* subspecies: *X. fastidiosa* subsp. [correction] *fastidiosa* [correction] subsp. nov., *X. fastidiosa* subsp. *multiplex* subsp. nov., and *X. fastidiosa* subsp. *pauca* subsp. nov. *Systematic and Applied Microbiology, 27*, 290–300.

Schoeler, G. B., Morón, C., Richards, A., Blair, P. J., & Olson, J. G. (2005). Human spotted fever rickettsial infections. *Emerging Infectious Diseases, 11*, 622–624.

Schwan, T. G. (1996). Ticks and *Borrelia*: Model systems for investigating pathogen-arthropod interactions. *Infectious Agents Disease, 5*, 167–181.

Schwanz, L. E., Voordouw, M. J., Brisson, D., & Ostfeld, R. S. (1989). *Borrelia burgdorferi* has minimal impact on the Lyme disease reservoir host *Peromyscus leucopus. Vector-Borne and Zoonotic Diseases, 11*, 117–124.

Shimelash, D., Alemu, T., Addis, T., Turyagyenda, F. L., & Blomme, G. (2008). Banana *Xanthomonas* wilt in Ethiopia: Occurrence and insect vector transmission. *African Crop Science Journal, 16*, 75–87.

Shiumi, T., & Sugiura, M. (1984). Differences among *Macrosteles orientalis*-transmitted MLO, potato purple-top wilt MLO in Japan and aster yellows MLO from USA). *Annals of the Phytopathological Society of Japan, 50*, 455–460. (In Japanese).

Silverman, D. J. (1991). Some contributions of electron microscopy to the study of the Rickettsiae. *European Journal of Epidemiology, 7*, 200–206.

Socolovschi, C., et al. (2009). Transovarial and trans-stadial transmission of *Rickettsiae africae* in *Amblyomma variegatum* ticks. *Clinical Microbiology and Infection, 15*(Suppl. 2), 317–318.

Sonenshine, D. E. (1994). *Ecological dynamics of tick-borne zoonoses*. New York: Oxford University Press, 464 pp.

Spencer, J. A., et al. (2003). Evaluation of permethrin and imidacloprid for prevention of *Borrelia burgdorferi* transmission from blacklegged ticks (*Ixodes scapularis*) to *Borrelia burgdorferi*-free dogs. *Parasitology Research, 90*, S106–S107.

Spinelli, F., Ciampolini, F., Cresti, M., Geider, K., & Costa, G. (2005). Influence of stigmatic morphology on flower colonization by Erwinia amylovora and Pantoea agglomerans. *European Journal of Plant Pathology, 113*, 395–405.

Stafford, K. C., et al. (1998). Temporal correlations between tick abundance and preva- lence of ticks infected with *Borrelia burgdorferi* and increasing incidence of Lyme disease. *Journal of Clinical Microbiology, 36*, 1240–1244.

Stavrinides, J., McCloskey, J. K., & Ochman, H. (2009). Pea aphid as both host and vector for the phytopathogenic bacterium *Pseudomonas syringae. Applied and Environmental Microbiology, 75*, 2230–2235.

Stavrinides, J., No, A., & Ochman, H. (2010). A single genetic locus in the phytopathogen Pantoea stewartii enables gut colonization and pathogenicity in an insect host. *Environmental Microbiology, 12*, 147–155.

Steiert, J. G., & Gilfoy, F. (2004). Infection Rates of *Amblyomma americanum* and *Dermacentor variabilis* by *Ehrlichia chaffeensis* and *Ehrlichia ewingii* in Southwest Missouri. *Vector-Borne and Zoonotic Diseases, 2*, 53–60.

Stenos, J., Graves, S. R., Popov, V. L., & Walker, D. H. (2003). *Aponomma hydrosauri*, the reptile-associated tick reservoir of *Rickettsia honei* on Flinders Island, Australia. *American Journal of Tropical Medicine and Hygiene, 69*, 314–317.

Stevenson, H. L., Labruna, M. B., Montenieri, J. A., Kosoy, M. Y., Gage, K. L., & Walker, D. H. (2005). Detection of *Rickettsia felis* in a new world flea species, *Anomiopsyllus nudata* (Siphonaptera: Ctenophthalmidae). *Journal of Medical Entomology, 42*, 163–167.

Subandiyah, S., Nikoh, N., Tsuyumu, S., Somowiyarjo, S., & Fukatsu, T. (2000). Complex endosymbiotic microbiota of the citrus psyllid *Diaphorina citri* (Homoptera: Psylloidea). *Zoological Science, 17*, 983–989.

Summer, E. J., et al. (2010). Genomic and biological analysis of phage Xfas53 and related prophages of *Xylella fastidiosa. Journal of Bacteriology, 192*, 179–190.

Svraka, S., Rolain, J. M., Bechah, Y., Gatabazi, J., & Raoult, D. (2006). *Rickettsia prowazekii* and real-time polymerase chain reaction. *Emerging Infectious Diseases, 12*, 428–432.

Tedeschi, R., & Alma, A. (2004). Transmission of apple proliferation phytoplasma by *Cacopsylla melanoneura* (Homoptera: Psyllidae). *Journal of Economic Entomology, 97*, 8–13.

Tedeschi, R., & Alma, A. (2006). *Fieberiella florii* (Homoptera: Auchenorrhyncha) as a vector of "*Candidatus* Phytoplasma mali". *Plant Disease, 90*, 284–290.

Teixeira, D. C., et al. (2005). '*Candidatus* Liberibacter americanus', associated with citrus huanglongbing (greening disease) in Sao Paulo State, Brazil. *International Journal of Systematic and Evolutonary Microbiology, 55*, 1857–1862.

Thomas, N. J., Bunikis, J., Barbour, A. G., & Wolcott, M. J. (2002). Fatal spirochetosis due to a relapsing fever-like *Borrelia* sp. in a Northern spotted owl. *Journal of Wildlife Diseases, 38*, 187–193.

Tinzaara, W., et al. (2006). Role of insects in the transmission of banana bacterial wilt. *African Crop Science Journal, 14*, 105–110.

Tissot-Dupont, H., Torres, S., Nezri, M., & Raoult, D. (1999). Hyperendemic focus of Q fever retlated to sheep and wind. *American Journal of Epidemiology, 150*, 67–74.

Tondella, M. L., et al. (1994). Isolation of *Haemophilus aegyptius* associated with Brazilian purpuric fever, of Chloropidae (Diptera) of the genera Hippelates and Liohippelates. *Revista do Instituto de Medicina Tropical de São Paulo, 36*, 105–109. (In Port.).

Toutoungi, L. N., & Gern, L. (1993). Ability of transovarially and subsequent transstadially infected *Ixodes hexagonus* ticks to maintain and transmit *Borrelia burgdorferi* in the laboratory. *Experimental & Applied Acarology, 17*, 581–586.

Trout Fryxell, R. T., & DeBruyn, J. M. (2016). The microbiome of *Ehrlichia*-infected and uninfected lone star ticks (*Amblyomma americanum*). *PLoS ONE, 11*, e0146651.

Trout Fryxell, R. T., et al. (2012). Survey of *Borreliae* in ticks, canines, and white-tailed deer from Arkansas, U.S.A. *Parasites and Vectors, 5*, 139.

Unsworth, N. B., et al. (2007). Flinders island spotted fever rickettsioses caused by "*marmionii*" strain of *Rickettsia honei*, eastern Australia. *Emerging Infectious Diseases, 13*, 566–573.

Vitorino, L., Chelo, I. M., Bacellar, F., & Zé-Zé, L. (2007). Rickettsiae phylogeny: A multigenic approach. *Microbiology, 153*, 160–168.

Wang, P., et al. (2014). Emergence of *Ixodes scapularis* and *Borrelia burgdorferi*, the Lyme disease vector and agent, in Ohio. *Frontiers in Cellular and Infection Microbiology, 4*, 70. doi:10.3389/fcimb.2014.00070.

Wei, W., Davis, R. E., Lee, I. M., & Zhao, Y. (2007). Computer-simulated RFLP analysis of 16S rRNA genes: Identification of ten new phytoplasma groups. *International Journal of Systematic and Evolutionary Microbiology, 57*, 1855–1867.

Weintraub, P. G., & Beanland, L. A. (2006). Insect vectors of phytoplasmas. *Annual Revue of Entomology, 51*, 91–111.

Weiss, E., & Dasch, G. A. (1991). Introduction to the Rickettsiales and other parasitic or mutualistic prokaryotes. In A. Balows, H. G. Trüper, M. Dworking, W. Harder, & K. H. Schleifer (Eds.), *The prokaryotes* (2nd ed., pp. 2402–2406). New York: Springer.

Wilson, M. D., Boakye, D. A., Mosi, L., & Asiedu, K. (2011). In the case of transmission of *Mycobacterium ulcerans* in Buruli ulcer disease *Acanthamoeba* species stand accused. *Ghana Medical Journal, 45*, 31–34.

Yuan, X., Morano, L., Bromley, R., Spring-Pearson, S., Stouthamer, R., & Nunney, L. (2010). Multilocus sequence typing of *Xylella fastidiosa* causing Pierce's disease and oleander leaf scorch in the United States. *Phytopathology, 100*, 601–611.

Zhao, Y., Wei, W., Davis, R. E., & Lee, I. M. (2010). Recent advances in 16S rRNA gene-based phytoplasma differentiation, classification and taxonomy. In P. G. Weintraub & P. Jones (Eds.), *Phytoplasmas: Genomes, plant hosts and vectors* (pp. 64–92). Wallingford: CABI.

Chapter 6
Travelling Bacteria: Phoresy

Abstract Examples of phoretic links with bacteria observed in some invertebrate groups are reviewed. They include insects transporting bacteria, symbionts horizontal transmission and the insect-killing bacteria associated to entomopathogenic nematodes. Other bacterial phoresis are reviewed for slug parasites, grass galling nematodes and microbiovorous species, and anellids.

Keywords *Anguina* • Corynebacteria • Entomopathogens • *Heterorhabditis* • Nematodes • *Phasmarhabditis* • *Photorhabdus* • Rhabditidae • *Steinernema* • Transport • *Xenorhabdus*

1 Introduction

The relationships linking invertebrates and bacteria often show uncertain, overlapping boundaries which make our classification attempts in some way arbitrary, although useful for research or systematic purposes. Many of these classifications, that may be considered as "anthropic" in origin, provide schemes simplifying the complexity of relationships among species and derived by their long term evolutionary patterns. When considering phoresy, these links can be assumed to involve the dispersal or targeted transport of an organism (the phoront) by a second one, which behaves as carrier and may (or may not) receive one or more benefits by this process.

Phoretic links are commonly found in many invertebrate lineages inhabiting a wide range of ecosystems, and may involve one or more species and different stages of the phoront cycle, involved in migration patters (Binns 2008). They are useful in the determination of the habits and properties of many invertebrates life-cycles and associations, and may be even exploited in the definition and timing of body decomposition processes, as applied in phorensic science (Perotti 2009).

Phoresy has a deep ecological, physiological and evolutionary impact, either in micro- and macro-habitats. It has been also considered as a necessary transition step in the evolution of animal parasitism among haematophagous invertebrates. This process was considered to have evolved from a passive association limited to transport and migration, towards a closer spatial contiguity (i.e. in nests), followed by the

A. Ciancio, *Invertebrate Bacteriology*, DOI 10.1007/978-94-024-0884-3_6

capacity to decompose excreta or body remnants (like blood drops or skin debris), until the last evolutionary adaptation providing the capacity to induce a damage by the direct feeding on blood or other body parts (Balashov 2006).

Mites, insects or nematodes provide many examples of phoretic links, often necessary for passive transport and life-cycle completion (Macchioni 2007). The many interactions of these invertebrates with bacteria present common evolutionary aspects (see other Chapters on symbiosys or pathogenicity). Studies carried out on the phylogenies of the bacteria phoretically associated to nematodes also showed that many links are highly specific, arising from evolutive adaptations possibly established around 250–500 Myr ago (Tailliez et al. 2006). A fossil record from Baltic amber showed an astigmatid mite attached on the carapace of a spider (Araneae: Dysderidae), indicating an Eocene association established around 44–49 Myr ago (Dunlop 2012).

When examining bacteria, the range of phoretic relationships spans from the simple transport and spread towards more favorable environments, to more specific endo-phoretic or necromenic links, involved in parasitic behaviours. This is shown for example by the entomopathogenic nematodes (EPNs) "waiting" for the host decomposition following death, which is caused by the infective bacteria they inoculate. Other possibilities include the association with plant symbionts, or the transport and dispersal of bacteria resident in different insect organs or in plants, soil or other niche microcosms (Sudhaus and Kühne 1990; Sudhaus 2008). These processes are often integrated with other symbiotic relationships, like commensalism (i.e. species inhabiting insects guts, in which no benefit or harm is provided) or mutualism, in which either the phoront and host receive a benefit (Dillon and Dillon 2004).

In this Chapter a number of phoretic relationships is examined, with examples provided for some invertebrate groups.

2 Insect-Associated Bacteria

Some common examples of phoresy include insects transporting bacteria which are parasitic or endosymbiotic in other insect hosts. This is the case of the horizontal transmission of endosymbiotic *Wolbachia* carried by the parasitoid wasp *Eretmocerus* sp. from infected to uninfected *Bemisia tabaci* whiteflies (Ahmed et al. 2015). Other cases of horizontal transmission of endosymbionts involve the transfer of *Hamiltonella defensa* or *Regiella insecticola* by parasitoid wasps, or the lateral transfer of male-killing *Spiroplama poulsonii* among *Drosophila* spp., carried by the ectoparasitic mite *Macrocheles subbadius* (Gehrer and Vorburger 2012; Jaenike et al. 2007).

The necromenic (see definition in previous Chapter) spread of insect-killing bacteria by EPNs plays a fundamental role in the natural regulation of many insect species, as well as in the practical and commercial exploitation of EPNs as biological control agents of crop pests. The phoretic relationships may be direct, or may involve a further organism which in turn transports the bacterial carriers. A direct phoresy occurs when the nematode inoculates the insect killing bacterium. A kind of "nested" phoresy, however, has been reported in the nematode *Heterorhabditis marelatus* that is patchily distributed in cultivated fields by the isopod *Porcellio scaber*. This occurs when the isopods excavate tunnels in the rhizosphere of lupine plants. In this microcosm, the nematode target host, the moth *Hepialus californicus*, is attacked by the EPNs that infect them with the entomopathogenic bacteria they carry and release, after penetration (Eng et al. 2005).

Similarly, the EPN *Steinernema feltiae* and associated bacteria have also a phoretic relationship with the earthworm *Eisenia fetida*. The nematodes were found to be still capable to attack and infect, with their mutualistic bacteria, the larvae of the piralid *Galleria mellonella*, after passing through the earthworm intestine (Campos-Herrera et al. 2006). Studies on the phoretic relationships of earthworm with *Steinernema* spp. highlighted the role of earthworms in EPNs dispersal in soil (Shapiro et al. 1995; Shapiro-Ilan and Brown 2013).

EPNs show many cases of coevolutive phoretic adaptations which may have several practical outcomes in modern agriculture, documented in a wide and growing literature (Kaya and Gaugler 1993; Grewal et al. 2005; Lacey and Georgis 2012; Lewis and Clarke 2012; Stock 2015; Lewis et al. 2015). Basically, two phylogenetically distant γ-Proteobacteria lineages converged to evolve a close association each with two genera of insect parasitic nematodes of the order Rhabditida. This relationship is necromenic and may be considered also as phoretic and mutualistic.

The main step in the association is the inoculation of the bacterial cells in the insect hosts, operated by the nematodes. The saepticemia eventually resulting by the infection, usually during the following 24 h, kills the insect. The cadavers become repleted with the bacterial cells digesting the insect and forming a kind of nutrient broth, on which the nematodes then feed. As the food source is depleted of nutrients the multiplying nematodes halt their development at the life stage of infective juvenile, acquiring and storing in their bodies the bacterial cells. Depending on the species, they may multiply or not in the nematodes, and will be used again as inoculum when starting a new infective process.

The two nematode groups, characterized by this sophisticated biological process, belong to the genera *Steinernema* and *Heterorhabditis*, which include several species. The motile and gram negative enterobacteria (γ-Proteobacteria) associated with each nematode lineage are grouped in the genera *Xenorhabdus* – of which the first recognized member was *X. nematophila*, now flanked by more than 20 species, associated to Steinernematidae – and *Photorhabdus* (Fig. 6.1). The latter genus is associated to Heterorhabditidae and includes *P. luminescens* (with subspp. *akhurstii*, *laumondii*, *luminescens*, *noenipuitensis* and *sonorensis*), and other species like *P. temperata*, *P. asymbiotica*, *P. heterorhabditis* and *P. zealandica* (Tailliez et al. 2006, 2010; Hincliffe et al. 2010; Orozco et al. 2013; Ferreira et al. 2014; Malan and Hatting 2015).

The genome of *P. asymbiotica* has been sequenced due to its pathogenicity to humans, and compared to that of *P. luminescens*. Data showed that the human pathogenic species has a smaller genome correlated to alower insect diversity in

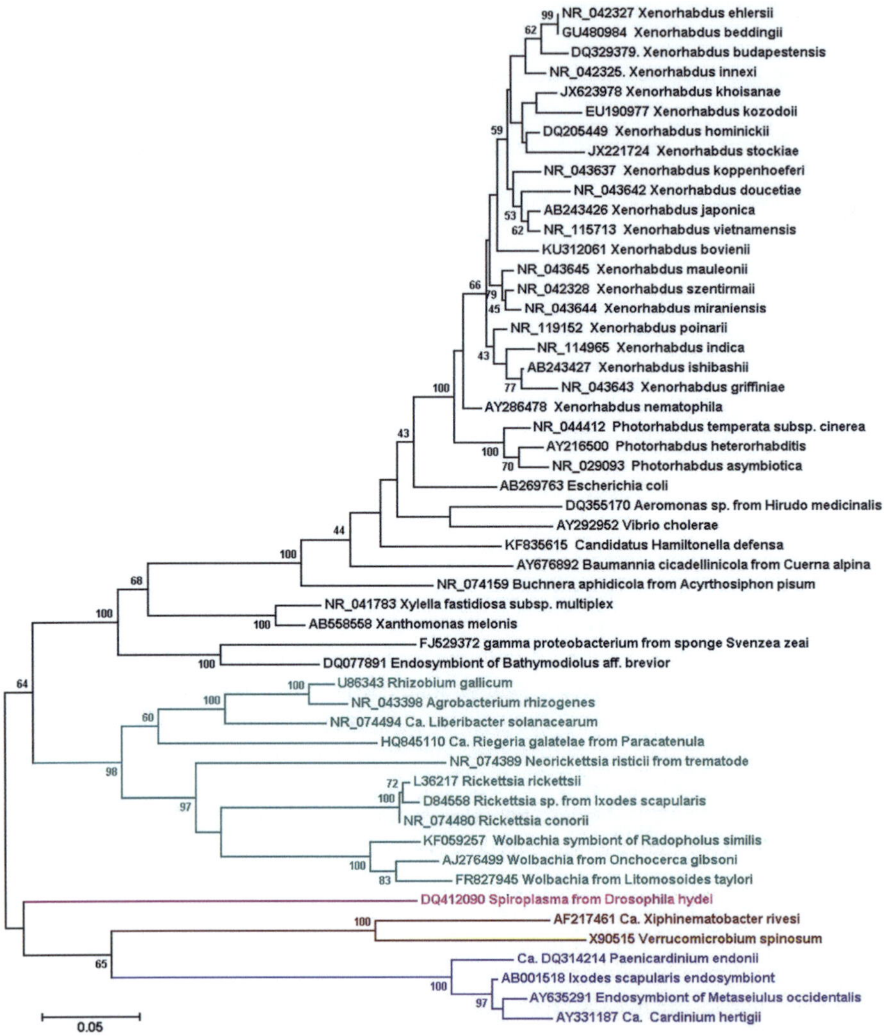

Fig. 6.1 Phylogenetic relationships of Xenorhabdus and *Photorhabdus spp.* and further *γ-Proteobacteria* with members of the *α-Proteobacteria* (*green*), *Mollicutes* (*red*), *Verrucomicrobia* (*brown*) and *Bacteriodetes* (*blue*). Dendrogram of aligned 16S rRNA ribosomial gene sequences, inferred by the Maximum Likelihood method based on the Jukes-Cantor model (For details of dendrogram construction see Fig. 3.4)

insecticidal genes, and that each species carries a specific DNA region of around 1 Mb. The *P. asymbiotica* genome also showed the presence of a *Yersinia pestis* related plasmid, and pathogenicity islands including a novel Type III SS (Wilkinson 2009).

The cells of *X. nematophila* are kept in the body of *Steinernema* spp. at the pharyngeal-intestinal junction, from where they are quickly released after insect

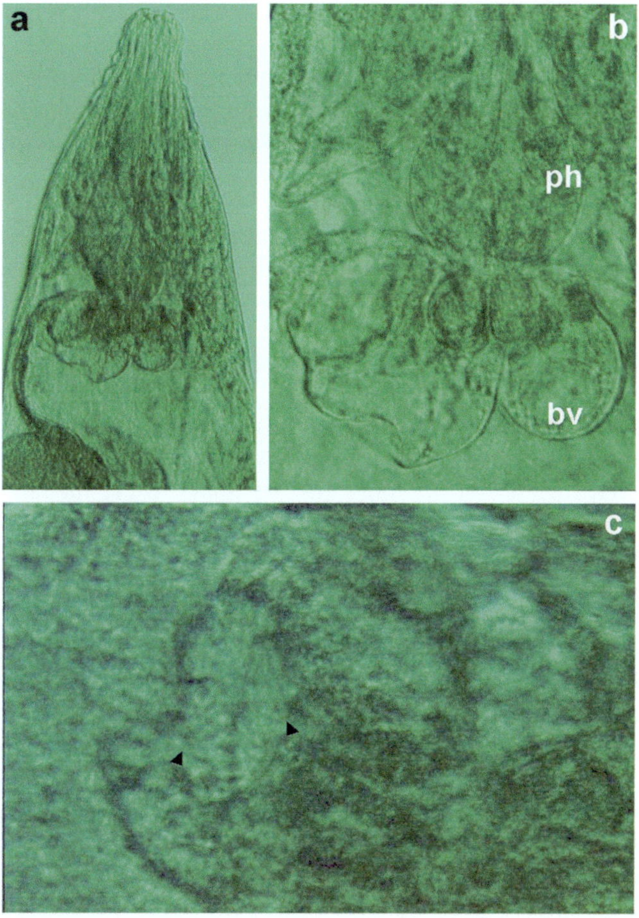

Fig. 6.2 Anterior region of an undescribed Steinernema sp. (**a**) showing (**b**, enlarged) the bacterial vesicles (*bv*) adiacent to the pharynx (*ph*) at the intestine junction, with bacterial cells stored inside (**c**, *arrowhead*) (Slide courtesy of A. Troccoli)

penetration and in which they also multiply (Bird and Akhurst 1983; Martens et al. 2003). The cells of *P. luminescens* are acquired by *Heterorhabditis* spp. during its matricidal intrauterine hatching and body evasion, when the juveniles breach the rectal gland epithelium. They are kept in the juvenile, in a pharyngeal intestinal valve cell or sac (Fig. 6.2) (Ciche et al. 2008).

The production of specific bacterial toxins is also involved in the death of the insect host, with practical implications for the identification of new, useful molecules with a highly specific mechanism of action (ffrench-Constant and Bowen 2000). Interestingly, the toxins introduced in the host haemolymph are lethal for the insect but not for the nematodes, indicating a consistent physiological and biochemical evolutive adaptation (Rajagopal and Bhatnagar 2002; Herbert and Goodrich-Blair 2007; Hinchliffe et al. 2010).

The routes of nematode penetration in the insect hosts usually follow the natural openings, including anus, mouth and possibly the tracheal system. The bacteria inoculation eventually occurs in the haemolymph. In infected *Spodoptera littoralis*, labelled cells of *X. nematophila* were found in the extracellular matrix of the connective tissues within the muscle layers of the insect midgut, where they were released by the vector *S. carpocapsae*. The bacteria eventually spread to the hemolymph and adiacent connective tissues. Similar observations were also reported for infected *Locusta migratoria*. In *G. mellonella*, the nematodes was observed to penetrate via the midgut wall after moulting, releasing the *X. nematophila* cells into the insect hemocoel (Sicard et al. 2004). The *Photorhabdus* spp. cells are released in the insect haemolymph through the nematode mouthpart and then migrate to the host extracellular matrix and the basal membrane of the midgut epithelium (Hincliffe et al. 2010).

The study of bacteria associated to EPNs increased sharply in the last decade due to the biocontrol potential of many species towards economically important pests. Several detailed reviews on the topic may be found in the literature, including Hincliffe et al. (2010), Stock (2015), and Lewis et al. (2015).

Phoretic associations of EPNs with other bacteria like *Bacillus* sp. have been also reported from *Heterorhabditis* sp. (Marti and Timper 1999). A *Paenibacillus* spp. was also reported from infective juveniles of *Steinernema diaprepesi*. The genus *Paenibacillus* includes entompathogenic species. However, the *S. diaprepesi* bacterium showed a specific phoretic association to the nematode, adhering only to its cuticular sheat. The biological process underpinning this association was not clear, but a reduction of the nematode migration and host penetration efficiency was observed (Campos-Herrera et al. 2012; El-Borai et al. 2005).

Entomopathogenic *Rhabditis* (*Oscheius*) spp. have been used in Kerala (India) to control *Mircarvalhoia arecae*, the areca nut spindle bug (Mohandas et al. 2007). The nematodes carried a number of bacteria which appeared lethal to several insect pests, including the rice yellow stem borer, *Scirpophaga incertulas*. Molecular data from isolates proceeding from cadavers of *G. mellonella* exposed to nematode infection showed occurrence of several γ- and β-Proteobacteria together with some *Bacillus* spp. (Sangeetha et al. 2016). In other *Rhabditis* spp. and bacteria associations a number of entomopathogenic *Serratia* spp. have been reported. Data suggest that the range of species and associations has yet to be fully explored and that new organisms will be discovered in the years to come, as far as the exploration of new micro-environments (like i.e. the dead bark of oak trees) will be carried out (Zhang et al. 2009; Tambong 2013).

Apart of entomopathogenic species, insect and bacteria phoretic relationships may involve one or more benefits for the carrier organism, including the phoront involvement in profitable metabolic or other biochemical pathways. In the mountain pine beetle *Dendroctonus ponderosae* a community of bacteria contributing to the metabolism of plant defensive terpenes is carried and dispersed by the insect (Adams et al. 2013). The bacteria may be carried in the insects gut and either be dispersed through the frass or regurgitated. In bark beetles, nutritional and fungicidal activities were reported for associated bacteria belonging to the genera *Microbacterium*,

Streptomyces, *Serratia*, *Rahnella*, *Pseudomonas* and *Brevundimonas* (Boone et al. 2013). The southern pine beetle *Dendroctonus frontalis* uses bacteria in oral secretions, including *Micrococcus luteus*, to antagonize noxious fungi (Cardoza et al. 2006). Similarly, the leaf-cutting ants *Acromyrmex* spp. transport a *Streptomyces* sp. characterized by antifungal properties, to control antagonistic fungi in their fungal gardens (Currie et al. 1999; Mercado et al. 2014).

Insects living in particular niches and exposed to sub-optimal or poor diets like termites rely on symbioses with a gut microbiota including several bacteria and protozoa. The termite gut is rich of unclassified bacterial lineages, including spirochetes, which provide either nutrients through symbiotic and metabolic processes and host protection from invasive, pathogenic species. Close links also occur with the microbiome of the feeding substrates through processes leading to the bacterial transfer and/or acquisition (Brune 1998; Dillon and Dillon 2004).

3 Other Invertebrate Bacteria

Further nematode phoretic relationships include the spatial spread of plant or rhizosphere bacteria. Metagenomic studies on meiofauna communities showed that these relationships often involve other invertebrates, and that the bacteria sequenced cannot be identified in a large number, being mostly unknown or unclassified (Kanzaki et al. 2012). Indeed, several lineages present in these processes still need to be described and characterized, and their phylogenies and functional role still remain to be fully determined. Nematode phoretic relationships involve species of Rhabditida and Anguininae, together with pathogenic bacteria, rhizobia and other functional groups.

3.1 Slug Parasitic Nematodes

The molluskicidal nematode *Phasmarhabditis hermaphrodita* kills slugs with a mechanism similar to those of other EPNs. In the slug *Deroceras reticulatum*, *Arion* and *Tandonia* spp., the nematode juveniles were found to penetrate the slug mantel and release the bacterium *Moraxella osloensis*. The cells of this species colonize the gut of the juvenile infective stages and are transported with them, as the nematodes move. The bacterium produces a molluskicidal polysaccharid lethal for the slug, and is associated to a complex bacterial community (Tan and Grewal 2001, 2003; Rae et al. 2010). The nematodes feed eventually on the bacteria. Cultures of *P. hermaphrodita* are commercially sold as biocontrol agents of garden slugs, and may be useful in eco-friendly applications aiming at reducing the slug densities (Wilson et al. 1993).

3.2 Grass Gall Nematodes

The genus *Corynebacterium* (Actinomycetes: *Microbacteriaceae*) include gram-positive, aerobic and pleomorphic bacteria with a cell wall composition lacking teichoic acids. These are polymers rich in peptidoglycan-linked phosphates, present in the cell wall of many gram-positive species (Mauël et al. 1989).

The taxonomy of this lineage has been reviewed, as it includes also animal and human pathogens (i.e. *C. diphtaeriae*, the causal agent of diphtheria) and other species producing systemic infections on plants of economic importance (Coyle and Lipsky 1990; Van den Velde et al. 2006). The diseases caused by plant pathogenic species arise after penetration of seeds, tubers, bulbs and/or other tissues through an infective route mainly associated to wounding. Major plant damage is produced through the action of toxins, hormones, extracellular polysaccharides and other metabolites (Vidaver 1981). The plant pathogenic forms have a high degree of host specificity, and associations of some species with nematodes have been observed. In these, some members of the nematode subfamily Anguininae (Tylenchida) play a fundamental role as vectors. The bacteria multiply inside the galls that are induced by the nematodes on a wide range of host grasses (Bird 1981; Riley and Ophel 1992; Riley and Reardon 1995; Evtushenko et al. 1994).

At the issue of the *Corynebacterium* group revision, based on sequencing and other taxonomic studies, the species have been initially classified in the genera *Curtobacterium* or *Clavibacter* (Davis et al. 1984). Some nematode associated species were later moved from the latter genus to *Aureobacterium* or *Rathayibacter* (Carlson and Vidaver 1982; Collins and Jones 1983; Zgurskaya et al. 1993; Sasaki et al. 1998).

The diseases caused by Anguininae-associated bacteria are unusual in that they may induce an animal pathology characterized by a neurotoxic reaction, called ryegrass toxicity (Bird 1981). In infected plants *R. toxicus* (syn. *Cl. toxicus*) induces a yellow bacterial slime or gumming, in particular evident in plant top parts and inflorescence, with dwarfing, distortion and formation of yellow seed galls. At maturity these contain, in the vessels and parenchyma, nematodes and bacteria. In sheep, horses or cattle grazing on the grasses the neurotoxic activity produces a number of reactions like staggering, collapse and violent convulsions, eventually followed by the animal death (Bird 1981; Riley and Ophel 1992; Ophel et al. 1993).

Species of *Rathayibacter* have been recognized in different continents and may represent a significant threat for livestock. An important disease of rye and other grasses in Australia is caused by *R. rathayi* (former *Cl. rathayi*) (Zgurskaya et al. 1993). This bacterium is vectored in natural conditions by the galling nematode *Anguina funesta* (syn.: *A. agrostis*, considered as a putative species complex) on the annual ryegrass, *Lolium rigidum*. *Rathayibacter toxicus* was experimentally found in Australia to be vectored on annual veltdgrass *Ehrharta longiflora* common in Western Australia, by adhering to juveniles of *Anguina australis*. The bacterium was capable to multiply in the galls induced by the nematode and produced toxins (Riley et al. 2001). Poisoning of livestock in Oregon in the 1940s–1960s has been attributed to the action of corynetoxins produced by *Rathayibacter* spp. in galls induced by anguinid nematodes on chewing fescue, *Festuca nigrescens* (Riley et al. 2003). Further *Rathayibacter* species, *R. caricis* and *R. festucae*, were described in East Europe from the phyllosphere of *Carex* sp. and in the leaf gall induced by the *A. graminis* on *Festuca rubra*, respectively (Dorofeeva et al. 2002).

Anguina tritici is responsible of the yellow slime or "tundu" disease of wheat, barley and other crops caused by *R. tritici*. The disease induces seed galls and curled leaves, with gummy exudates. The symptoms include the presence of leaf veins with pale, parallel striations. The bacteria are vectored by the nematodes and produce slime and gumming leaves, but no toxicity was reported (Vidaver 1981). Two other grass-associated bacteria were also reported from root galls induced on *Poa annua* by *Subanguina radicicola*, described as *Leifsonia poae* and *L. aquatica* (Evtushenko et al. 2000).

3.3 Microbiovorous Nematodes

In class Secernentea the order Rhabditida includes highly diversified lineages of free-living and bacteria-feeding nematodes. They are ubiquitous and colonized almost any terrestrial habitat and niche, including extreme environments like caves (Curcić et al. 2008; Sudhaus 2008). Many species can be easily cultured for weeks on water agar with organic debris proceeding from almost any soil. The order include also specialized lineages like EPNs (see Sect. 6.1). They have established many phoretic links with soil bacteria, including species of plant pathogens or useful ones like rhizobia (Sudhaus 2008).

Multitrophic interactions in soil involve many organisms, organized in complex systems of relationships also called "food-webs". These include for example roots, bacteria (either beneficial or plant pathogenic) and invertebrates like bacteriovorous nematodes. A number of phoretic relationships have been identified in these systems, in which nematodes may play different roles, either beneficial or detrimental for plants.

Bacteria-feeding nematodes living in soil or on various types of decaying organic matter rely on chemotaxis and on gradients of volatiles that can be perceived even at long (some meters) distances (Troemel et al. 1997; Allaire et al. 2002). Since roots exudates are known to diffuse only at a short distance, Horiuchi et al. (2005) studied the interactions of rhizobia with *Caenorhabditis elegans*. In particular they aimed at explaining how the bacterium may reach, over a long distance, a suitable legume root to establish a nitrogen fixing symbiosis, involving the majority of plants as often observed in the field. Experimental observations showed that *C. elegans* was capable of feeding on, and actively transporting, cells of *Sinorhizobium meliloti*. The bacteria survived and multiplyed into the nematode alimentary canal or were attached onto its cuticle. The rhizobia were transported and then released from the cuticle or through the faeces, as the nematode moved along a gradient towards the roots of *Medicago truncatula*, attracted by the volatile dimethyl sulfide they produced (Horiuchi et al. 2005).

In a different, plant detrimental association, *Pectobacterium* and *Dickeya* spp. were shown to resist grazing by bacterial feeders like *C. elegans* and *Pristionchus* spp. The bacteria cause soft-rot on potato tubers in Finland. Their cells remained viable in the nematodes intestine and once released they could be dispersed on experimentally exposed potato tubers. This phoretic links appeared involved in the disease ecology and partially responsible for the bacterial spatial spreading (Nykyri et al. 2014).

Bacteriophagous nematodes may ingest, among other bacteria, also human enteropathogenic species. They may thus potentially act as bacterial reservoirs, contributing to their protection, or even behave as vectors, eventually spreading the pathogens on fruits and vegetable (Caldwell et al. 2003; Anderson et al. 2006; Lacharme-Lora et al. 2009). *In vitro* assays also showed that *C. elegans* could act as a passive carrier of *Salmonella enterica* and that commercial cleaners and sanitizers could only partially remove the cells of the pathogen from the bodies of dried nematodes (Kenney et al. 2004).

Nematode interactions with soil bacteria also involve the Diplogasteridae, a lineage of insect or nematode predators and bacterial feeders. Due to their biology they can be multiplied *in vitro* on artificial cultures (Bilgrami 2009). Many bacteria have been found in associations with Diplogasteridae, which may be also responsible for their dispersal. The sequencing the 16S ribosomal genes from the intestinal microflora of *Koerneria sudhausi* allowed identification of several species, including *Pseudomonas putida*, *Pigmentiphaga kullae*, *Acidovorax avenae*, *Bosea thioxidans*, *Xenophilus* and *Bradyrhizobium* spp. (Colagiero et al. 2011). In beetle parasitic *Pristionchus* spp., the bacterial lineages found (Pseudomonadales, Burkholderiales, Flavobacteria and Xanthomonadales), included plant (*Erwinia* and *Agrobacterium* spp.) and human pathogens (*Bordetella*, *Burkholderia* or *Microbacterium* spp.). A range of interactions was reported for these associations, spanning from dissemination to mortality, including the nematode capacity to evade pathogenic species as *Serratia marcescens* or *Bacillus thuringiensis* (Rae et al. 2008).

Predators like *K. sudhausi* have a potential as biocontrol agents, in the management of other plant parasitic nematodes. Their *in vitro* multiplication on bacterial cultures facilitates their application as massive inocula. Diplogasteridae are selective predators and a prey preference (including plant parasitic nematodes) was reported for *Mononchoides fortidens*, *M. longicaudatus* and *M. gaugleri* (Bilgrami and Jairajpuri 1988; Chitambar and Noffsinger 1989; Bilgrami et al. 2005). Studies on predation activity also showed prey chemoattraction (Bilgrami 2009).

The eggs of some Diplogasteridae like *K. sudhausi* are provided with irregular and external swellings, that persist during the embryo development until hatching (Fig. 6.3). Once placed on agar, the eggs were gradually covered by a dense layer of bacterial cells, and freshly hatched juveniles were often observed around eggs, touching the irregularities with their lips and feeding on the bacteria (Colagiero et al. 2011). Although circumstantial, this observation suggests that the relationship of Diplogasteridae with bacteria may be more sophisticated than a simple and direct predatory behaviour, with evolutionary implications worth further investigations.

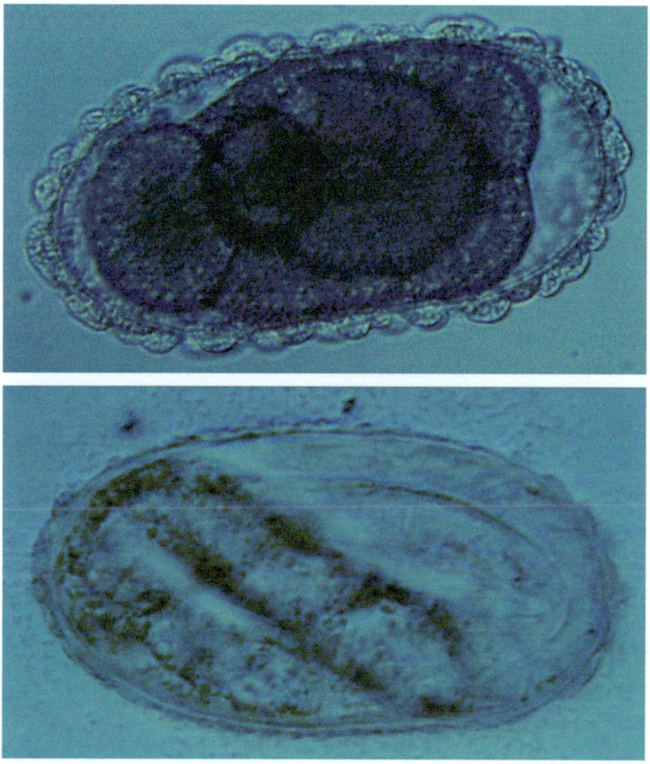

Fig. 6.3 Eggs of Koherneria sudhausi showing surface swellings (*top*) and bacteria surrounding the external egg surface, at a more advanced embryo developmental stage. Emerging larvae were observd to feed on the bacteria, soon after hatching

3.4 Anellids

The contribution of earthworms to the fertility of soils is a well documented process, known since the first observations carried out by Charles Darwin in the nineteenth century. Several reviews are available in the literature to explore many aspects of earthworms ecology and biology (Edwards and Bohlen 1996; Blanchart et al. 1999; Drake and Horn 2007; Bertrand et al. 2015). Their role in the phoresy of EPNs and associated bacteria has been previously reported in this Chapter.

Earthworms actively transport and disperse several groups of functional bacterial species, including useful rhizobia or plant growth promoting bacteria. The mechanisms on which these processes depend are related to the burial of organic matter, the continuous detritivore activity, the dispersal of egg capsules or the decomposition and release of organic matter and other excreta (Madsen and Alexander 1982; Stephens et al. 1994; Hendriksen 1995; Thorpe et al. 1996; Daane and Häggblom 1999).

Earthworms play a fundamental role in the conservation and enhancement of many soil physico-chemical properties, in the dispersal of other soil microarthropods through their burrowing tunnels, in the aeration of soil or in the production of N_2O through their activated gut microbiota. Further activities concern the preservation of the soil microbial diversity, enriched by a number of associated and specific bacterial sub-populations (Horn et al. 2003; Cameron et al. 2013; Frisli et al. 2013).

Anellids and ostracods (Crustacea) exploit phoretic relationships also for their own dispersal in some specific environments like the epiphytic bromeliad plants in tropical forests. They showed a chemically oriented behaviour, attaching to frogs or other small animals for transport (Lopez et al. 1999, 2005). In both organisms, this behaviour implies carrying with them their whole gut and intestinal microbiota. Similarly, also mites exploit phoresy, as shown for *Hemisarcoptes cooremani* hosted by *Chilocorus cacti* (Coleoptera: Coccinellidae). The phoretic link was considered as a mean for a transition among habitats, in an evolutionary perspective related to a transition towards insect parasitism (Houck and O'Connor 1991). Several phoretic relationships also occur among anellids, nematodes, crustaceans and other arthropods inhabiting cavernicolous habitats (Vandel 1965).

References

Adams, A. S., et al. (2013). Mountain pine beetles colonizing historical and naïve host trees are associated with a bacterial community highly enriched in genes contributing to terpene metabolism. *Applied and Environmental Microbiology, 79*, 3468–3475.

Ahmed, M. Z., et al. (2015). The intracellular bacterium *Wolbachia* uses parasitoid wasps as phoretic vectors for efficient horizontal transmission. *PLoS Pathogens, 11*, e1004672.

Allaire, S. E., Yates, S. R., Ernst, F. F., & Gan, J. (2002). A dynamic twodimensional system for measuring volatile organic compound volatilization and movement in soils. *Journal of Environmental Quality, 31*, 1079–1087.

Anderson, G. L., Kenney, S. J., Millner, P. D., Beuchat, L. R., & Williams, P. L. (2006). Shedding of foodborne pathogens by *Caenorhabditis elegans* in compost-amended and unamended soil. *Food Microbiology, 23*, 146–153.

Balashov, I. (2006). The origin and evolution of parasitism on terrestrial vertebrates in insects, mites, and ticks. *Parazitologiia, 40*, 409–424. [in Russian].

Bertrand, M., et al. (2015). Earthworm services for cropping systems. A review. *Agronomy for Sustainable Development, 35*, 553–567.

Bilgrami, A. L. (2009). Biological control potentials of predatory nematodes. In A. Ciancio & K. G. Mukerji (Eds.), *Integrated management and biocontrol of vegetable and grain crops nematodes* (pp. 3–28). Dordrecht: Springer.

Bilgrami, A. L., & Jairajpuri, M. S. (1988). Attraction of *Mononchoides longicaudatus* and *M. fortidens* (Nematoda: Diplogasteridae) towards prey and factors influencing attraction. *Revue de Nematologie, 11*, 195–202.

Bilgrami, A. L., Gaugler, R., & Brey, C. (2005). Prey preference and feeding behaviour of the diplogastrid predator *Mononchoides gaugleri* (Nematoda: Diplogastridae). *Nematology, 7*, 333–342.

Binns, E. S. (2008). Phoresy as migration – Some functional aspects of phoresy in mites. *Biological Reviews, 57*, 561–620.

Bird, A. F. (1981). In B. M. Zuckerman & R. A. Rohde (Eds.), *Plant parasitic nematodes*. New York: Academic, 508 pp.

Bird, A. F., & Akhurst, R. J. (1983). The nature of the intestinal vesicle in nematodes of the family Steinernematidae. *International Journal of Parasitology, 13*, 599–606.

Blanchart, E., et al. (1999). Effects of earthworms on soil structure and physical properties. In P. Lavelle, L. Brussaard, & P. Hendrix (Eds.), *Earthworm management in tropical agroecosystems* (pp. 149–172). Wallingford: CAB International.

Boone, C. K., et al. (2013). Bacteria associated with a tree-killing insect reduce concentrations of plant defense compounds. *Journal of Chemical Ecology, 39*, 1003–1006.

Brune, A. (1998). Termite guts: The world's smallest bioreactors. *Trends in Biotechnology, 16*, 16–21.

Caldwell, K. N., Anderson, G. L., Williams, P. L., & Beuchat, L. R. (2003). Attraction of a free-living nematode, *Caenorhabditis elegans*, to foodborne pathogenic bacteria and its potential as a vector of *Salmonella* Poona for preharvest contamination of cantaloupe. *Journal of Food Protection, 66*, 1964–1971.

Cameron, E. K., Proctor, H. C., & Bayne, E. M. (2013). Effects of an ecosystem engineer on belowground movement of microarthropods. *PLoS ONE, 8*, e62796.

Campos-Herrera, R., Trigo, D., & Gutiérrez, C. (2006). Phoresy of the entomopathogenic nematode *Steinernema feltiae* by the earthworm *Eisenia fetida*. *Journal of Invertebrate Pathology, 92*, 50–54.

Campos-Herrera, R., El-Borai, F. E., & Duncan, L. W. (2012). Real-time PCR as an effective technique to assess the impact of phoresy by *Paenibacillus* sp. bacteria on *Steinernema diaprepesi* nematodes in nature. *Molecular Ecology Resources, 12*, 885–893.

Cardoza, Y. J., Klepzig, K. D., & Raffa, K. F. (2006). Bacteria in oral secretions of an endophytic insect inhibit antagonistic fungi. *Ecological Entomology, 31*, 636–645.

Carlson, R. R., & Vidaver, A. K. (1982). Taxonomy of *Corynebacterium* plant pathogens, including a new pathogen of wheat, based on polyacrylamide gel electrophoresis of cellular proteins. *International Journal of Systematic Bacteriology, 32*, 315–326.

Chitambar, J. J., & Noffsinger, M. (1989). Predaceous behaviour and life history of *Odontopharynx longicaudata* (Diplogasteridae). *Journal of Nematology, 21*, 284–291.

Ciche, T. A., Kim, K., Kaufmann-Daszczuk, B., Nguyen, K. C. Q., & Hall, D. H. (2008). Cell invasion and matricide during *Photorhabdus luminescens* transmission by *Heterorhabditis bacteriophora* nematodes. *Applied and Environmental Microbiology, 74*, 2275–2287.

Colagiero, M., Rosso, L. C., Ciancio, A., & Murga Gutierrez, S. N. (2011). Observations on the biology of a predatory nematode belonging to Diplogasteridae. *Redia, 93*, 133–135.

Collins, M. D., & Jones, D. (1983). Reclassification of *Corynebacterium flaccumfaciens*, *Corynebacterium betae*, *Corynebacterium oortii* and *Corynebacterium poinsettiae* in the genus *Curtobacterium*, as *Curtobacteiium flaccumfaciens* comb. nov. *Journal of General Microbiology, 129*, 3545–3548.

Coyle, M. B., & Lipsky, B. A. (1990). Coryneform bacteria in infectious diseases: Clinical and laboratory aspects. *Clinical Microbiology Reviews, 3*, 227–246.

Curcić, B. P., Sudhaus, W., Dimitrijević, R. N., Makarov, S. E., & Tomić, V. T. (2008). *Rhabditophanes schneideri* (Rhabditida) phoretic on a cave pseudoscorpion. *Journal of Invertebrate Pathology, 99*, 254–256.

Currie, C. R., Scott, J. A., Summerbell, R. C., & Malloch, D. (1999). Fungus-growing ants use antibiotic-producing bacteria to control garden parasites. *Nature, 398*, 701–704.

Daane, L. L., & Häggblom, M. M. (1999). Earthworm egg capsules as vectors for the environmental introduction of biodegradative bacteria. *Applied and Environmental Microbiology, 65*, 2376–2381.

Davis, M. I., Gillaspie, A. G., Vidaver, A. K., & Harris, R. W. (1984). *Clavibacter*: A new genus containing some phytopathogenic coryneform bacteria, including *Clavibacter xyli* subsp. *xyli* sp. nov., subsp. nov. and *Clavibacterxyli* subsp. *cynodontis* subsp. nov., pathogens that cause ratoon stunting disease of sugarcane and bermudagrass stunting disease. *International Journal of Systematic Bacteriology, 34*, 107–117.

Dillon, R. J., & Dillon, V. M. (2004). The gut bacteria of insects: Nonpathogenic interactions. *Annual Reviews of Entomology, 49*, 71–92.
Dorofeeva, L. V., et al. (2002). *Rathayibacter caricis* sp. nov. and *Rathayibacter festucae* sp. nov., isolated from the phyllosphere of *Carex* sp. and the leaf gall induced by the nematode *Anguina graminis* on *Festuca rubra* L., respectively. *International Journal of Systematic and Evolutionary Microbiology, 52*, 1917–1923.
Drake, H. L., & Horn, M. A. (2007). As the worm turns: The earthworm gut as a transient habitat for soil microbial biomes. *Annual Review of Microbiology, 61*, 169–189.
Dunlop, J. A. (2012). A minute fossil phoretic mite recovered by phase-contrast X-ray computed tomography. *Biology Letters, 8*, 457–460.
Edwards, C. A., & Bohlen, P. J. (1996). *Biology and ecology of earthworms* (Vol. 3). London: Chapman & Hall, 433 pp.
El-Borai, F. E., Duncan, L. W., & Preston, J. F. (2005). Bionomics of a phoretic association between *Paenibacillus* sp. and the entomopathogenic nematode *Steinernema diaprepesi. Journal of Nematology, 37*, 18–25.
Eng, M. S., Preisser, E. L., & Strong, D. R. (2005). Phoresy of the entomopathogenic nematode *Heterorhabditis marelatus* by a non-host organism, the isopod *Porcellio scaber. Journal of Invertebrate Pathology, 88*, 173–176.
Evtushenko, L. I., Dorofeeva, L. V., Dobrovolskaya, T. G., & Subbotin, S. (1994). Coryneform bacteria from plant galls induced by nematodes of the subfamily Anguininae. *Russian Journal of Nematology, 2*, 99–104.
Evtushenko, L. I., Dorofeeva, L. V., Subbotin, S. A., Cole, J. R., & Tiedje, J. M. (2000). *Leifsonia poae* gen. nov., sp. nov., isolated from nematode galls on *Poa annua*, and reclassification of '*Corynebacterium aquaticum*' Leifson 1962 as *Leifsonia aquatica* (ex Leifson 1962) gen. nov., nom. rev., comb. nov. and *Clavibacter xyli* Davis et al. 1984 with two subspecies as *Leifsonia xyli* (Davis et al. 1984) gen. nov., comb. nov. *International Journal of Systematic and Evolutionary Microbiology, 50*, 371–380.
Ferreira, T., et al. (2014). *Photorhabdus heterorhabditis* sp. nov., a symbiont of the entomopathogenic nematode *Heterorhabditis zealandica. International Journal of Systematic and Evolutionary Microbiology, 64*, 1540–1545.
ffrench-Constant, R. H., & Bowen, D. J. (2000). Novel insecticidal toxins from nematode-symbiotic bacteria. *Cell and Molecular Life Sciences, 57*, 828–833.
Frisli, T., Haverkamp, T. H., Jakobsen, K. S., Stenseth, N. C., & Rudi, K. (2013). Estimation of metagenome size and structure in an experimental soil microbiota from low coverage next-generation sequence data. *Journal of Applied Microbiology, 114*, 141–151.
Gehrer, L., & Vorburger, C. (2012). Parasitoids as vectors of facultative bacterial endosymbionts in aphids. *Biology Letters, 8*, 613–615.
Grewal, P. S., Ehlers, R. U., & Shapiro-Ilan, D. I. (2005). *Nematodes as biological control agents*. Wallingford: CABI Publishing.
Hendriksen, N. B. (1995). Effects of detritivore earthworms on dispersal and survival of the bacterium *Aeromonas hydrophila. Acta Zoologica Fennica, 196*, 115–119.
Herbert, E. E., & Goodrich-Blair, H. (2007). Friend and foe: The two faces of *Xenorhabdus nematophila. Nature Reviews Microbiology, 5*, 634–646.
Hinchliffe, S. J., Hares, M. C., Dowling, A. J., & ffrench-Constant, R. H. (2010). Insecticidal toxins from the *Photorhabdus* and *Xenorhabdus* bacteria. *The Open Toxinology Journal, 3*, 83–100.
Horiuchi, J., Prithiviraj, B., Bais, H. P., Kimball, B. A., & Vivanco, J. M. (2005). Soil nematodes mediate positive interactions between legume plants and rhizobium bacteria. *Planta, 222*, 848–857.
Horn, M. A., Schramm, A., & Drake, H. L. (2003). The earthworm gut: An ideal habitat for ingested N_2O-producing microorganisms. *Applied and Environmental Microbiology, 69*, 1662–1669.
Houck, M. A., & O'Connor, B. M. (1991). Ecological and evolutionary significance of phoresy in the Astigmata. *Annual Review of Entomology, 36*, 611–636.

Jaenike, J., Polak, M., Fiskin, A., Helou, M., & Minhas, M. (2007). Interspecific transmission of endosymbiotic *Spiroplasma* by mites. *Biology Letters, 3*, 23–25.

Kanzaki, N., et al. (2012). Reverse taxonomy for elucidating diversity of insect-associated nematodes: A case study with termites. *PLoS ONE, 7*, e43865.

Kaya, H. K., & Gaugler, R. (1993). Entomopathogenic nematodes. *Annual Review of Entomology, 38*, 181–206.

Kenney, S. J., Anderson, G. L., Williams, P. L., Millner, P. D., & Beuchat, L. R. (2004). Effectiveness of cleaners and sanitizers in killing *Salmonella* Newport in the gut of a free-living nematode, *Caenorhabditis elegans*. *Journal of Food Protection, 67*, 2151–2157.

Lacey, L. A., & Georgis, R. (2012). Entomopathogenic nematodes for control of insect pests above and below ground with comments on commercial production. *Journal of Nematology, 44*, 218–225.

Lacharme-Lora, L., Perkins, S. E., Humphrey, T. J., Hudson, P. J., & Salisbury, V. (2009). Use of bioluminescent bacterial biosensors to investigate the role of free-living helminths as reservoirs and vectors of *Salmonella*. *Environmental Microbiology Reports, 1*, 198–207.

Lewis, E. E., & Clarke, D. J. (2012). Nematode parasites and entomopathogens. In F. E. Vega & H. K. Kaya (Eds.), *Insect pathology* (Vol. II, pp. 395–443). San Diego: Academic.

Lewis, E. E., Hazir, S., Hodson, A., & Gulcu, B. (2015). Trophic relationships of entomopathogenic nematodes in agricultural habitats. In R. Campos-Herrera (Ed.), *Nematode pathogenesis of insects and other pests* (pp. 139–163). Switzerland: Springer International Publishing.

Lopez, L. C. S., Rodrigues, P. P., & Rios, R. I. (1999). Frogs and snakes as phoretic dispersal agents of bromeliad ostracods (*Elpidium*) and Annelids (*Dero*). *Biotropica, 31*, 705–708.

Lopez, L. C. S., Filizola, B., Deiss, I., & Rios, R. I. (2005). Phoretic behaviour of bromeliad annelids (*Dero*) and ostracods (*Elpidium*) using frogs and lizards as dispersal vectors. *Hydrobiologia, 549*, 15–22.

Macchioni, F. (2007). Importance of phoresy in the transmission of Acarina. *Parassitologia, 49*, 17–22.

Madsen, E. L., & Alexander, M. (1982). Transport of *Rhizobium* and *Pseudomonas* through soil. *Soil Science Society of America Journal, 46*, 557–560.

Malan, A., & Hatting, J. L. (2015). Entomopathogenic nematode exploitation: Case study in laboratoryand field applications from South Africa. In R. Campos-Herrera (Ed.), *Nematode pathogenesis of insects and other pests* (pp. 477–508). Switzerland: Springer International Publishing.

Martens, E. C., Heungens, K., & Goodrich-Blair, H. (2003). Early colonization events in the mutualistic association between *Steinernema carpocapsae* nematodes and *Xenorhabdus nematophila* bacteria. *Journal of Bacteriology, 185*, 3147–3154.

Marti, O. G., & Timper, P. (1999). Phoretic relationship between a *Bacillus* sp. and the entomopathogenic nematode, *Heterorhabditis* sp. *Journal of Nematology, 31*, 553.

Mauël, C., Young, M., Margot, P., & Karamata, D. (1989). The essential nature of teichoic acids in *Bacillus subtilis* as revealed by insertional mutagenesis. *Molecular & General Genetics, 215*, 388–394.

Mercado, J. E., Hofstetter, R. W., Reboletti, D. M., & Negrón, J. F. (2014). Phoretic symbionts of the Mountain Pine Beetle (*Dendroctonus ponderosae* Hopkins). *Forest Science, 60*, 512–526.

Mohandas, C., Sheeba, M., Firoza, A. J., & Rajamma, P. (2007). Bacteria associated with *Rhabditis* (*Oscheius*) spp. (Rhabditidae: Nematoda) for the biocontrol of insect pests. In *Proceedings of national seminar on achievements and opportunities in post harvest management and value addition in root and tuber crops* (NSRTC–2), Trivandrum, Kerala, India, pp. 195–198.

Nykyri, J., et al. (2014). Evidence that nematodes may vector the soft rot-causing enterobacterial phytopathogens. *Plant Pathology, 63*, 747–757.

Ophel, K. M., Bird, A. F., & Kerr, A. (1993). Association of bacteriophage particles with toxin production by *Clavibacter toxicus*, the causal agent of annual ryegrass toxicity. *Phytopathology, 83*, 676–681.

Orozco, R. A., Hill, T., & Stock, S. P. (2013). Characterization and phylogenetic relationships of *Photorhabdus luminescens* subsp. *sonorensis* (c-Proteobacteria: Enterobacteriaceae), the bacterial symbiont of the entomopathogenic nematode *Heterorhabditis sonorensis* (Nematoda: Heterorhabditidae). *Current Microbiology, 66*, 30–39.

Perotti, M. A. (2009). Mégnin re-analysed: The case of the newborn baby girl, Paris, 1878. *Experimental and Applied Acarology, 49*, 37–44.

Rae, R., et al. (2008). Isolation of naturally associated bacteria of necromenic *Pristionchus* nematodes and fitness consequences. *Journal of Experimental Biology, 211*, 1927–1936.

Rae, R. G., Tourna, M., & Wilson, J. M. (2010). The slug parasitic nematode *Phasmarhabditis hermaphrodita* associates with complex and variable bacterial assemblages that do not affect its virulence. *Journal of Invertebrate Pathology, 104*, 222–226.

Rajagopal, R., & Bhatnagar, R. K. (2002). Insecticidal toxic proteins produced by *Photorhabdus luminescens akhurstii,* a symbiont of *Heterorhabditis indica. Journal of Nematology, 34*, 23–27.

Riley, I. T., & Ophel, K. M. (1992). *Clavibacter toxicus* sp. nov. the bacterium responsible for annual ryegrass toxicity in Australia. *International Journal of Systematic Bacteriology, 42*, 64–68.

Riley, I. T., & Reardon, T. B. (1995). Isolation and characterisation of *Clavibacter tritici* associated with *Anguina tritici* in Western Australia. *Plant Pathology, 44*, 805–810.

Riley, I. T., Schmitz, A., & de Silva, P. (2001). *Anguina australis*, a vector for *Rathayibacter toxicus* in *Ehrharta longiflora. Australasian Plant Pathology, 30*, 171–175.

Riley, I. T., Gregory, A. R., Allen, J. G., & Edgar, J. A. (2003). Poisoning of livestock in Oregon in the 1940s to 1960s attributed to corynetoxins produced by *Rathayibacter* in nematode galls in chewing fescue (*Festuca nigrescens*). *Veterinary and Human Toxicology, 45*, 160–162.

Sangeetha, B. G., et al. (2016). Molecular characterization and amplified ribosomal DNA restriction analysis of entomopathogenic bacteria associated with *Rhabditis* (*Oscheius*) spp. *3 Biotech, 6*, 32.

Sasaki, J., Chijimatsu, M., & Suzuki, K. I. (1998). Taxonomic significance of 2,4-diaminobutyric acid isomers in the cell wall peptidoglycan of actinomycetes and reclassification of *Clavibacter toxicus* as *Rathayibacter toxicus* comb. nov. *International Journal of Systematic Bacteriology, 48*, 403–410.

Shapiro, D. I., Tylka, G. L., Berry, E. C., & Lewis, L. C. (1995). Effects of earthworms on the dispersal of *Steinernema* spp. *Journal of Nematology, 27*, 21–28.

Shapiro-Ilan, D. I., & Brown, I. (2013). Earthworms as phoretic hosts for *Steinernema carpocapsae* and *Beauveria bassiana*: Implications for enhanced biological control. *Biological Control, 66*, 41–48.

Sicard, M., et al. (2004). Stages of infection during the tripartite interaction between *Xenorhabdus nematophila*, its nematode vector, and insect hosts. *Applied and Environmental Microbiology, 70*, 6473–6480.

Stephens, P. M., Davoren, C. W., Ryder, M. H., & Doube, B. M. (1994). Influence of the earthworm *Aporrectodea trapezoides* (Lumbricidae) on the colonization of alfalfa (*Medicago sativa* L.) roots by *Rhizobium meliloti* L5-30R and the survival of *R. meliloti* L5-30R in soil. *Biology and Fertility of Soils, 18*, 63–70.

Stock, S. P. (2015). Diversity, biology and evolutionary relationships. In R. Campos-Herrera (Ed.), *Nematode pathogenesis of insects and other pests* (pp. 3–27). Switzerland: Springer International Publishing.

Sudhaus, W. (2008). Evolution of insect parasitism in rhabditid and diplogastrid nematodes. In S. E. Makarov & R. N. Dimitrijević (Eds.), *Advances in arachnology and developmental biology* (pp. 143–161). Vienna: SASA, Belgrade & UNESCO MAB Serbia.

Sudhaus, W., & Kühne, R. (1990). Nematodes associated with Psychodidae: Description of *Rhabditis berolina* sp. n. and redescription of *R. dubia* Bovien, 1937 (Nematoda: Rhabditidae), with biological and ecological notes, and a phylogenetic discussion. *Nematologica, 35*, 305–320.

Tailliez, P., Pagès, S., Ginibre, N., & Boemare, N. (2006). New insight into diversity in the genus *Xenorhabdus*, including the description of ten novel species. *International Journal of Systematic and Evolutionary Microbiology, 56*, 2805–2818.

Tailliez, P., et al. (2010). Phylogeny of *Photorhabdus* and *Xenorhabdus* based on universally conserved protein-coding sequences and implications for the taxonomy of these two genera. Proposal of new taxa: *X. vietnamensis* sp. nov., *P. luminescens* subsp. *caribbeanensis* subsp. nov., *P. luminescens* subsp. *hainanensis* subsp. nov., *P. temperata* subsp. *khanii* subsp. nov., *P. temperata* subsp. *tasmaniensis* subsp. nov., and the reclassification of *P. luminescens* subsp. *thracensis* as *P. temperata* subsp. *thracensis* comb. nov. *International Journal of Systematic and Evolutionary Microbiology, 60*, 1921–1937.

Tambong, J. T. (2013). Phylogeny of bacteria isolated from *Rhabditis* sp. (Nematoda) and identification of novel entomopathogenic *Serratia marcescens* strains. *Current Microbiology, 66*, 138–144.

Tan, L., & Grewal, P. S. (2001). Pathogenicity of *Moraxella osloensis*, a bacterium associated with the nematode *Phasmarhabditis hermaphrodita*, to the slug *Deroceras reticulatum. Applied and Environmental Microbiology, 67*, 5010–5016.

Tan, L., & Grewal, P. S. (2003). Characterization of the first molluscicidal lipopolysaccharide from *Moraxella osloensis. Applied and Environmenatal Microbiology, 69*, 3646–3649.

Thorpe, I. S., Killham, K., Prosser, J. I., & Glover, L. A. (1996). The role of the earthworm *Lumbricus terrestris* in the transport of bacterial inocula through soil. *Biology and Fertility of Soils, 23*, 132–139.

Troemel, E. R., Kimmel, B. E., & Bargmann, C. I. (1997). Reprogramming chemotaxis responses: Sensory neurons define olfactory preferences in *C. elegans. Cell, 91*, 161–169.

Van den Velde, S., et al. (2006). Species identification of corynebacteria by cellular fatty acid analysis. *Diagnostic Microbiology and Infectious Disease, 54*, 99–104.

Vandel, A. (1965). *Biospeleology: The biology of cavernicolous animals*. London: Pergamon Press, 524 pp.

Vidaver, A. K. (1981). The plant pathogenic Corynebacteria. *Annual Reviews of Microbiology, 36*, 491–517.

Wilkinson, P. (2009). Comparative genomics of the emerging human pathogen *Photorhabdus asymbiotica* with the insect pathogen *Photorhabdus luminescens. BMC Genomics, 10*, 302.

Wilson, M. J., Glen, D. M., & George, S. K. (1993). The rhabditid nematode *Phasmarhabditis hermaphrodita* as a potential biological-control agent for slugs. *Biocontrol Science and Technology, 3*, 503–511.

Zgurskaya, H. I., Evtushenko, L. I., Akimov, V. N., & Kalakoutskii, L. V. (1993). *Rathayibacter* gen, nov., including the species *Rathayibacter rathayi* comb. nov., *Rathayibacter tritici* comb. nov., *Rathayibacter iranicus* comb. nov., and six strains from annual grasses. *International Journal of Systematic Bacteriology, 43*, 143–149.

Zhang, C. X., et al. (2009). *Serratia nematodiphila* sp. nov., associated symbiotically with the entomopathogenic nematode *Heterorhabditidoides chongmingensis* (Rhabditida: Rhabditidae). *International Journal of Systematic and Evolutionary Microbiology, 59*, 1603–1608.

Part II
Molecular Processes

Chapter 7
Defense and Immune Systems

Abstract The complex of defensive strategies characterizing invertebrate immune systems and response is reviewed. These include the role of pathogen recognition molecules and evolutionary aspects concerning a number of molecular components. Effective humoral responses to invasive pathogens rely on a first level based on melanization and phenoloxidase activity and on specific pathogen-associated molecular patterns. Other invertebrate defensive pathways reviewed include lectin-mediated complement and activation, antimicrobial peptides and heat shock proteins, with pathways for nitric oxide and reactive oxygen species, and lysozymes. Cellular defense processes include phagocytosis by immunocytes, lysosomes, encapsulation and hemolymph coagulation. Some aspects concerning selectivity, specificity and evasion are also reviewed, with host resistance and tolerance.

Keywords Antimicrobial peptides • Biochemical pathways • Defensins • Hemolymph • Humoral response • Immune system • Lysozyme • Melanin • Pathogenicity • Pattern recognition receptors • Phenoloxidase • Selectivity

1 Introduction

Invertebrates live in their environment with a plethora of microorganisms. To overcome the attacks of pathogenic species, they developed a wide range of defensive strategies. The responses deployed include the capacity to recognize pathogenic organisms and their variants and to discriminate them from the species living in non-pathogenic associations, like the endosymbiotic bacteria. Pathogens, and bacteria among them, represent indeed a potent selective pressure that gave shape to speciations paths and elicited the evolution of several fundamental diversification traits, including the insurgence of sexual recombination (see Chap. 4). In most invertebrate lineages a wide repertoire of sophisticated defense mechanisms is integrated in an innate immune system relying on cells and molecular products, supporting a complex array of pathways active against invasive pathogens.

Vertebrates benefit of a further, specialized and adaptive immune system characterized by a complex of systemic and selective responses, including the activity of lymphocytes and the capacity to produce immunoglobulins, that react to new anti-

A. Ciancio, *Invertebrate Bacteriology*, DOI 10.1007/978-94-024-0884-3_7

gens keeping their "memory". Invertebrates instead only rely on what appears to be a complex of innate defensive strategies, in many phyla based on cells called hemocytes and phagocytes, involved in phagocytosis.

These systems are flanked by the expression of a repertoire of genetically controlled humoral responses. All these components, in conjunction with the presence of specific defensive products, originate the innate immunity system (Iwanaga and Lee 2005). Its main traits are the presence of biochemical processes leading to phagocytosis or to the formation of physical barriers, and/or the integration of complex cell and humoral responses leading to the release of antimicrobial or other active molecules (Rinkevich and Müller 1996; Iwanaga and Lee 2005; Nappi and Christensen 2005; Jiravanichpaisal et al. 2006).

Several bahavioral and physical factors also contribute to the evasion of pathogens, like the grinding capacity to mechanically disrupt bacterial cells during feeding by means of schlerotized chitin-rich gut components (shown by many insects and nematodes feeding on bacteria), the presence of physical cuticular barriers present in the gut or around other natural body openings, or the sensing capacity for potentially harmful species, leading to their avoidance (Schulenburg et al. 2004).

The invertebrate immune system has been mostly studied in insects, molluscs and crustaceans, whose immune-competent hemocytes are active in phagocytosis, cell adhesion, encapsulation and melanization, as well as in the release of several defense reactive species. However, the number of studies and the eventual discoveries concerning properties and structural details of the invertebrate immune systems, and the data concerning their variation among phyla as well, are still restricted to a low number of cases when compared to the number of invertebrate phyla. These mainly involve *C. elegans*, *Drosophila* and Crustacea, with a few further additional lineages. Any generalization hence may appear premature, given the high rate of evolutive divergence and adaptation shown by the systems studied. The actual coverage of these processes is considered in fact partial, when compared to the large phylogenetic distances separating the different invertebrate evolutive radiations, and the complexity of the reactions involved (Hibino et al. 2006). Furthermore, substantial differences exist among various invertebrate lineages as concerns the presence and role of one or more components of the innate immune system, as shown by the limited relevance found in *C. elegans* for some of the most common defensive processes active in other phyla (Schulenburg et al. 2004).

In spite of the lack of a complex similar to the vertebrate adaptive system, the invertebrate immune reactions are, however, highly selective and capable to discriminate among closely related pathogens or to distinguish them from mutualistic bacteria or other cells, i.e. those present in the gut as food. Since a few phyla and groups have been explored in detail as concerns their immune systems, the variations in phylogenetically close lineages is still underexplored, and it is possible that even in the same phylum, significant differences might be found. This is the case for example of the pea aphid *Acyrthoshiphon pisum*, whose sequenced genome revealed the lack of several defence-related genes, widespread among insects, for which a possible substitution by the activity of endosymbionts was hypothesized (Gerardo et al. 2010).

The "universe" of invertebrate immune systems to explore is indeed very large, and relatively few data are still available on the evolution of these fundamental processes among i.e. the species or groups living in extreme environments. It is hence possible that the range of immune adaptations and mechanisms may result wider and more complex, as far as the number of case studies is enlarged. In this chapter some basic traits of invertebrate immune systems and responses are examined, with an overview on the most fundamental molecular mechanisms and products associated.

2 Origins and Evolution

The innate immune systems relies on three levels of response, the first one being formed by a range of pathogen recognition molecules, which activate a second complex front of biochemical pathways and genes, whose cascade may finally yield a variety of effector molecules and cell activation steps (Roeder et al. 2010). Many invertebrate immune systems include epitope recognition pathways like the Toll-like receptors (TLRs) present in *Drosophila* (also present among vertebrates), indicating a common and ancient evolutionary origin (Aderem and Ulevitch 2000; Medzhitov and Janeway 2000). However, the effector molecules which ultimately kill or block the intruding pathogens are considered as the most ancient and likely the first that have been recruited for defense, being common and widespread among different Kingdoms (Roeder et al. 2010). The widespread protective action exerted by products like melanin is indicative of the ancient evolutionary origin of the immune system, being this polymer one of the most common protective pigments present across Kingdoms. The melanin production pathways persisted in fact during eons of life evolution on earth, as shown by the residual traces that have been found in fossils of squids, dinosaurs, birds and other organisms (Solano 2014).

The evolutionary changes related to the more complex physiology of vertebrates (in particular the advent of adipocytes) are considered as the steps that ignited the evolution of their adaptive immune system, which has no homolog among invertebrates (van Niekerk and Engelbrecht 2015). The invertebrate innate immune systems, however, show a wide range of receptors and pathways that sustain an expanded repertoire of responses which are equally specific and effective, whose evolution independently proceeded among different phyla.

Invertebrates evolved a common set of immune molecular components as sophisticated as the vertebrates adaptive ones, by affording similar evolutionary challenges (Bailey et al. 2013). This is also shown by recent discoveries concerning the presence of a kind of invertebrate immune "memory", by the Rag1/2-like gene cluster (that was considered unique for vertebrates) and by the high incidence of genes of the TLR type covering around 1 % of the genome, which were found in the sea urchin *Strongylocentrotus purpuratus* (Echinodermata) (van Niekerk and Engelbrecht 2015). The presence in *Drosophila* of homologs of the mammalian

TLR family, active in the insect immune response against fungal infections, also confirms the ancient origin of this pathway (Ferrandon et al. 2007; Lemaitre and Hoffmann 2007).

Overall data indicate that many invertebrate lineages developed an immune system that evolved, more than 600 Myrs ago, from that of their common ancestor with other phyla like vertebrates. All along their subsequent evolutive radiations, this process proceeded through a series of enrichment steps rather than by constructing a second adaptive system, as occurred for vertebrates.

The acquired immunity system of vertebrates is considered to have arosed at the time of divergence between the lamprey cyclostomes and cartilaginous fishes (Hoffmann et al. 1999; Fujita 2002). Apart of its basic significance in the study of natural evolutionary history, a deeper understanding of the invertebrate immune systems is indeed important, given the possibility to induce a sort of "modulation" of the immunitary protection in species with an economic impact, or due to the severe effects exerted by some invertebrates on the health of humans, animals and plants (Bailey et al. 2013).

3 Humoral Defense

3.1 Melanization and Phenoloxidase Activity

Many invertebrates deploy a rapid and effective humoral response to invasive pathogens, including bacteria, through a process called melanization, culminating in the production of melanin. Studies carried out on insects and crustaceans highlighted the role of phenoloxidase (PO, also known as tyrosinase), a key enzyme involved in the reaction, which is regulated by a proteolytic activation cascade named proPO. This system has been found in many lineages, including insects, crustaceans, molluscs and others (Iwanaga and Lee 2005). In the hemocytes of the fresh water crayfish, *Pacifastacus leniusculus* (Arthropoda, subphylum Crustacea), a serine protease activates proPO in the hemocytic granules in which it is synthesized and stored, to be eventually released through exocytosis induced by pattern recognition receptors (PRR). These are binding proteins that recognize components of the bacterial cell wall like lipopolysaccharides (LPS), peptidoglycans and the β-1,3-glucans of fungi (Cerenius et al. 1994). After binding to microbial cell wall components, the PRRs proteins activate the serine protease zymogens[1] culminating in the enzyme activation and PO release (Lee et al. 2000).

The substrate of PO is the amino acid L-tyrosine, its main monophenol precursors. Through its (and other monophenol substrates) hydroxylation and oxidation, PO catalyzes two reactions, starting in the first one the production of diphenols and in the second one the release of toxic quinones. These products eventually result in the polymerization of melanin and reactive intermediates, which are involved in several

[1] *Zymogens* (proenzymes) are inactive enzyme precursors that can be activated by a biochemical or other structural molecular change.

defense mechanisms, including a direct action against invasive pathogens (Sugumaran 2002; Nappi and Christensen 2005; Kanost and Gorman 2008; Amparyup et al. 2013; Solano 2014; Jearaphunt et al. 2015).

The ProPO protective role and efficacy have been indirectly demonstrated by the susceptibility of hosts in which it was artificially suppressed, as shown by *P. leniusculus* infected by *Aeromonas hydrophila*, by the higher mortality observed for the kuruma shrimp *Marsupenaeus japonicus* after bacterial infections, or by the infection of the shrimp *Penaeus monodon* by *Vibrio harveyi* and a further fungal pathogen (Liu et al. 2007; Amparyup et al. 2009; Fagutao et al. 2009; Charoensapsri et al. 2014).

Given the toxicity of its intermediate products, the ProPO system is tightly controlled and localized (Jearaphunt et al. 2015). In *P. leniusculus* it is regulated by inactive zymogens (in turn activated by proteolysis) which act as serine protease inhibitors, such as a 150 kDa pacifastin, an α2-macroglobulin and a Kazal inhibitor (Johansson et al. 1994; Iwanaga and Lee 2005). Studies on gene expression and RNA interference[2] (RNAi) in *P. monodon* showed that several serine proteases and related homologs (SPH) are involved in the ProPO regulatory pathways. The fundamental role of masquerade-like proteins and SPHs in the ProPO cascade is integrated by their activity also as pathogen binding proteins, and by their interactions with other immune related genes and antimicrobial products (Jearaphunt et al. 2015). A direct antibacterial activity has also been observed for serine protease *in vivo* (Iwanaga and Lee 2005). The masquerade-like proteins and SPHs of *P. leniusculus* or the mud crab *Scylla paramamosain* have been observed binding to LPS, β-1,3-glucans and components of the Gram-negative bacteria cell walls (Lee and Söderhäll 2001; Liu et al. 2011; Zhang et al. 2013).

The production of melanin is widespread among animals and is also present in bacteria and plants (Solano 2014). Melanin and PO pathways are similarly widespread among invertebrates, having been found in many distant lineages such as corals, anellids, echinoderms or cephalopods (Porchet-Henneré and Vernet 1992; Roch et al. 1992; Beschin et al. 1998; Mydlarz and Palmer 2011; Novoa et al. 2011). In Arthropoda, melanin is also found during the hardening and darkening of the cuticle and in response to wounding (Boucias and Pendland 1998). Melanization stimulates further defensive reactions like the mobilization and attachment of hemocytes, with phagocytosis and encapsulation (Cerenius and Söderhäll 2004; Iwanaga and Lee 2005).

3.2 Receptors and Recognition

The Pattern Recognition Receptors (PRRs) are heterogeous proteins that recognize specific pathogen-associated molecular patterns (PAMPs). PRRs can be present in the cell cytoplasm, in the body fluid as secreted proteins or as cell surface receptors.

---

[2] An experimental approach based on the post-transcriptional silencing of a gene, induced by short RNA fragments.

Most important PRRs are characterized by the presence of domains like the C-type lectin, the scavenger receptor cysteine-rich and/or the leucine-rich repeats (LRR) (Medzhitov and Janeway 2000). Their binding with PAMPs represents one of the first steps in the activation of the invertebrate immune response, followed by a number of humoral or immune cell reactions. Humoral responses include the PO patterns leading to melanization, or the activation of cytokines[3] with eventual expression of genes coding for antimicrobial peptides (AMPs).

The common traits of targeted PAMPs are: (i) their unique production by pathogens, since these molecules are not found in the host body (thus allowing an immediate recognition of self/non-self molecules and cells, a property in common with homolog vertebrates cells); (ii) their essential role and conservation among large groups of pathogens, being fundamental for their biology and survival, and (iii) their function as molecular identifiers of the pathogen(s) lineage (Medzhitov and Janeway 2000). Some examples of PAMPs are the peptidoglycans and lipoteichoic acids (LTA) from the cell walls of Gram-positive bacteria, the LPS found in Gram-negative species or lipoarabinomannan, a molecular signature of mycobacteria.

The production and release of AMPs is mediated by TLRs and represent a critical component of the invertebrate immune system (Lemaitre et al. 1996; Imler and Hoffmann 2000; Krutzik et al. 2001; Underhill and Orinsky 2002). Advances in this reserach field have been achieved through the study of the *Drosophila* immune system and, as shown by its (and other insects') sequenced genome, of their homologs in the human immune system. By studying mutations in *Drosophila* the TLR 18-Wheeler was shown to be involved in the specific recognition of Gram-negative bacteria and in the activation of a defensive pathway secreting attacin, an antibacterial peptide which acts specifically against Gram-negative species (Williams et al. 1997).

These studies also showed that several members of the TLRs family found in insects have vertebrate homologs. They act through the specific identification of different groups of microbial targets, yielding a recognition capability that is mediated by a number of PAMPs and their corresponding reactive peptides (Medzhitov and Janeway 2000). *Drosophila* TLRs activate the nuclear factor-kB (NF-kB), which induces expression of various AMPs (Schulenburg et al. 2004).

There is evidence that many receptors are present among the invertebrate lineages examined. Annotations from the sequenced genome of the tobacco hornworm *Manduca sexta* (Lepidoptera) revealed genes coding for a large number of PRRs including, apart of known C-type lectin-domain proteins (4 types) and TLR (16 types), the following additional PRRs: LRR proteins (76 types), peptidoglycan recognition proteins (14 types), EGF/Nim-domain proteins (6 types), β-1,3-glucanase-related proteins (5 types), immunoglobulin-domain (5 types), galectins and fibrinogen-related proteins (4 types each), thioester proteins (3 types), hemocytins (2 types) and one Reeler (Zhang et al. 2015). Similarly, several TLR genes were also found in the sequenced genome of the sea urchin *S. purpuratus* (Smith et al. 2006). In *Drosophila* and *Anopheles gambiae* the hypervariable PRR Dscam (Down syn-

[3] *Cytokines* = a group of small proteins released by cells, with a signalling function towards other cells or receptors.

drome) cell adhesion molecule gene has a complex organization with 101 exons yielding more than 18,000 (*Drosophila*) to 31,000 (*A. gambiae*) alternative splice isoforms. This broad range of PRRs shows a large set of different combinations of pathogen-specific adhesive domains, involved in the *Drosophila* binding to bacteria and in the mosquito resistance to bacteria and *Plasmodium*. These findings suggested a previously unsuspected complexity of the insect innate immune system, sustaining a broad capacity to face a wide range of pathogenic invasions (Christophides et al. 2004; Watson et al. 2005; Dong et al. 2006).

Assays with *Drosophila* infected by *Micrococcus luteus* or fungi showed that the Toll pathway regulates the antifungal drosomycin response (Lemaitre et al. 1996), whereas the locus identified as the immunodeficiency (IMD) pathway regulates the response to Gram-negative or Gram-positive bacteria, as shown by mutation studies (Kleino and Silverman 2014). In *D. melanogaster* either the Toll and IMD pathways activate the expression of AMPs producing genes, and may act in synergy. IMD recognizes the bacterial diaminopimelic acid (DAP)-type peptidoglycans through the recognition proteins LC (PGRP-LC) and LE. These induce expression of the adaptor protein IMD and of kinases[4] that activate the transcription factor Relish. After cleveage the Relish N-terminal portion is translocated to the nucleus, inducing the expression of the AMPs diptericin and attacin-A (Tanji et al. 2007). The activation of the IMD pathway occurs in minutes, whereas the Toll pathway responds to fungi and Gram-positive bacteria through an activation process lasting several hours (Kleino and Silverman 2014).

Further recognition pathways present in the innate immune response involve the C-jun-N-terminal kinase (JNK) and the Janus kinase/signal transducers and activators of transcription (JAK/STAT). JNKs are a group of conserved MAP kinases identified in insect and mammal signaling pathways, activated by cytokines or environmental stress factors (Botella et al. 2001). JAK/STAT is an evolutionary conserved pathway whose signal transduction cascade is active in developmental, immune response, hematopoiesis and cancer processes. In adult stages of *D. melanogaster* the JAK/STAT pathway appeared involved in the expression of AMPs, whereas in *A. gambiae* it translocated into the nucleus of fat body cells, after bacterial infection (Barillas-Mury et al. 1999).

3.3 Lectin-Mediated Complement and Activation

Lectins are proteins bearing at least one binding site and carbohydrate-binding domain, reversibly interacting in a specific way with a sugar moiety (Grubhoffer et al. 2004). They are involved in several immune reactions, and originated as part of a very ancient defensive system. This is shown by their diffuse presence either in invertebrate and vertebrate immune systems, and by the wide repertoire of structural motifs present and functions deployed (Vasta et al. 2004). The term was coined

[4] *Kinases* = phosphotransferase enzymes catalyzing the transfer of a phosphate group from a high-energy donor to a specific substrate.

by Boyd and Shapleigh in 1954 from the latin *legere* (to gather), due to the property of these proteins to aggregate red blood cells *in vitro* (from which the term "hemagglutinins" also used for these proteins) (Grubhoffer et al. 2004).

Lectins are active in many evolutionary ancient, defense processes including the early immune reaction steps leading to the elimination of pathogens, through processes like agglutination, immobilization, complement-mediated opsonization[5] and lysis, integrated by multiple cell-based and humoral recognition processes and interactions (Renwrantz 1983; Vasta et al. 2004).

The superfamily of highly conserved C-type lectins (Ca-dependent, CTLs) is found in many invertebrate lineages, including porifera, nematodes, molluscs, planariae or arthropoda (Maizels et al. 2006). CTLs show a conserved 115–130 amino acid carbohydrate-recognition domain (CRD) with several Ca^{++} binding sites, determining their C-type ligation, with four cysteine residues involved in sugar binding. In the genome of *C. elegans* at least 278 CTL genes have been found, a large part involved in antibacterial defensive functions (Schulenburg et al. 2008). *In vitro* assays with the pathogenic *Microbacterium nematophilum* showed that the most abundant group among the 68 *C. elegans* genes up-regulated 6 h post-infection encoded products with a CTL domain (O'Rourke et al. 2006). Further nematode genomic and transcriptomic data showed that CTLs are widely distributed and also occur in the plant parasitic nematodes *Meloidogyne incognita*, *Rotylenchulus reniformis* and *Heterodera glycines* (Abad et al. 2008; Ganji et al. 2013). CTLs are also present in animal parasitic nematodes, in which they are likely involved in interactions with their hosts' immune systems (Harcus et al. 2009).

In crustaceans, CTLs are involved in defense mechanisms from pathogenic bacteria. The CTLs of the shrimp *Litopenaeus vannamei* agglutinate Gram-negative species like *Vibrio alginolyticus* and *V. parahaemolyticus* and the Gram-positive *Bacillus subtilis*, in presence of Ca ions (Li et al. 2014). The CTLs genes of the giant freshwater prawn *Macrobrachium rosenbergii* were found to be mainly expressed in the hepatopancreas, gills and stomach, after challenges with *Vibrio anguillarum* and the White spot syndrome virus (Ren et al. 2012).

Lectins may also perform other, non-defensive functions like i.e. mediating the adhesion of bacterial ectosymbionts, as observed for the external cuticular layers of the marine nematode *Laxus oneistus* (Bulgheresi et al. 2006). Apart of CTLs, other lectin families have been found in invertebrates like galectins (found in *C. elegans*, *D. melanogaster* and the marine sponge *Geodia cydonium*), F-type lectins (present in the purple sea urchin, planariae, molluscs, horseshoe crabs, insects, echinoderms) or pentraxins (Vasta et al. 2004).

Lectins are also involved in other, more evolutionary recent, defense pathways. In the innate immune system, the activation of the thermal-labile Complement proteins[6] yields a cascade of proteolysis and assembly reactions (called the "activation pathway") leading to the cleveage of a key component, the third complement protein (C3). This product (identified among invertebrates in the purple sea urchin *S.*

[5] A mechanism in which the bacterial cell surface is covered by small proteins (opsonins) that facilitate phagocytosis by host macrophages like cells, bearing opsonin-specific receptors.

[6] So called in early immunology studies since they complemented the thermally stable, pathogen-specific antibodies.

purpuratus), originates a further lytic pathway with the formation of the membrane-attack complex (MAC) (not found, however, in the sequenced sea urchin genome) (Hibino et al. 2006). In vertebrates, the activation may be induced by a classical antibody-based pathway, by lectins or by an alternative pathway. In the invertebrate immune systems only the two latter pathways are present, and converge towards the C3 activation, that eventually starts the final lytic process. In the case of lectins, a carbohydrate is recognized by a PRR such as the mannose-binding lectin (MBL) or ficolins (in mammals), which mediate bacteria and other pathogens recognition, eventually activating specific enzymes known as MBL-associated serine proteases (MASPs). In the alternative pathway, the C3 is covalently linked to hydroxyl or amine groups present on the surface of the microbial invaders (Fujita 2002).

In ticks, lectins and other humoral components of the immune system are produced by haemocytes and fat body cells and then secreted into the hemolymph, the extracellular fluid of the body cavity (called hemocel or celome), containing several types of proteins and hemocytes. They show Ca-dependent hemagglutinin properties and binding to sialic acid (recognizing Gram negative bacteria), N-acetyl-D-glucosamine and D-galactose. A 85 kDa structural lectin of the sheep tick *Ixodes ricinus*, localized in hemocytes, midgut cells and granular inclusions of nephrocytes, is a recognition molecule active in the tick immune system (Grubhoffer et al. 2004).

In *I. ricinus*, sialic acid—specific lectins secreted in the hemolymph—mediate the particular coiling phagocytosis of the spirochete *Borrelia burgdorferi*, a severe human pathogen transmitted by the tick's bite and etiologic agent of Lyme disease (see Chap. 5). In this process the phagocytes extend a pedicellate protrusion of the cell membrane, that turns around the bacterium in a spiral coil, eventually introgressing the bacterium for its final lysis. The 85 kDa structural lectin of *I. ricinus*, produced and stored in hemocytes, recognizes the sialic acid from the wall of *B. burgdorferi* and likey acts either in the phagocytosis of the blood cell in the tick meal and in the defense reaction of the vector to the bacterium. Only a small fration (less than 5 %) of *B. burgdorferi* cells can in fact migrate from the tick midgut and hemocel to the salivary glands, from where they are later transmitted (Grubhoffer et al. 2004).

Genes coding for TLR and members of the alternative and lectin complement pathways have been found in invertebrates, as shown by the sequenced genome of the purple sea urchin (Hibino et al. 2006; Smith et al. 2006). In these animals, celomic fluid cells (celomocytes) mediate a wide range of immune defense responses, including important functions like opsonization and phagocytosis by macrophage-like cells, encapsulation or production of antibacterial compounds like echinochrome A (a naphthoquinone).

Genome data analysis showed several genes active in the sea urchin immune system, coding for TLRs, C-type lectins, Complement homologs of the vertebrate C3/4/5 genes and opsonins. Although no vertebrate MASP homolog was found in the *S. purpuratus* genome, members of the terminal pathway included proteins with similarities to the Membrane Attack Complex/Perforin (MACPF) whose vertebrate homologs induce the formation of pores on the bacterial cell membrane (Hibino et al. 2006; Smith et al. 2006). C3 homologs were also found in the horseshoe crab *Carcinoscorpius rotundicauda* and in the gorgonian coral *Swiftia exerta* (Zhu et al. 2005; Dishaw et al. 2005).

3.4 Antimicrobial Peptides (AMPs)

The first observations of a kind of antibacterial inducible defense in insects was provided by Boman et al. (1972) who showed that fruit flies survived to a lethal dose of pathogenic *Pseudomonas aeruginosa* if they had been previously challenged with an attenuated cell suspension of *Enterobacter cloacae* (Kleino and Silverman 2014). This study was followed by the discovery of the first AMP (cecropin) in *Cecropia* moths, and by the subsequent identification of new AMPs, either in invertebrate and vertebrate species (Kleino and Silverman 2014).

AMPs are small and ubiquitous molecules (<10 kDa) formed by 15 to less than 200 amino acids, acting as multifunctional components of the innate immune system of many invertebrates, modulating signal transductions or cytokine production and release, with additional antitumor and mitogenic activities. AMPs are evolutionary conserved since they can be found across all Kingdoms, including plants or vertebrates. The number of discovered AMPs is still growing, being actually in the order of >2500, as shown by their online database http://aps.unmc.edu/AP/main.php (Wang and Wang 2004; Thomas et al. 2010). AMPs show a high degree of sequence diversity, mirrored by a wide range of functional and evolutive adaptations, conserving their basic antimicrobial activity for several hundred Myrs (Peschel and Sahl 2006). These molecules elicited a wide attention not only for their meaning in the biology and evolution of the immune defense in many species, but also for their application potential in the human biomedical field (Wiesner and Vilcinskas 2010; Sperstad et al. 2011; Vilcinskas 2011; Yi et al. 2014).

AMPs are cationic and amphipathic[7] peptides mostly encoded by a single gene, which can target and disrupt the bacterial cell membrane. Due to their cationic and amphipathic properties, AMPs damage the bacterial plasma membranes through the formation of pores or by blocking, after their introgression in the target bacterium, one or more metabolic reactions or products. This results in the pathogen inhability to outperform its invasive strategy and in its eventual death (Rosa and Barracco 2010). The AMPs biochemical properties facilitate their insertion in the anionic bacterial cell wall and phospholipid membranes through a number of mechanisms, depending on their amino acid sequence and their target species (Bulet et al. 2004).

Insects AMPs can be classified in three main groups: linear amphipathic and hydrophobic α-helices, cysteine-stabilized cyclic peptides—forming β-sheets or α-helix/β-sheets—and Gly/Pro-rich peptides (Lorenzini et al. 2003; Reddy et al. 2004; Yi et al. 2014). Crustaceans (Decapoda, i.e. the penaeid shrimps) also show

[7] *Amphipathic* = the property of a molecule containing either polar (water soluble) and nonpolar (water insoluble) domains (i.e. the phospholipids forming the lipid bilayer of the cell membrane).

further multi-domain, chimeric and other unconventional AMPs like histones and derived fragments, together with hemocyanin-derived peptides (Rosa and Barracco 2010).

AMPs may be induced by infection via TLR, IMD or other pathways (Lemaitre and Hoffmann 2007) or may be constitutively present (i.e. in holometabolous insects, mussels, shrimps or crabs). In insects, the induction of AMPs occurs as a response to sepsis, with a several hundred-fold concentration increase in hemocytes or in the fat body (an organ whose function is similar to the vertebrate liver) in which they are mostly produced, as well as in epithelial cells (Bulet et al. 2004; Lemaitre and Hoffmann 2007).

Insects AMPs are produced and released in the hemolymph within a few hours from infection. When constitutively present, AMPs are stored in hemocytes and released, after infection, in the hemolymph or phagocytic vesicles (Lorenzini et al. 2003). In the Japanese horseshoe crab (*Tachypleus tridentatus*) a number of defense molecules are stored in secretory granules of two types (large or small), from which they are released through exocytosis after bacterial invasion. Larger granules contain clotting factors involved in hemolymph coagulation, with protease inhibitors (serpins and cystatins), anti-lipopolysaccharide (LPS) factors and tachylectins, with LPS binding and bacterial agglutinating activities. Cysteine-rich basic proteins such as tachyplesins, defensin, tachycitin and tachystatins with antimicrobial or agglutination activities are stored in smaller granules. Both groups of defense proteins act in synergy in the hemolymph, after release (Iwanaga et al. 1998).

Many AMPs adapted to the evolutive history of their producers and are likely the result of G × G selective races. For example, the honeybee apidaecins are specifically targeted and immediately effective against a number of bacteria encountered by the bees during their feeding activity, like enteric or plant pathogenic bacteria (Table 7.1). *In vitro* tests showed, however, that Bt and other insect bacterial parasites were resistant to these products (Casteels et al. 1989).

Several AMP genes occur in the same invertebrate species and many related variants may be found in the same lineage. In the genome of *M. sexta*, 86 AMP genes were identified, coding for a wide repertoire of functional products and clusters including attacins, cecropins, defensins, diapausins, gallerimycins, gloverin, lebocins, lysozyme-related proteins, moricins, X-tox splicing variants, whey acidic protein homologs and transferrins (He et al. 2015).

3.5 Heat Shock Proteins (HSPs)

HSPs are members of a multifunctional protein family present either in Prokaryotes and Eukaryotes, activated in response to high temperatures or other stress factors, as well as functional in the immune response. They vary from 27 to 110 kDa and may act as chaperones, involved in the folding and stabilization of other proteins, or as proteases that degrade other damaged proteins (Gerardo et al. 2010). In presence of stress, HSPs sustain refolding of misfolded proteins or prevent their aggregation.

Table 7.1 Properties of main goups of antimicrobial peptides

AMPs	Targets/activity	Producer	References
Acanthoscurrin	*Escherichia coli*, *Candida albicans*	Arachnida (*Acanthoscurria gomesiana*)	Lorenzini et al. (2003)
Anti-LPS factor	Gram-negative (R-types i.e. *Salmonella minnesota*); Gram-positive. Inhibits LPS-mediated hemolymph coagulation	Crustacea (*Limulus polyphemus*, *Tachypleus tridentatus*, *Litopenaeus setiferus*, *Penaeus monodon*, *Scylla paramamosain*, *Eriocheir sinensis*)	Tanaka et al. (1982), Gross et al. (2001), Supungul et al. (2002), Cuthbertson et al. (2004), Imjongjirak et al. (2007), Iwanaga (2007), Li et al. (2008), and Tassanakajon et al. (2015)
Apidaecins	Gram-negative (*Agrobacterium*, *Erwinia*, *Rhizobium*, *Pseudomonas*, *Salmonella*, *Shigella*)	Insecta (Hymenoptera: *Apis mellifera*)	Casteels et al. (1989), Evans et al. (2006)
Arasin	Antibacterial (Gram positive/negative); antifungal	Crustacea, Decapoda (*Hyas araneus*)	Stensvåg et al. (2008) and Paulsen et al. (2013)
Armadillidin	Gram-positive (*Bacillus megaterium*)	Isopoda (*Armadillidium vulgare*), insects, spiders	Lorenzini et al. (2003) and Herbinière et al. (2005)
Astacidins	Gram-positive and Gram-negative	Crustacea, Decapoda (*Procambarus clarkii*, *Pacifastacus leniusculus*)	Lee et al. (2003), Jiravanichpaisal et al. (2007), and Shi et al. (2014)
Attacins	Gram-negative (*E. coli* growth inhibitor; involved in *Glossina* trypanosome resistance)	Insecta (*Hyalophora cecropia*, *M. sexta*, *Drosophila melanogaster*, *Glossina morsitans*; *Harmonia axyridis*)	Hultmark et al. (1983), Hu and Aksoy (2006), Vilcinskas et al. (2013), and He et al. (2015)
Bac-like	Gram-positive (*Micrococcus luteus*); Gram-negative (*Psychrobacter immobilis*)	Crustacea (Decapoda, *Carcinus maenas*)	Schnapp et al. (1996)
Caenacins, antimicrobial neuropeptide-like proteins (NLP), with YGGYG motif	Epidermal response to nematophagous fungus (*Drechmeria coniospora*)	Nematoda: *Anisakis simplex*, *Bursaphelenchus xylophilus*, *B. mucronatus*, *C. elegans*, *C. remanei*, *Globodera rostochiensis*, *Meloidogyne incognita*, *Pristonchus pacificus*, *Strongyloides ratti*	McVeigh et al. (2008) and Tarr (2012)

(continued)

Table 7.1 (continued)

AMPs	Targets/activity	Producer	References
Caenopores	*E. coli*	Nematoda (*C. elegans*)	Roeder et al. (2010)
Callinectin	Gram-positive and Gram-negative	Crustacea (Decapoda, *Callinectes sapidus*)	Khoo et al. (1999)
Cecropins	Gram-positive and Gram-negative	Insecta (Lepidoptera, *M. sexta*; *Hyalophora cecropia*); Nematoda (*A. suum*)	Hultmark et al. (1980) and He et al. (2015)
Coleoptericins	Gram-negative	Insecta (Coleoptera: *Allomyrina dichotoma, Nicrophorus vespilloides, H. axyridis*)	Vilcinskas et al. (2013)
Crustins	Gram-positive and Gram-negative; proteinase inhibitors	Crustacea (Decapoda, *Homarus americanus, Pacifastacus leniusculus*); Insecta (Hymenoptera)	Jiravanichpaisal et al. (2007), Battison et al. (2008), and Tassanakajon et al. (2015)
Defensins	Gram-positive and Gram negative; fungi	Insecta (Coleoptera, *Zophobas atratus*; (Coleoptera, *H. axyridis*; Lepidoptera, *M. sexta*); Scorpions (*Androctonus australis, Leiurus quinquestriatus*); Acarina (tick, *Ixodes scapularis*); Crustacea (*Panulirus japonicus*); Nematoda (*Ascaris suum, Toxascaris leonina, Toxocara canis, C. elegans, Xiphinema index, Meloidogyne javanica, M. hapla*); Mollusca (*Mytilus edulis, M. galloprovincialis*)	Bulet et al. (1999), Hubert et al. (1996), Pisuttharachai et al. (2009), Gerdol et al. (2012), Vilcinskas et al. (2013), Tonk et al. (2014), Tarr (2012), and He et al. (2015)
Drosocin	Gram-negative	Insecta (*Drosophila*)	Bulet et al. (1996)
Gallerimycins	Fungi	Insecta (*M. sexta*)	He et al. (2015)
Gloverin	Gram-negative	Insecta (*Helicoverpa armigera, M. sexta*)	Mackintosh et al. (1998) and He et al. (2015)
Gomesin	Gram-positive and Gram-negative; fungi	Arachnida (*Acanthoscurria gomesiana*)	Silva et al. (2000)

(continued)

Table 7.1 (continued)

AMPs	Targets/activity	Producer	References
Homarin	*Vibrio* sp., ciliates	Decapoda (*Homarus americanus*)	Battison et al. (2008)
Hyastatin	Gram-positive and Gram-negative; yeasts	Crustacea (Decapoda: *Hyas araneus*)	Sperstad et al. (2009)
Lebocin	*E. coli*	Insecta (*M. sexta*, *Bombyx mori*)	Chowdhury et al. (1995) and He et al. (2015)
Moricin	*E. coli*, *S. aureus*	Insecta (*M. sexta*, *B. mori*)	Hara and Yamakawa (1995) and He et al. (2015)
Nemapore	Gram-positive (*B. megaterium*, *B. thuringiensis*); Gram-negative (*E. coli*)	Nematoda (*C. elegans*, *Pristionchus pacificus*; ascarids, filariae)	Hoeckendorf et al. (2012) and Tarr (2012)
Penaeidin	Gram-positive; fungi	Crustacea (*Litopenaeus setiferus*)	Cuthbertson et al. (1994)
Pyrrhocoricin	Gram-negative (*E. coli*). Gram-positive (*M. luteus*, *B. subtilis*); inhibits chaperone DnaK	Insecta (*Pyrrhocoris apterus*)	Cociancich et al. (1994)
Saposin	*E. coli*	Nematoda (*C. elegans*)	Banyai and Patthy (1998)
Scygonadin	Gram-positive (*M. luteus* *Streptococcus aureus*, *S. pyogenes*)	Crustacea (*Scylla serrata*)	Huang et al. (2006)
Strongylocins	Gram-positive, Gram-negative	Echinodermata (*Strongylocentrotus purpuratus*)	Li et al. (2010)
Stylicin	*Vibrio penaeicida*; *Fusarium oxysporum*	Crustacea (*Litopenaeus stylirostris*)	Rolland et al. (2010)
Tachyplesin	Gram-positive and negative; fungi	Crustacea (horseshoe crab, *Tachypleus tridentatus*); spiders	Iwanaga et al. (1998)
Tenecin	Gram-negative	Insecta (*Tenebrio molitor*)	Chae et al. (2012)
Thaumatins	Antifungal, suppress germination	Nematoda (*C. elegans*); Insecta (*A. pisi*)	Shatters et al. (2006) and Gerardo et al. (2010)

The HSP family members are classified in five groups depending on their size (HSP20-40, 60, 70, 90 and 110). HSP70, the most conserved and studied, has been found in invertebrates including nematodes, crustacea (including hydrothermal vent shrimps), insects, molluscs and other lineages (Ravaux et al. 2007).

The HSPs production is activated after injury and sepsis. In the malaria vector *Anopheles gambiae* several HSPs were found to be up regulated in experimental assays using heat inactivated cells of *Salmonella typhimurium* and *Staphylococcus aureus*, and were considered as involved in defense reactions (Aguilar et al. 2005).

In *C. elegans*, the hsp-60 gene was up regulated in the intestine upon infection by *Enterococcus faecalis*. Its product was considered to protect proteins by other reactive species produced by the host, sustaining their homeostasis (Mohri-Shiomi and Garsin 2008). HSP70 was also produced in the haemocytes of the bay scallop *Argopecten irradians* up to 16 h after experimental infection by *Vibrio anguillarum*, in a time-dependent expression pattern (Song et al. 2006; 2015).

Proteomic studies on the hemolymph proteins of the model species *D. melanogaster* showed that, 24 h after immune challenge with a feeding medium containing a microbial membrane suspension, the HSP70 gene was upregulated together with other immune-related proteins and antioxidant system components (De Morais Guedes et al. 2005). HSP70 was also upregulated in the midgut of *Rhodnius prolixus* (Insecta: Reduviidae), vector of the protozoan parasite *Trypanosoma cruzi* responsible of human trypanosomiasis (Chagas disease), after immunization by microinjection with *E. coli* and *M. luteus* (Ursic-Bedoya and Lowenberger 2007).

3.6 Nitric Oxide (NO) and Reactive Oxygen Species (ROS)

NO is a very ancient and highly conserved signalling and defense molecule, found in all Kingdoms. It is a highly reactive, unstable and diffusible gas, participating in basal cell metabolism, produced by the enzyme NO synthase (NOS) using as substrate L-arginine, which is oxidized to citrullin. Constitutive (neuronal or endothelial) and induced NO forms may act as neurotransmitters (in chemosensory signalling) and as defensive molecules, respectively. After infection, induced and highly toxic NO forms are rapidly produced at very high concentrations in cells by NOS, until complete depletion of their substrate.

NO was identified in brain cells of *Apis mellifera*, *D. melanogaster* and *Schistocerca gregaria* (Müller 1997). Its release was also observed in hemocytes of the horseshoe crab *L. polyphemus* (Radomski et al. 1991) and later reported from many invertebrates (Ottaviani et al. 1993; Müller 1997). Production and release of pathogens-killing NO forms in insects is part of a general, broad and non-specific humoral immune response (Rivero 2006). NO expression was reported as a defense mechanism in vector *Anopheles stephensi* mosquitoes after infection by the eukaryotic parasite *Plasmodium berghei* (Rivero 2006). NO-related genes have been also found in *Mytilus galloprovincialis* and in the genome of the purple sea urchin (Gourdon et al. 2001; Hibino et al. 2006).

The production of ROS (hydrogen peroxide, H_2O_2 and superoxide, O_2^- radicals), also known as respiratory burst, concerns the release of toxic free oxygen radicals via an oxidative metabolism, with a direct oxidative stress effect induced on target invasive cells. NO and ROS function in innate immunity including the modulation of the cytokine response of lymphocytes and the regulation of apoptosis (programmed death) in immune cells. Reactive N and O intermediates are intracellular targets for members of several signal transduction pathways (Bogdan et al. 2000).

In *D. melanogaster*, ROS and intermediate molecules are released in the gut after their pathways have been activated by a bacterial infection. This is one of the first defensive responses of the insect, in which ROS are produced by the gut epithelial cells through synthesis by the NADPH oxidase enzyme Duox, and detoxified by a catalase regulated by the immune system. Surviving bacterial cells are then controlled when PGRPs eventually detect their peptidoglycan fragments and/or elicit IMD pathways, for subsequent AMP release (Vallet-Gely et al. 2008).

ROS are produced in *C. elegans* via a NADPH oxidase in response to the Gram-positive and pathogenic *Enterococcus faecalis*, with upregulation of stress responses after exposure to the pathogen (Chávez et al. 2009). In molluscs, the production of toxic oxygen radicals was histochemically demonstrated for the hemocytes of the scallop *Patinopecten yessoensis* (Nakamura et al. 1985). ROS have been later reported from hemocytes of the snail *Lymnaea stagnalis* (Dikkeboom et al. 1987), the oyster *Crassostea virginica* (Austin and Paynter 1995), the bivalve *Pecten maximus*, as a reaction to an infection by a *Rickettsia*-like pathogen (Le Gall et al. 1991), in the mussel *Mytilus edulis* and in other molluscs (Adema et al. 1991; Pipe 1992).

ROS are also involved in the expulsion of chromatin from the nucleus, in what appears as an evolutionary ancient defense reaction of vertebrates that was recently found among invertebrates. These include the shore crab *Carcinus maenas*, the blue mussel *Mytilus edulis* and the sea anemone *Actinia equina* (Cnidaria). The chromatin extrusion process, termed ETosis, is induced by various factors like protein kinase C activator, oxidative bursts, LPS, bacteria and others. The chromatin forms meshes and networks which entrap bacteria or other pathogens, which eventually undergo encapsulation and degradation by hemocytes (Robb et al. 2014).

3.7 Lysozyme and Other Pathways

Lysozymes (also known as muramidases or N-acetylmuramide glycanhydrolases) are a large family of bacteria-degrading enzymes, contributing to the immune response in vertebrates and invertebrates. The first lysozyme was discovered by Alexander Fleming in 1921. These enzymes are grouped in four main types, all encountered among invertebrates. The c-type (also indicated as LysC, chicken or conventional-type) targets and degrades the cell wall peptidoglycans of Gram positive bacteria, by hydrolysing the bonds between the N-acetylglucosamine and the N-acetylmuramic acid residues. A minor activity against Gram negative species and fungi was also reported (Vilcinskas et al. 2013).

A second group of lysozymes, the i-type (LysI or invertebrate-type), present only in invertebrates, was initially found in the starfish *Asterias rubens*, in the bivalve *Venerupis philippinarum* (formerly *Tapes japonica*) and then in several other invertebrate lineages, including members of Anellida, Arthropoda, Echinodermata, Mollusca and Nematoda (Van Herreweghe and Michiels 2012; Bathige et al. 2013). The g-type (LysG or goose-type), originally found in eggs of the Embden goose, was later identified in other mammals, fish and amphibians. Among invertebrates, it has been found thus far only in molluscs. Finally, the ch-type (chalaropsis) was initially identified as an extracellular bacteriolytic enzyme produced by the fungus *Chalaropsis*, and targets N,6-O-diacetylmuramic acid (Callewaert and Michiels 2010).

The functional role and main activity of the different lysozymes are mainly targeted towards hydrolysis of the bacterial cell wall and eventual cell degradation. Some lysozymes, however, also exibit a chitinase activity (i.e. the *Venerupis philippinarum* lysozyme), likely due to the similarity between the β-1,4 N-acetylglucosamine chitin homopolymer bond and that between acetylmuramic acid and β-1,4 N-acetylglucosamine, forming the peptidoglycan heteropolymer of the bacterial cell wall.

Studies on Lys-I from bivalves and other invertebrates showed the presence, in the same species, of lysozyme variants with different patterns of temporal and tissue or cell expressions (i.e. digestive organs, muscles, tubules, hemocytes), suggesting a number of functional roles in the animal metabolism likely involved, apart of defense from invading bacteria, in further digestive functions (Van Herreweghe and Michiels 2012).

In *C. elegans* 15 genes, classified in three clades, code for lysozymes, the first five of the i-type. They have an affinity to the peptidoglycan of Gram-positive bacteria like *Staphylococcus aureus* and *Neisseria gonorrhoeae*, which act as virulence factors. The remaining ch-type lysozymes, also present in other *Caenorhabditis* species, are more similar to homolog genes of *Entamoeba histolytica* rather than to those of insects. Several nematode lysozymes act in the intestine as antimicrobial compounds, leading to the bacterial cell wall disruption. Cel-Lys1, 7 and 8 are also expressed in the nervous system, in the juvenile muscle cells and in the terminal pharyngeal bulb, respectively (Schulenburg and Boehnisch 2008).

After infection with pathogenic *Serratia marcescens* Db11, *C. elegans* expressed genes included those encoding lysozymes and lectins, in part controlled by a transforming growth factor-b (TGF-b)-like pathway (Mallo et al. 2002). Further pathways involved in *C. elegans* innate response to bacterial pathogens were the p38 mitogen-activated protein kinase (MAPK) pathway, the programmed cell death (PCD) involved in resistance to *S. enterica*, and the insulin-like receptor pathway, conferring resistance to Gram negative and many Gram positive pathogens. All these pathways are considered to interact in a general *C. elegans* biotic stress response (Schulenburg et al. 2008). TGF-b and MAPK pathways are also active in insects innate response, i.e. in the expression of anti-*Plasmodium* effector NOS genes in *Anopheles* (Surachetpong et al. 2009).

Four c-type and six i-type lysozyme genes were found in the invasive ladybird beetle *Harmonia axyridis*, two of which showed different functional expression patterns and body part locations (Beckert et al. 2015). Two c-type lysozymes (an acidic c-lys3 and a basic c-lys4) were expressed in *H. axyridis* after bacterial injection, boosting the insect defense response. *Harmonia axyridis* beetles show a two layered immune system (a first layer based on the low-molecular-mass antimicrobial compound harmonine and a second one based on several AMPs), considered to provide the beetle a superior advantage by underpinning its spatial spreading in new areas and colonization of new environments (Beckert et al. 2015).

Further adaptations of the innate defense reactions may also follow different patterns, as shown by the reduced response observed in *A. pisi*. After experimental piercing with bacteria-loaded needles, the aphid hemolymph showed elicited activity of lysozyme flanked by a limited production of antimicrobial compounds, indicative of a major aphid investment in the production of viviparous offsprings rather than in the activation of direct defense processes (Altincicek et al. 2008).

4 Cellular Defense Processes

4.1 Phagocytosis

The activations of one or more humoral responses, like the production of ROS or agglutinins-lectins, mobilizes the phagocytes, a group of invertebrate cells characterized by similarities to the vertebrate macrophages and neutrophils. Several cell lines, identified by different names, have been identified in invertebrate phyla, all involved in similar phagocytic functions, and a general term "immunocytes" was proposed for them (Ottaviani 2011).

In general, the immunocytes get in contact with the pathogens cells or any other non-self material by random contacts or by attractve processes, in some cases mediated by finger-like protrusions or pseudopodia (Fig. 7.1). The particles or intruders are then internalized as small vesicles called phagosomes and ultimately killed by the reactive molecules and enzymes therein present (Greenberg and Grinstein 2002). The process is rapid, and occurs in the order of minutes. In experimental observations with hemocytes of the edible ivory snail, *Babylonia areolata*, pathogenic cells of *Vibrio parahaemolyticus* were phagocytosed after 30 min *in vitro* exposure, and the process activation increased up to the followin 90 min (Di et al. 2013).

Phagocytosis is a very ancient process, frequently encountered among animals, and deriving by a primitive form of cell nourishment behaviour. The invertebrate body structural organization has been considered as a valid parameter to classify them in three major groups: the acelomates (lacking a body cavity), the pseudocelomates (with a pseudocel, a body cavity filled with fluid but deprived of own walls and cells), and the celomates, showing a body cavity (celome) with outer and inner surfaces of mesodermic origins, lined by an epithelium and

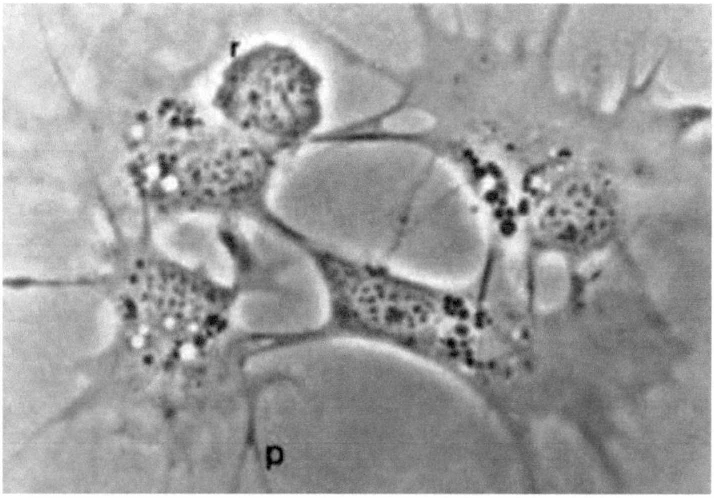

Fig. 7.1 Phagocytes of the pond snail *Lymnaea stagnalis* showing pseudopodia (p) and internal granules (Adapted from van der Knaap et al. 1993)

filled with a body fluid and cells. In some groups of the latter type (i.e. molluscs, arthropods) the fluid (also called hemolymph) is the equivalent of the blood found in some other lineages (anellids, cephalopods and echinoderms), in which a vascular system with blood vessels is also present (Ottaviani 2011).

Although common among invertebrate lineages, phagocytosis is not ubiquitous, and i.e. has been observed in *C. elegans* only as an engulfment process carried out, after a cell programmed death (apoptosis), by adiacent cells in a scavenger-recycling behaviour (Gumienny and Hengartner 2001). In invertebrates deprived of a celomic cavities (i.e. sponges, corals, anemones) the immunocytes (often also called phagocytic amebocytes) wander through their tissues devouring foreign material and particles, as well as self cell debris and derived decaying material (Bayne 1990). In celomate invertebrates, provided with a vascular system (like insects or crustaceans), phagocytosis is the final outcome of many immune responses mediated by the recognition of the bacterial cell wall or other pathogens' components. These responses include hemolymph coagulation, melanization, production of one or more AMPs or activation of lectin and complement pathways (Bayne 1990; Iwanaga and Lee 2005; Buchmann 2014). In insects like the pea aphid, showing a reduced set of immune defenses, up to five differentiated cell categories were observed in the hemolymph of adults (Laughton et al. 2011; Schmitz et al. 2012).

In vertebrates, phagocytosis is mediated by lectin-based molecular recognition mechanisms in which lectins present on the surface of phagocytes recognize and bind to sugars present on the bacterial cell surface as well as viceversa, with a third possibility given by lectins with two sugar recognizing domains or glycoconjugates, that act as bridges between the sugar residues exposed on both surfaces (Ofek and Sharon 1988). Similarly, these mechanisms play a fundamental role also in the invertebrate phagocytosis. Lectins binding to different types of hemocytes were found in the snail *Biomphalaria glabrata* (Joky et al. 1983), in the mussel *M. edulis* (Pipe 1992) and, with some differences among geographycally distant populations, in the oyster *Crassostrea virginica* (Cheng et al. 1995). However, different defensive strategies based on variations in the lectin types and/or on their occurrence was found even within the same invertebrate lineage, as shown by molluscs (Hooper et al. 2007). Several lectins present in the hemolymph of the cockroach, *Blaberus discoidalis* are responsible for the non-self recognition and eventual enhanced phagocytotic reactions, based on their sugar specificities allowing bacterial cells recognition. As per *Bombyx mori* and other insects, they enhance the activation of the ProPO cascade, followed by melanization and other defensive reactions (Wilson et al. 1999).

In *Oligochaeta* (Anellida), cytotoxic cellular reactions were observed after close contacts of celomocytes with target cells, likely mediated by cell surface glycoproteins (Cossarizza et al. 1995). Phagocytosis is followed by encapsulation by large celomocytes, showing the involvement in the process of two distinct cell types, which support hypothesis of early divergence between lytic and phagocytosic reactions (Quaglino et al. 1996; Salzet et al. 2006).

In anellids, the development of advanced structures like the celomic cavity and the vascular system, coupled to the metameric body organization, represented significant evolutive steps mirrored by an immune system characterized by more advanced defense strategies, cell diversification and organization. The two classes of cells—those proceeding from the mesoderm (celome) and those originated in the vascular system—allowed the evolution of two separated cellular lines, both fundamental in the immune response and functionally integrated. Given the structural and phylogenetic distances separating the anellids, these cell groups have been often indicated as amoebocytes (present in all three anellid classes), eleocytes (in *Polychaeta* and *Oligochaeta*), erythrocytes and hemocytes (*Polychaeta*), celomocytes, vascular lymphocytes, macrophages (*Oligochaeta*) and chloragocytes (*Hirudinea*) (Vetvicka and Sima 2009). The five subgroups of the Polychaeta show immunocytes (called amebocytes or granulocytes) involved in phagocytosis, with further additional functions like waste removal, pathogens elimination, antibacterial defense and encapsulation. Their hemoglobin-rich erythrocytes are also involved in phagocytosis.

The immune systems of the Oligochaeta has been the object of extensive cytological studies as an invertebrate immunity and defense model. It also provides an example of the complexity of the cells classification task, as cells showing different developmental stages may belong to the same type or can proceed from distinct lines, as shown for molluscs (*Mytilus* spp.), insects, anellids or sea urchins (Ottaviani 2011).

In the earthworms *Lumbricus terrestris* and *Eisenia foetida* the free cells are classified as celomocytes (amoebocytes and eleocytes) and hemocytes (blood cells) with further sub-classifications (Fig. 7.2). Some groups from both lineages are involved in immune reaction processes, like recognition of non-self and reaction to transplants, cytotoxicity, encapsulation, endocytosis with enzymatic digestion of introgressed material, wound healing and tissue regeneration. Cells identified as large celomocytes are involved in phagocytosis and encapsulation, whereas the smaller ones display a cytotoxic activity (Vetvicka and Sima 2009).

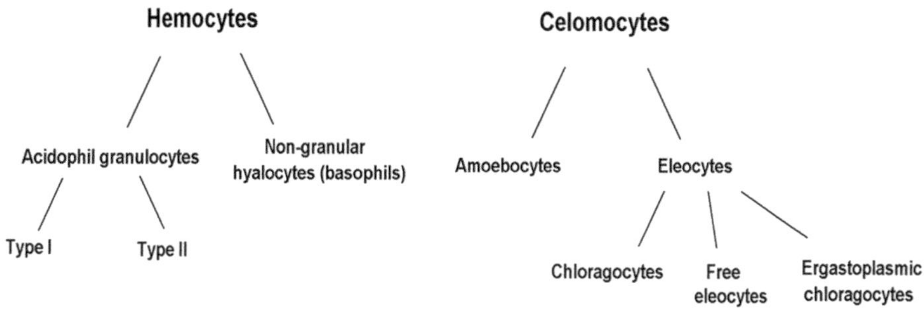

Fig. 7.2 A schematic classification of the different types of earthworm cell lines

In Hirudinea, which are characterized by the fusion of the celome with the vascular system (an evolutive process that originated the hemocel), the blood cell populations are classified as amoebocytes and chloragogens. The former (also referred to as leukocytes or lymphocytes) have a main role in phagocytosis, whereas the latter are classified as macrophage-like, NK-like and granular cells. They are involved in encapsulation (being the macrophages provided with recognition surface molecules) and bacteria phagocytosis (De Eguileor et al. 2000; Vetvicka and Sima 2009).

4.1.1 Lysosomes

Lysosome are eukaryotic cell organelles in charge of the final destruction and lysis of the material introgressed through phagocytosis. In the lysosome, a number of hydrolytic enzymes, active at low pH, decompose the introgressed materials to their basal components, for their eventual re-cycling in the cellular environment or final elimination. The strict acidic affinity and requirement of these enzymes (i.e. acid phosphatase, lysozyme, cysteine protease cathepsin C) present as zymogens, prevent the risk of an enzymatic cell degradation in case of lysozyme rupture and enzymes diffusion in the cell compartment. The lysozyme function is hence the confinement of potentially dangerous enzymes needed for the pathogens and invaders final degradation and lysis (Tryselius and Hultmark 1997; Garcia-Carreño et al. 2014).

4.2 Encapsulation and Hemolymph Coagulation

This process is a particular outcome of phagocytosis, in which the phagocytic cell is uncapable of completely engulfing the foreign material, stretching itself along its surface and inducing the arrival and aggregation of further phagocytes. The result is a cellular clotting around the foreign body or particle, which becomes completely surrounded and isolated (Bayne 1990). In many invertebrates the cellular layers are composed by dead, melanized hemocytes. Being a generic and aspecific response of cells to wounding, encapsulation and melanization also occur as part of a wound healing reaction (Lavine and Strand 2002). This was also experimentally shown by i.e. the injection of artificial bodies like nylon implants in insects or microbeads in snails (van der Knaap et al. 1993; Baer et al. 2006).

Encapsulation may also result as a reaction mediated by common invertebrate recognition pathways, flanked by specific cellular and biochemical products. In the more than 200 Myrs old defense system of the horseshoe crab, the hemocytes circulating in the hemolymph are sensitive to minimal amounts of LPS or to foreign granular components. LPS activates three serine protease zymogens (factors C, B and a proclotting enzyme) and a clottable protein (coagulogen), forming an insoluble coaguling gel that embeds the pathogens, for subsequent killing by one or more AMPs produced. Factor C is responsible for the detection of Gram-negative bacteria, whereas a further zymogen, the factor G, senses the β-(1,3) glucan of fungi. Other encapsulation mechanisms include agglutinins, present either in the hemolymph or in cells, and a cytolytic activity of the hemolymph targeting the intruder cells (Muta and Iwanaga 1996).

In *Themiste petricola*, a celomate marine worm (phylum Sipunchula), clotting is a reaction of specific celomic cells to contact with sea water. The cells aggregate forming a first clot mass that acts as a barrier, finalized to the repair of lesions and injuries of their muscular body wall. Due to their hydroskeleton and intracelomic pressure, whose preservation is critical for the worm feeding and survival, a rupture could determine the rapid loss of the worm body content and integrity. The clot also entraps bacteria and other organic particles, that later are attacked and eliminated by neighbour phagocytes (Lombardo and Blanco 2012).

Hemolymph coagulation is a general defense response in Arthropoda aiming at wound sealing and avoidance of bacteria entrance in the hemocoel. It is mediated by soluble and cell factors and actively takes part in the innate immunity (Muta and Iwanaga 1996; Theopold et al. 2004). Differing from the vertebrate blood clotting process, arthropods clotting relies on a set of protein domains differently assembled, and on the recruitment of additional proteins.

In the arthropods open circulatory system (in which hemolymph infiltrates tissues from the hemocoel), clotting is originated by microbial elicitors. In the horseshoe crab, hemolymph coagulation is initiated when even minimal amounts of LPS are detected, converting the soluble protein coagulogen into insoluble coagulin at the end of a proteolytic cascade in which LPS and β-1,3 glucan react with Factors B, C and G, with production of further small cleavage molecules which act as AMPs (Theopold et al. 2004). The clot kills bacteria in conjunction with a plasma factor. The components of the clot system are stored in different types of hemocyte granules, and released after clotting induction (Iwanaga et al. 1998). The crab coagulogen is the functionally equivalent of the vertebrate fibrinogen but evolved independently. It is instead a plasma lectin that has similarity to the fibrinogen domain, and acts as a pattern recognition protein mediating polymerization of fibrin monomers. A further process involves hemocyanin, that is converted to PO by reaction with either the clotting factor or the Factor B (Theopold et al. 2004).

In the crayfish *Pacifastacus leniusculus*, the hemolymph clotting is based on a clotting protein (a homodimer of two 200 kDa monomers linked by disulphide bridges) and a hemocyte-released transglutaminase, also found in the horseshoe crab clotting. In crayfish also the components of the proPO system are activated a few minutes after microbial polysaccharide PRR recognition, and released in the hemolymph from granules. The proPO signal is then amplified by an activated peroxinectin with a cell adhesive and opsonic function, and spreads to the adiacent cells. Crustacean clotting proteins are mainly lipoproteins, and their lipophilic lysine and glutamine residues, cross-linked through transglutaminase, contribute to its precipitation during the process. In crayfish the proteinase-scavenger α_2-macroglobulin exposes free lysine and glutamine residues, and is integrated by transminase cross-linking in the clotting, favouring its action as inhibitor of the pathogens proteinases (Hall and Söderhäll 1994; Theopold et al. 2004).

ProPO activation and transaminases are also involved in insect hemolymph clotting, in which a role in PPO activation was found for a serine protease inhibitor (serpin27A), which also regulates dorso-ventral polarity in *Drosophila* (Rushlow 2004). Further proteins like hemolectin (*Drosophila*) and hemocytin (*Bombyx mori*) are also involved in hemolymph clotting. Genomic data showed presence of fibrin-like genes, some of which induced on immune challenge, that likely evolved independently by recent gene duplication events (Theopold et al. 2004).

5 Selectivity, Specificity and Evasion

Selectivity is an obligate requirement of any defensive system. It is necessary either for the distinction of "self" cells from those of the intruders and to recognize a pathogenic species from a mutualist or from bacteria used as food. Selectivity works through a number of signals translated to cells in order to trigger a subsequent specific cascade of negative regulators or other defense-related genes. This trait and the detailed mechanisms involved are difficult to demonstrate experimentally, since a comparative scheme is needed for this purpose, in which either the pathogen and non-pathogenic bacteria must be provided to the host, in controlled conditions. Furthermore, actual knowledge on this process is still incomplete, and most informations produced mainly refer to insects and associated endosymbionts.

Tolerance and control of endosymbionts is achieved in insects by a number of mechanisms including bacterial adaptations, modulation of the host immune response or bacteriocyte compartmentalization, resulting is a sort of simplification of the host immune system and functionning (Ratzka et al. 2012). Other strategies may involve gene expression and regulation. In the light organ of the squid *Euprymna scolopes*, members of the thioester-containing proteins superfamily like Es-CD109, involved in the antimicrobial immune reaction, are down regulated or expressed at reduced levels, in order to allow colonization by the bioluminescent symbiotic bacteria which help *E. scolopes* to evade predation (Yazzie et al. 2015).

The adaptations of endosymbiotic bacteria mostly concern polymorphisms of specific receptors like the amino acids of the outer membrane protein (OmpA), as shown by *Sodalis glossinidius* present in the gut, hemolymph and phagocytes of its host, the tsetse fly *Glossina morsitans morsitans*. The differences in the exposed aminoacidic patterns allow evasion or resistance to the AMPs-based reaction of the host immune system, which is instead elicited in presence of pathogenic *E. coli* strain K12. This mechanisms was confirmed by assays carried out recombining the

OmpAs between the two bacteria (Weiss et al. 2008). Other mechanisms selectively active on the insect immune system include a sort of control or block of the host reaction, as shown by *Wolbachia* in *Drosophila simulans* Riverside. The bacterium appears to evade the host reaction after detection, likely through the expression of genes encoding ankyrin repeat (ANK) proteins that mediate protein-protein interactions (McGraw and O'Neill 2004). A suppression of the host immune reaction was also observed in *D. melanogaster* infected by *Spiroplasma* endosymbiont strains MSRO or NSRO (Ratzka et al. 2012).

Genes acting as negative regulators of the immune system response have been found to be expressed in bacteriomes, including a homolog of a mammal TLR activator and a weevil homolog of the *Drosophila pgrp-lb* gene, whose product down-regulates the immune response towards commensal gut bacteria (Zhang and Ghosh 2002; Gottar et al. 2006). Since the weevils bacteriome originated from the gut tissue, it was suggested that the mechanisms sustaining gut commensal bacteria may have been adopted to estabilish and control the bacteriome symbiosis (Anselme et al. 2008). PGRP-LB is also active in the down regulation of the immune response towards symbionts and in their control through autophagocytic processes (Ratzka et al. 2012).

Studies involving the application of RNAi showed that in the midgut of adult *Drosophila* the IMD pathway is selectively active towards the Gram-negative species susceptible to the AMPs released, thus affecting the abundance and species composition of the gut microbiota. Similarly, silencing the *duox* gene involved in ROS production increased the amounts of gut bacteria in *Aedes aegypti*. This system, however, showed different effects in other insects, i.e. displaying, in *Anopheles gambiae*, a protective effect on the gut microbiota. This effect likely occurs at the level of the gut peritrophic membrane, by reducing its permeability and thus lowering the traffic of microbial elicitors of the immune reaction and/or of effective AMPs. The Duox system was instead found to be activated in the gut of *Drosophila* in presence of uracil released by ingested pathogenic *Erwinia*, a molecule not produced by the other symbiotic bacteria (Douglas 2014).

The direct competition or the indirect negative regulation of a bacterium by an endosymbiont may also account for selective, indirect host protection from pathogens. This is shown by symbiotic *Wigglesworthia* whose presence protects its *Glossina* host from opportunistic infection by *E. coli*, an outcome achieved through the induction of effectors that are mainly active against the pathogenic bacterium (Weiss et al. 2012; Douglas 2014).

A selective capacity to discriminate between pathogenic and non pathogenic bacteria has been shown for the pea aphid *Acyrthosiphon pisum* exposed to cultured isolates of the mutualistic bacterium *Serratia symbiotica*—which provides the host with resistance traits towards heat stress and parasitoids (see Chap. 3) – or the close pathogenic species *S. marcescens* (see Chap. 4), capable of killing the host a few days after ingestion (Renoz et al. 2015). A reduced defense capacity was reported for the aphid, which lacks many genes related to an effective response towards pathogens, including AMPs and IMD pathways. The aphid is also chracterized by a low number of hemocytes, with reduced lysozyme and PO activities (Altincicek et al. 2011; Laughton et al. 2011). This association was investigated to identify mechanisms related to the different reactions to colonization by either pathogenic and symbiotic *Serratia* spp. In controlled assays Renoz et al. (2015) showed that, differing from the pathogenic *S. marcescens*, the *S. symbiotica* isolate did not elicit an immune response by the host, and that the bacterium was able to estabilish and multiply in the aphid digestive tract, aggregating in the gut without inducing any pathogenic effect. The innate immune response is regulated in aphids by Macrophage Inhibitory Factors (ApMIFs), which were not expressed in presence of facultative symbionts or during colonization, but were up-regulated during the invasion of pathogenic bacteria (Dubreuil et al. 2014).

Phagocytosis has been considered as the possible source of specificity in the innate immune response, as shown by the response of the woodlouse *Porcellio scaber* (Crustacea: Isopoda), challenged by a previously encountered bacterial strain (Roth and Kurtz 2009). This immunological priming is considered as favoured by selection in the case of repeated encounters with the same parasite, reducing the energy and cellular "investment" needed for defense. A further advantage appears related to the parasites co-evolution and adaptation to a host general defence (Schulenburg et al. 2007). Some evidence indicative of a kind of immune specificity was provided by the waterflea *Daphnia magna*, whose offsprings showed increased resistance against a specific strain of the microparasite *Pasteuria ramosa* but not against a second tested strain (Little et al. 2003) and by the species-specific immune response of the bumblebee *Bombus terrestris*, towards two closely related *Paenibacillus* spp. (Sadd and Schmid-Hempel 2006). Although the molecular mechanisms of specificity are yet only in part understood, a factor considered as fundamental in this process is the genetic diversity of the pathogen recognition receptors or defensive effectors, sustaining a simple key-lock parasite detection system. However, since the numbers achieved by these molecules do not still appear sufficient to warrant specificity, other factors like the synergistic interactions acting among receptor molecules and in different reaction steps (like recognition, signal transduction, pathogen elimination), and possibly the dosage effects associated with immunological priming, were proposed as involved in specificity (Schulenburg et al. 2007).

Differing from resistance, which implies the destruction of the invading pathogens, tolerance is the capability to withstand the attack and the direct or indirect (defense cost) damage induced, surviving and reproducing even in presence of a high pathogen cell loads. In invertebrates, tolerance requires the capability to maintain the homeostasis of major physiological functions, including the replacement of the cells (i.e. intestinal or gut cells) lost as a consequence of the pathogen's attack (Ferrandon 2009).

Since bacteria are the causal agents of many invertebrate diseases, it is evident that they developed a number of processes underpinning pathogenesis and virulence, thus counteracting the immune defenses set in place by their hosts. These evasion processes received thus far less attention than the study of the innate immune system on the host's side, including the medical implications and the vertebrate immune system. New discoveries and perspectives are hence expected by the progress in the study of the ecology and evolution of invertebrate immune systems as by the study of evasion strategies (Schmid-Hempel 2005; 2009).

References

Abad, P., et al. (2008). Genome sequence of the metazoan plant-parasitic nematode *Meloidogyne incognita*. *Nature Biotechnology, 26*, 909–915.

Adema, C. M., van der Knaap, W. P. W., & Sminia, T. (1991). Molluscan haemocyte-mediated cytotoxicity: The role of reactive oxygen intermediates. *Reviews in Aquatic Sciences, 4*, 210–223.

Aderem, A., & Ulevitch, R. (2000). Toll-like receptors in the induction of the innate immune response. *Nature, 406*, 782–787.

Aguilar, R., et al. (2005). Global gene expression analysis of *Anopheles gambiae* responses to microbial challenge. *Insect Biochemistry and Molecular Biology, 35*, 709–719.

Altincicek, B., Gross, J., & Vilcinskas, A. (2008). Wounding-mediated gene expression and accelerated viviparous reproduction of the pea aphid *Acyrthosiphon pisum*. *Insect Molecular Biology, 17*, 711–716.

Altincicek, B., Ter Braak, B., Laughton, A. M., Udekwu, K. I., & Gerardo, N. M. (2011). *Escherichia coli* K-12 pathogenicity in the pea aphid, *Acyrthosiphon pisum*, reveals reduced antibacterial defense in aphids. *Developmental & Comparative Immunology, 35*, 1091–1097.

Amparyup, P., Charoensapsri, W., & Tassanakajon, A. (2009). Two prophenoloxidases are important for the survival of *Vibrio harveyi* challenged shrimp *Penaeus monodon*. *Developmental and Comparative Immunology, 33*, 247–256.

Amparyup, P., Charoensapsri, W., & Tassanakajon, A. (2013). Prophenoloxidase system and its role in shrimp immune responses against major pathogens. *Fish Shellfish Immunology, 34*, 990–1001.

Anselme, C., et al. (2008). Identification of the weevil immune genes and their expression in the bacteriome tissue. *BMC Biology, 6*, 43.

Austin, K. A., & Paynter, K. T. (1995). Characterization of the chemiluminescence measured in hemocytes of the eastern oyster, *Crassostrea virginica*. *Journal of Experimental Zoology, 273*, 461–471.

Baer, B., Armitage, S. A. O., & Boomsma, J. J. (2006). Sperm storage induces an immunity cost in ants. *Nature, 441*, 872–875.

Bailey, M., Christoforidou, Z., & Lewis, M. (2013). Evolution of immune systems: Specificity and autoreactivity. *Autoimmunity Reviews, 12*, 643–647.

Banyai, L., & Patthy, L. (1998). Amoebapore homologs of *Caenorhabditis elegans*. *Biochimica et Biophysica Acta, 1429*, 259–264.

Barillas-Mury, C., Han, Y. S., Seeley, D., & Kafatos, F. C. (1999). *Anopheles gambiae* Ag-STAT, a new insect member of the STAT family, is activated in response to bacterial infection. *The EMBO Journal, 18*, 959–967.

Bathige, S. D. N. K., et al. (2013). A bifunctional invertebrate-type lysozyme from the disk abalone, *Haliotis discus discus*: Genome organization, transcriptional profiling and biological activities of recombinant protein. *Developmental and Comparative Immunology, 41*, 282–294.

Battison, A. L., Summerfield, R., & Patrzykat, A. (2008). Isolation and characterisation of two antimicrobial peptides from haemocytes of the American lobster *Homarus americanus*. *Fish and Shellfish Immunology, 25*, 181–187.

Bayne, C. J. (1990). Phagocytosis and non-self recognition in invertebrates. *Bioscience, 40*, 723–731.

Beckert, A., et al. (2015). Two c-type lysozymes boost the innate immune system of the invasive ladybird *Harmonia axyridis*. *Developmental and Comparative Immunology, 49*, 303–312.

Beschin, A., et al. (1998). Identification and cloning of a glucan- and lipopolysaccharide- binding protein from *Eisenia foetida* earthworm involved in the activation of prophenoloxidase cascade. *Journal of Biological Chemistry, 273*, 24948–24954.

Bogdan, C., Rollinghoff, M., & Diefenbach, A. (2000). Reactive oxygen and reactive nitrogen intermediates in innate and specific immunity. *Current Opinion in Immunology, 12*, 64–76.

Boman, H. G., Nilsson, I., & Rasmuson, B. (1972). Inducible antibacterial defence system in *Drosophila. Nature, 237*, 232–235.
Botella, J. A., Baines, I. A., Williams, D. D., Goberdhan, D. C. I., Proud, C. G., & Wilson, C. (2001). The Drosophila cell shape regulator c-Jun N-terminal kinase also functions as a stress-activated protein kinase. *Insect Biochemistry and Molecular Biology, 31*, 839–847.
Boucias, D. G., & Pendland, J. C. (1998). *Principles of insect pathology*. New York: Springer. 550 pp.
Buchmann, K. (2014). Evolution of innate immunity: Clues from invertebrates via fish to mammals. *Frontiers in Immunology, 5*, 459.
Bulet, P., Urge, L., Ohresser, S., Hetru, C., & Otvos, L., Jr. (1996). Enlarged scale chemical synthesis and range of activity of drosocin, an O-glycosylated antibacterial peptide of *Drosophila. European Journal of Biochemistry, 238*, 64–69.
Bulet, P., Hetru, C., Dimarcq, J. L., & Hoffmann, D. (1999). Antimicrobial peptides in insects; structure and function. *Developmental and Comparative Immunology, 23*, 329–344.
Bulet, P., Stocklin, R., & Menin, L. (2004). Antimicrobial peptides: From invertebrates to vertebrates. *Immunological Reviews, 198*, 169e184.
Bulgheresi, S., et al. (2006). A new C-type lectin similar to the human immunoreceptor DC-SIGN mediates symbiont acquisition by a marine nematode. *Applied and Environmental Microbiology, 72*, 2950–2956.
Callewaert, L., & Michiels, C. W. (2010). Lysozymes in the animal kingdom. *Journal of Biosciences, 35*, 127–160.
Casteels, P., Ampe, C., Jacobs, F., Vaeck, M., & Tempst, P. (1989). Apidaecins: Antibacterial peptides from honeybees. *EMBO Journal, 8*, 2387–2391.
Cerenius, L., et al. (1994). Structure and biological activity of a 1,3-b-D-glucan-binding protein in crustacean blood. *Journal of Biological Chemistry, 269*, 29462–29467.
Cerenius, L., & Söderhäll, K. (2004). The prophenoloxidase-activating system in invertebrates. *Immunological Reviews, 198*, 116–126.
Charoensapsri, W., Amparyup, P., Suriyachan, C., & Tassanakajon, A. (2014). Melanization reaction products of shrimp display antimicrobial properties against their major bacterial and fungal pathogens. *Developmental and Comparative Immunology, 47*, 150–159.
Chávez, V., Mohri-Shiomi, A., & Garsin, D. A. (2009). Ce-Duox1/BLI-3 generates reactive oxygen species as a protective innate immune mechanism in *Caenorhabditis elegans. Infection and Immunity, 77*, 4983–4989.
Cheng, T. C., Manzi, J. J., & Burrell, V. G. (1995). Differences in lectin binding by haemocytes of oysters (*Crassostrea virginica*) from three regions and further evidence for the correlation between the presence of lathyrose and the absence of *Haplosporidium nelsoni. Journal of Shellfish Research, 14*, 477e81.
Chowdhury, S., et al. (1995). cDNA cloning and gene expression of lebocin, a novel member of antibacterial peptides from the silkworm, *Bombyx mory. Biochemical and Biophysical Research Communications, 214*, 271–278.
Christophides, G. K., Vlachou, D., & Kafatos, F. C. (2004). Comparative and functional genomics of the innate immune system in the malaria vector *Anopheles gambiae. Immunology Reviews, 198*, 127–148.
Cociancich, S., et al. (1994). Novel inducible antibacterial peptides from a hemipteran insect, the sap-sucking bug *Pyrrhocoris apterus. Biochemical Journal, 300*(2), 567–575.
Cossarizza, A., et al. (1995). Mitochondrial mass and membrane potential in coelomocytes from the earthworm *eisenia foetida*: Studies with fluorescent probes in single intact cells. *Biochemical and Biophysical Research Communications, 214*, 503–510.
Cuthbertson, B. J., Bullesbach, E. E., Fievet, J., Bachere, E., & Gross, P. S. (2004). A new class (penaeidin class 4) of antimicrobial peptides from the Atlantic white shrimp (*Litopenaeus setiferus*) exhibits target specificity and an independent proline- rich-domain function. *Biochemical Journal, 381*, 79–86.

De Eguileor, M., et al. (2000). Different types of response against foreign antigens by leech leukocytes. *Tissue and Cell, 32*, 40–48.
De Morais Guedes, S., et al. (2005). Proteomics of immune-challenged *Drosophila melanogaster* larvae hemolymph. *Biochemical and Biophysical Research Communications, 328*, 106–115.
Di, G., Zhang, Z., & Ke, C. (2013). Phagocytosis and respiratory burst activity of haemocytes from the ivory snail, *Babylonia areolata. Fish & Shellfish Immunology, 35*, 366e374.
Dikkeboom, R., Tijnagel, J. M. G. H., Mulder, E. C., & van der Knaap, W. P. W. (1987). Hemocytes of the pond snail *Lymnaea stagnalis* generate reactive forms of oxygen. *Journal of Invertebrate Pathology, 49*, 321–331.
Dishaw, L. J., Smith, S. L., & Bigger, C. H. (2005). Characterization of a C3-like cDNA in a coral: Phylogenetic implications. *Immunogenetics, 57*, 535–548.
Dong, Y., Taylor, H. E., & Dimopoulos, G. (2006). AgDscam, a hypervariable immunoglobulin domain-containing receptor of the *Anopheles gambiae* innate immune system. *PLoS Biology, 4*, e229.
Douglas, A. E. (2014). The molecular basis of bacterial–insect symbiosis. *Journal of Molecular Biology, 426*, 3830–3837.
Dubreuil, G., Deleury, E., Crochard, D., Simon, J. C., & Coustau, C. (2014). Diversification of MIF immune regulators in aphids: Link with agonistic and antagonistic interactions. *BMC Genomics, 15*, 762.
Evans, J. D., et al. (2006). Immune pathways and defence mechanisms in honey bees *Apis mellifera. Insect Molecular Biology, 15*, 645–656.
Fagutao, F. F., et al. (2009). Increased bacterial load in shrimp hemolymph in the absence of prophenoloxidase. *FEBS Journal, 276*, 5298–5306.
Ferrandon, D. (2009). Host tolerance versus resistance and microbial virulence in the host-pathogen equation. *Cell Host & Microbe, 6*, 203–205.
Ferrandon, D., Imler, J. L., Hetru, C., & Hoffmann, J. A. (2007). The *Drosophila* systemic immune response: Sensing and signalling during bacterial and fungal infections. *Nature Reviews Immunology, 7*, 862–874.
Fujita, T. (2002). Evolution of the lectin-complement pathway and its role in innate immunity. *Nature Reviews Immunology, 2*, 346–353.
Ganji, S., Jenkins, J. N., & Wubben, M. J. (2013). Molecular characterization of the reniform nematode C-type lectin gene family reveals a likely role in mitigating environmental stresses during plant parasitism. *Gene, 537*, 269–278.
Garcia-Carreño, F. L., Navarrete del Toro, M. A., & Muhlia-Almazan, A. (2014). The role of lysosomal cysteine proteases in crustacean immune response. *Invertebrate Survival Journal, 11*, 109–118.
Gerardo, N. M., et al. (2010). Immunity and other defenses in pea aphids, *Acyrthosiphon pisum. Genome Biology, 11*, R21.
Gerdol, M., De Moro, G., Manfrin, C., Venier, P., & Pallavicini, A. (2012). Big defensins and mytimacins, new AMP families of the Mediterranean mussel *Mytilus galloprovincialis. Developmental and Comparative Immunology, 36*, 390–399.
Gottar, M., et al. (2006). Dual detection of fungal infections in *Drosophila* via recognition of glucans and sensing of virulence factors. *Cell, 127*, 1425–1437.
Gourdon, I. L., Guerin, M. C., Torreilles, J., & Roch, P. (2001). Nitric oxide generation by hemocytes of the mussel Mytilus galloprovincialis. *Nitric Oxide, 5*, 1–6.
Greenberg, S., & Grinstein, S. (2002). Phagocytosis and innate immunity. *Current Opinions in Immunology, 14*, 136–145.
Gross, P. S., Bartlett, T. C., Browdy, C. L., Chapman, R. W., & Warr, G. W. (2001). Immune gene discovery by expressed sequence tag analysis of hemocytes and hepatopancreas in the Pacific white shrimp, Litopenaeus vannamei, and the Atlantic white shrimp, L. setiferus. *Developmental and Comparative Immunology, 25*, 565–577.
Grubhoffer, L., Kovář, V., & Rudenko, N. (2004). Tick lectins: Structural and functional properties. *Parasitology, 129*, S113–S125.

Gumienny, T. L., & Hengartner, M. O. (2001). How the worm removes corpses: The nematode *C. elegans* as a model system to study engulfment. *Cell Death and Differentiation, 8*, 564–568.

Hall, M., & Söderhäll, K. (1994). Crayfish α-macroglobulin as a substrate for transglutaminases. *Comparative Biochemistry and Physiology Part B: Comparative Biochemistry, 108*, 65–72.

Hara, S., & Yamakawa, M. (1995). Moricin, a novel type of antibacterial peptide isolated from the silkworm, *Bombyx mori. Journal of Biological Chemistry, 270*, 29923–29927.

Harcus, Y., et al. (2009). C-type lectins from the nematode parasites *Heligmosomoides polygyrus* and *Nippostrongylus brasiliensis. Parasitology International, 58*, 461–470.

He, Y., et al. (2015). A genome-wide analysis of antimicrobial effector genes and their transcription patterns in *Manduca sexta. Insect Biochemistry and Molecular Biology, 63*, 23–37.

Herbinière, J., et al. (2005). Armadillidin: A novel glycine-rich antibacterial peptide directed against gram-positive bacteria in the woodlouse *Armadillidium vulgare* (Terrestrial Isopod, Crustacean). *Developmental and Comparative Immunology, 29*, 489–499.

Hibino, T., et al. (2006). The immune gene repertoire encoded in the purple sea urchin genome. *Developmental Biology, 300*, 349–365.

Hoeckendorf, A., Stanisak, M., & Leippe, M. (2012). The saposin-like protein SPP-12 is an antimicrobial polypeptide in pharyngeal neurons of *Caenorhabditis elegans* and participates in defence against a natural bacterial pathogen. *Biochemical Journal, 445*, 205–212.

Hoffmann, J. A., Kafatos, F. C., Janeway, C. A., & Ezekowitz, R. A. (1999). Phylogenetic perspectives in innate immunity. *Science, 284*, 1313–1318.

Hooper, C., Day, R., Slocombe, R., Handlinger, J., & Benkendorff, K. (2007). Stress and immune responses in abalone: Limitations in current knowledge and investigative methods based on other models. *Fish & Shellfish Immunology, 22*, 363e379.

Hu, C., & Aksoy, S. (2006). Innate immune responses regulate trypanosome parasite infection of the tsetse fly *Glossina morsitans morsitans. Molecular Microbiology, 60*, 1194–1204.

Hubert, F., Noël, T., & Roch, P. (1996). A member of the arthropod defensin family from edible Mediterranean mussels (*Mytilus galloprovincialis*). *European Journal of Biochemistry, 240*, 302–306.

Hultmark, D., Steiner, H., Rasmuson, T., & Boman, H. G. (1980). Insect immunity. Purification and properties of three inducible bactericidal proteins from hemolymph of immunized pupae of Hyalophora cecropia. *European Journal of Biochemistry, 106*, 7–16.

Hultmark, D., et al. (1983). Insect immunity. Attacins, a family of antibacterial proteins from *Hyalophora cecropia. The EMBO Journal, 2*, 571–576.

Imjongjirak, C., Amparyup, P., Tassanakajon, A., & Sittipraneed, S. (2007). Anti-lipopolysaccharide factor (ALF) of mud crab Scylla paramamosain: Molecular cloning, genomic organization and the antimicrobial activity of its synthetic LPS binding domain. *Molecular Immunology, 44*, 3195–3203.

Imler, J. L., & Hoffmann, J. A. (2000). Signaling mechanisms in the antimicrobial host defense of Drosophila. *Current Opinion in Microbiology, 3*, 16–22.

Iwanaga, S. (2007). Biochemical principle of *Limulus* test for detecting bacterial endotoxins. *Proceedings of the Japan Academy Series B Physical and Biological Sciences, 83*, 110–119.

Iwanaga, S., & Lee, B. L. (2005). Recent advances in the innate immunity of invertebrate animals. *Journal of Biochemistry and Molecular Biology, 38*, 128–150.

Iwanaga, S., Kawabata, S., & Muta, T. (1998). New types of clotting factors and defense molecules found in horseshoe crab hemolymph: Their structures and functions. *Journal of Biochemistry, 123*, 1–15.

Jearaphunt, M., et al. (2015). Shrimp serine proteinase homologues PmMasSPH-1 and -2 play a role in the activation of the prophenoloxidase system. *PLoS ONE, 10*, e0121073.

Jiravanichpaisal, P., Lee, B. L., & Söderhäll, K. (2006). Cell-mediated immunity in arthropods: Hematopoiesis, coagulation, melanization and opsonization. *Immunobiology, 211*, 213–236.

Jiravanichpaisal, P., Lee, S. Y., Kim, Y. A., Andren, T., & Söderhäll, I. (2007). Antibacterial peptides in hemocytes and hematopoietic tissue from freshwater crayfish *Pacifastacus leniusculus*: Characterization and expression pattern. *Developmental and Comparative Immunology, 31*, 441–455.

Johansson, M. W., Keyser, P., & Söderhäll, K. (1994). Purification and cDNA cloning of a four-domain Kazal proteinase inhibitor from crayfish blood cells. *European Journal of Biochemistry, 223*, 389–394.

Joky, A., Matricon-Gondran, M., & Benex, J. (1983). Fine structural differences in the amoebocytes of *Biomphalaria glabrata. Developmental and Comparative Immunology, 7*, 669–672.

Kanost, M. R., & Gorman, M. J. (2008). Phenoloxidase in insect immunity. In N. E. Beckage (Ed.), *Insect immunology* (pp. 69–96). San Diego: Academic.

Khoo, L., Robinette, D. W., & Noga, E. J. (1999). Callinectin, an antibacterial peptide from blue crab, *Callinectes sapidus*, hemocytes. *Marine Biotechnology, 1*, 44–51.

Kleino, A., & Silverman, N. (2014). The Drosophila IMD pathway in the activation of the humoral immune response. *Developmental and Comparative Immunology, 42*, 25–35.

Krutzik, S. R., Sieling, P. A., & Modlin, R. L. (2001). The role of Toll-like receptors in host defense against microbial infection. *Current Opinions in Immunology, 13*, 104–108.

Laughton, A. M., Garcia, J. R., Altincicek, B., Strand, M. R., & Gerardo, N. M. (2011). Characterisation of immune responses in the pea aphid, *Acyrthosiphon pisum. Journal of Insect Physiology, 57*, 830–839.

Lavine, M. D., & Strand, M. R. (2002). Insect hemocytes and their role in immunity. *Insect Biochemistry and Molecular Biology, 32*, 1295–1309.

Le Gall, G., Bachère, E., & Mialhe, E. (1991). Chemiluminescence analysis of the activity of *Pecten maximus* hemocytes stimulated with zymosan and host-specifie Rickettsiales-like organisms. *Diseases of Aquatic Organisms, 11*, 181–186.

Lee, S. Y., Wang, R., & Soderhall, K. (2000). A lipopolysaccharide-and beta-1,3-glucan-binding protein from hemocytes of the freshwater crayfish Pacifastacus leniusculus. Purification, characterization, and cDNA cloning. *Journal of Biological Chemistry, 275*, 1337–1343.

Lee, S. Y., & Söderhäll, K. (2001). Characterization of a pattern recognition protein, a masquerade-like protein, in the freshwater crayfish *Pacifastacus leniusculus. Journal of Immunology, 166*, 7319–7326.

Lee, S. Y., Lee, B. L., & Soderhall, K. (2003). Processing of an antibacterial peptide from hemocyanin of the freshwater crayfish *Pacifastacus leniusculus. Journal of Biological Chemistry, 278*, 7927–7933.

Lemaitre, B., & Hoffmann, J. (2007). The host defense of *Drosophila melanogaster. Annual Review of Immunology, 25*, 697–743.

Lemaitre, B., Nicolas, E., Michaut, L., Reichhart, J. M., & Hoffmann, J. A. (1996). The dorsoventral regulatory gene cassette spatzle/Toll/cactus controls the potent antifungal response in *Drosophila* adults. *Cell, 86*, 973–983.

Li, C., et al. (2008). Molecular cloning, genomic organization and functional analysis of an anti-lipopolysaccharide factor from Chinese mitten crab Eriocheir sinensis. *Developmental and Comparative Immunology, 32*, 784–794.

Li, C., Blencke, H. M., Smith, L. C., Karp, M. T., & Stensvåg, K. (2010). Two recombinant peptides, SpStrongylocins 1 and 2, from *Strongylocentrotus purpuratus*, show antimicrobial activity against Gram-positive and Gram-negative bacteria. *Developmental and Comparative Immunology, 34*, 286–292.

Li, M., et al. (2014). Identification of a C-type lectin with antiviral and antibacterial activity from pacific white shrimp *Litopenaeus vannamei. Developmental & Comparative Immunology, 46*, 231–240.

Little, T. J., O'Connor, B., Colegrave, N., Watt, K., & Read, A. F. (2003). Maternal transfer of strain-specific immunity in an invertebrate. *Current Biology, 13*, 489–492.

Liu, H., et al. (2007). Phenoloxidase is an important component of the defense against *Aeromonas hydrophila* infection in a crustacean, *Pacifastacus leniusculus. Journal of Biological Chemistry, 282*, 33593–33598.

Liu, H., et al. (2011). Peptidoglycan activation of the proPO-system without a peptidoglycan receptor protein (PGRP)? *Developmental and Comparative Immunology, 35*, 51–61.

Lombardo, T., & Blanco, G. A. (2012). Clot formation in the sipunculid worm *Themiste petricola*: A haemostatic and immune cellular response. *International Journal of Cell Biology, 2012*, 280675.

Lorenzini, D. M., da Silva, P. I., Jr., Fogaça, A. C., Bulet, P., & Daffre, S. (2003). Acanthoscurrin: A novel glycine-rich antimicrobial peptide constitutively expressed in the hemocytes of the spider *Acanthoscurria gomesiana. Developmental and Comparative Immunology, 27*, 781–791.

Mackintosh, J. A., et al. (1998). A gloverin-like antibacterial protein is synthesized in *Helicoverpa armigera* following bacterial challenge. *Developmental and Comparative Immunology, 22*, 387–399.

Maizels, R. M., Schabussova, I., Callister, D. M., & Nicoll, G. (2006). Molecular biology and immunology of *Toxocara canis*. In C. V. Holland & H. V. Smith (Eds.), *Toxocara: The enigmatic parasite* (pp. 3–17). Wallingford: CAB International.

Mallo, G. V., et al. (2002). Inducible antibacterial defence system in *C. elegans. Current Biology, 12*, 1209–1214.

McGraw, E. A., & O'Neill, S. L. (2004). *Wolbachia pipientis*: Intracellular infection and pathogenesis in *Drosophila. Current Opinions in Microbiology, 7*, 67–70.

McVeigh, P., et al. (2008). Neuropeptide-like protein diversity in phylum Nematoda. *International Journal of Parasitology, 38*, 1493–1503.

Medzhitov, R., & Janeway, C. (2000). Innate immune recognition: Mechanisms and pathways. *Immunological Reviews, 173*, 89–97.

Mohri-Shiomi, A., & Garsin, D. A. (2008). Insulin signaling and the heat shock response modulate protein homeostasis in the *Caenorhabditis elegans* intestine during infection. *Journal of Biological Chemistry, 283*, 194–201.

Müller, U. (1997). The nitric oxide system in insects. *Progress in Neurobiology, 51*, 363–381.

Muta, T., & Iwanaga, S. (1996). The role of hemolymph coagulation in innate immunity. *Current Opinions in Immunology, 8*, 41–47.

Mydlarz, L. D., & Palmer, C. V. (2011). The presence of multiple phenoloxidases in Caribbean reef-building corals. *Comparative Biochemistry and Physiology A, 159*, 372–378.

Nakamura, M., Mori, K., Inooka, S., & Nomura, T. (1985). *In vitro* production of hydrogen peroxide by the amoebocytes of the scallop, *Patinopecten yessoensis* (Jay). *Developmental and Comparative Immunology, 9*, 407–417.

Nappi, A. J., & Christensen, B. M. (2005). Melanogenesis and associated cytotoxic reactions: Applications to insect innate immunity. *Insect Biochemistry and Molecular Biology, 35*, 443–459.

Novoa, B., Roch, P., Figueras, A., & Pallavicini, A. (2011). Insights into the innate immunity of the Mediterranean mussel *Mytilus galloprovincialis. Genomics, 12*, 69–88.

O'Rourke, D., Baban, D., Demidova, M., Mott, R., & Hodgkin, J. (2006). Genomic clusters, putative pathogen recognition molecules, and antimicrobial genes are induced by infection of *C. elegans* with *M. nematophilum. Genome Research, 16*, 1005–1016.

Ofek, I., & Sharon, N. (1988). Lectinophagocytosis: A molecular mechanism of recognition between cell surface sugars and lectin in the phagocytosis of bacteria. *Infection and Immunity, 56*, 539–547.

Ottaviani, E. (2011). Immunocyte: The invertebrate counterpart of the vertebrate macrophage. *Invertebrate Survival Journal, 8*, 1–4.

Ottaviani, E., Paemen, L. R., & Stefano, G. B. (1993). Evidence for nitric oxide production and utilization as a bactericidal agent by invertebrate immunocytes. *European Journal of Pharmacology, 248*, 319–324.

Paulsen, V. S., et al. (2013). Structure-activity relationships of the antimicrobial peptide Arasin 1—and mode of action studies of the N-Terminal, proline-rich region. *PLoS ONE, 8*, e53326.

Peschel, A., & Sahl, H. G. (2006). The co-evolution of host cationic antimicrobial peptides and microbial resistance. *Nature Reviews Microbiology, 4*, 529–536.
Pipe, R. K. (1992). Generation of reactive oxygen metabolites by the haemocytes of the mussel *Mytilus edulis. Developmental and Comparative Immunology, 16*, 111–122.
Porchet-Henneré, E., & Vernet, G. (1992). Cellular immunity in an annelid (*Nereis diversicolor*, Polychaeta): Production of melanin by a subpopulation of granulocytes. *Cell and Tissue Research, 269*, 167–174.
Quaglino, et al. (1996). Earthworm coelomocytes in vitro: Cellular features and "granuloma" formation during cytotoxic activity against the mammalian tumor cell target K562. *European Journal of Cell Biology, 70*, 278–288.
Radomski, M. W., Martin, J. F., & Moncada, S. (1991). Synthesis of nitric oxide by haemocytes of the American borseshoe crab (*Limulus polyphemus*). *Philosophical Transactions of the Royal Society of London (Biology), 334*, 129–133.
Ratzka, C., Gross, R., & Feldhaar, H. (2012). Endosymbiont tolerance and control within insect hosts. *Insects, 3*, 553–572.
Ravaux, J., et al. (2007). First hsp70 from two hydrothermal vent shrimps, mirocaris fortunata and rimicaris exoculata: Characterization and sequence analysis. *Gene, 386*, 162–172.
Reddy, K. V., Yedery, R. D., & Aranha, C. (2004). Antimicrobial peptides: Premises and promises. *International Journal of Antimicrobial Agents, 24*, 536e547.
Ren, Q., Li, M., Du, J., Zhang, C. Y., & Wang, W. (2012). Immune response of four dual-CRD C-type lectins to microbial challenges in giant freshwater prawn *Macrobrachium rosenbergii. Fish & Shellfish Immunology, 33*, 155–167.
Renoz, F., Noël, C., Errachid, A., Foray, V., & Hance, T. (2015). Infection dynamic of symbiotic bacteria in the pea aphid *Acyrthosiphon pisum* gut and host immune response at the early steps in the infection process. *PLoS ONE, 10*, e0122099.
Renwrantz, L. (1983). Involvement of agglutinins (lectins) in invertebrate defense reactions: The immuno-biological importance of carbohydrate-specific binding molecules. *Developmental and Comparative Immunology, 7*, 603–608.
Rinkevich, B., & Müller, W. E. G. (Eds.). (1996). *Invertebrate immunology*. Dordrecht: Springer.
Rivero, A. (2006). Nitric oxide: An antiparasitic molecule of invertebrates. *Trends in Parasitology, 22*, 219–225.
Robb, C. T., Dyrynda, E. A., Gray, R. D., Rossi, A. G., & Smith, V. J. (2014). Invertebrate extracellular phagocyte traps show that chromatin is an ancient defence weapon. *Nature Communications, 5*, 4627.
Roch, P., Canicattì, C., & Sammarco, S. (1992). *Tetrameric* structure of the active phenoloxidase evidenced in the coelomocytes of the echinoderm *Holothuria tubulosa. Comparative Biochemistry and Physiology, 102B*, 349–355.
Roeder, T., et al. (2010). Caenopores are antimicrobial peptides in the nematode *Caenorhabditis elegans* instrumental in nutrition and immunity. *Developmental and Comparative Immunology, 34*, 203–209.
Rolland, J. L., et al. (2010). Stylicins, a new family of antimicrobial peptides from the Pacific blue shrimp *Litopenaeus stylirostris. Molecular Immunology, 47*, 1269–1277.
Rosa, R. D., & Barracco, M. A. (2010). Antimicrobial peptides in crustaceans. *Invertebrate Survival Journal, 7*, 262–284.
Roth, O., & Kurtz, J. (2009). Phagocytosis mediates specificity in the immune defence of an invertebrate, the woodlouse *Porcellio scaber* (Crustacea: Isopoda). *Developmental and Comparative Immunology, 33*, 1151–1155.
Rushlow, C. (2004). Dorsoventral patterning: A serpin pinned down at last. *Current Biology, 14*, R16–R18.
Sadd, B. M., & Schmid-Hempel, P. (2006). Insect immunity shows specificity in protection upon secondary pathogen exposure. *Current Biology, 16*, 1206–1210.
Salzet, M., Tasiemski, A., & Cooper, E. (2006). Innate immunity in lophotrochozoans: The annelids. *Current Pharmaceutical Design, 12*, 3043–3050.

Schmid-Hempel, P. (2005). Evolutionary ecology of insect immune defenses. *Annual Review of Entomology, 50*, 529–551.

Schmid-Hempel, P. (2009). Immune defence, parasite evasion strategies and their relevance for 'macroscopic phenomena' such as virulence. *Philosophical Transactions of the Royal Society of London B Biological Sciences, 364*(1513), 85–98.

Schmitz, A., et al. (2012). The cellular immune response of the pea aphid to foreign intrusion and symbiotic challenge. *PLoS ONE, 7*, e42114.

Schnapp, D., Kemp, G. D., & Smith, V. J. (1996). Purification and characterization of a proline-rich antibacterial peptide, with sequence similarity to bactenecin-7, from the haemocytes of the shore crab, *Carcinus maenas. European Journal of Biochemistry, 240*, 532–539.

Schulenburg, H., & Boehnisch, C. (2008). Diversification and adaptive sequence evolution of *Caenorhabditis* lysozymes (Nematoda: Rhabditidae). *BMC Evolutionary Biology, 8*, 114.

Schulenburg, H., Kurz, C. L., & Ewbank, J. J. (2004). Evolution of the innate immunesystem: The worm perspective. *Immunological Reviews, 198*, 36–58.

Schulenburg, H., Boehnisch, C., & Michiels, N. K. (2007). How do invertebrates generate a highly specific innate immune response? *Molecular Immunology, 44*, 3338–3344.

Schulenburg, H., Hoeppner, M. P., Weiner, J., 3rd, & Bornberg-Bauer, E. (2008). Specificity of the innate immune system and diversity of C-type lectin domain (CTLD) proteins in the nematode *Caenorhabditis elegans. Immunobiology, 213*, 237–250.

Shatters, R. G., Boykin, L. M., Lapointe, S. L., Hunter, W. B., & Weathersbee, A. A. (2006). Phylogenetic and structural relationships of the PR5 gene family reveal an ancient multigene family conserved in plants and select animal taxa. *Journal of Molecular Evolution, 63*, 12–29.

Shi, X. Z., Zhao, X. F., & Wang, J. X. (2014). A new type antimicrobial peptide astacidin functions in antibacterial immune response in red swamp crayfish *Procambarus clarkii. Developmental and Comparative Immunology, 43*, 121–128.

Silva, P. I., Jr., Daffre, S., & Bulet, P. (2000). Isolation and characterization of gomesin, an 18-residue cysteine-rich defense peptide from the spider *Acanthoscurria gomesiana* hemocytes with sequence similarities to horseshoe crab antimicrobial peptides of the tachyplesin family. *Journal of Biological Chemistry, 275*, 33464–33470.

Smith, L. C., et al. (2006). The sea urchin immune system. *Invertebrate Survival Journal, 3*, 25–39.

Solano, F. (2014). Melanins: Skin pigments and much more—types, structural models, biological functions, and formation routes. *New Journal of Science, 2014*, 498276.

Song, L., Wu, L., Ni, D., Chang, Y., Xu, W., & Xing, K. (2006). The cDNA cloning and mRNA expression of heat shock protein 70 gene in the haemocytes of bay scallop (Argopecten irradians, Lamarck 1819) responding to bacteria challenge and naphthalin stress. *Fish & Shellfish Immunology, 21*, 335–345.

Song, L., Wang, L., Zhang, H., & Wang, M. (2015). The immune system and its modulation mechanism in scallop. *Fish & Shellfish Immunology, 46*, 65–78.

Sperstad, S. V., Haug, T., Vasskog, T., & Stensvåg, K. (2009). Hyastatin, a glycine-rich multi-domain antimicrobial peptide isolated from the spider crab (*Hyas araneus*) hemocytes. *Molecular Immunology, 46*, 2604–2612.

Sperstad, S. V., Haug, T., Blencke, H. M., Styrvold, O. B., Li, C., & Stensvåg, K. (2011). Antimicrobial peptides from marine invertebrates: Challenges and perspectives in marine antimicrobial peptide discovery. *Biotechnology Advances, 29*, 519–530.

Stensvåg, K., et al. (2008). Arasin 1, a proline- arginine-rich antimicrobial peptide isolated from the spider crab, *Hyas araneus. Developmental & Comparative Immunology, 32*, 275–285.

Sugumaran, M. (2002). Comparative biochemistry of eumelanogenesis and the protective roles of phenoloxidase and melanin in insects. *Pigment Cell Research, 15*, 2–9.

Supungul, P., Klinbunga, S., Pichyangkura, R., Jitrapakdee, S., Hirono, I., Aoki, T., et al. (2002). Identification of immune-related genes in hemocytes of black tiger shrimp (Penaeus monodon). *Marine Biotechnology, 4*, 487–494.

Surachetpong, W., Singh, N., Cheung, K. W., & Luckhart, S. (2009). MAPK ERK signaling regulates the TGF-β1-dependent mosquito response to *Plasmodium falciparum. PLoS Pathogens, 5*, e1000366.

Tanaka, S., Nakamura, T., Morita, T., & Iwanaga, S. (1982). Limulus anti-LPS factor: An anticoagulant which inhibits the endotoxin-mediated activation of Limulus coagulation system. *Biochemical and Biophysical Research Communications, 105*, 717–723.

Tanji, T., Hu, X., Weber, A. N. R., & Ip, Y. T. (2007). Toll and IMD pathways synergistically activate an innate immune response in *Drosophila melanogaster. Molecular and Cellular Biology, 27*, 4578–4588.

Tarr, D. E. K. (2012). Distribution and characteristics of ABFs, cecropins, nemapores, and lysozymes in nematodes. *Developmental and Comparative Immunology, 36*, 502–520.

Tassanakajon, A., Somboonwiwat, K., & Amparyup, P. (2015). Sequence diversity and evolution of antimicrobial peptides in invertebrates. *Developmental and Comparative Immunology, 48*, 324–341.

Theopold, U., Schmidt, O., Söderhäll, K., & Dushay, M. S. (2004). Coagulation in arthropods: Defense, wound closure and healing. *Trends in Immunology, 25*, 289–294.

Thomas, S., Karnik, S., Barai, R. S., Jayaraman, V. K., & Idicula-Thomas, S. (2010). CAMP: A useful resource for research on antimicrobial peptides. *Nucleic Acids Research, 38*, D774–D780.

Tonk, M., et al. (2014). Defensins from the tick *Ixodes scapularis* are effective against phytopathogenic fungi and the human bacterial pathogen *Listeria grayi. Parasites & Vectors, 7*, 554.

Tryselius, Y., & Hultmark, D. (1997). Cysteine proteinase 1 (CP1), a cathepsin L-like enzyme expressed in the *Drosophila melanogaster* haemocyte cell line mbn-2. *Insect Molecular Biology, 6*, 173–181.

Underhill, D. M., & Orinsky, A. (2002). Toll-like receptors: Key mediators of microbe detection. *Current Opinions in Immunology, 14*, 103–110.

Ursic-Bedoya, R. J., & Lowenberger, C. A. (2007). *Rhodnius prolixus*: Identification of immune-related genes up-regulated in response to pathogens and parasites using suppressive subtractive hybridization. *Developmental and Comparative Immunology, 31*, 109–120.

Vallet-Gely, I., Lemaitre, B., & Boccard, F. (2008). Bacterial strategies to overcome insect defences. *Nature Reviews Microbiology, 6*, 302–313.

van der Knaap, W. P. W., Adema, C. M., & Sminia, T. (1993). Invertebrate blood cells: Morphological and functional aspects of the haemocytes in the pond snail *Lymnaea stagnalis. Comparative Haematology International, 3*, 20–26.

Van Herreweghe, J. M., & Michiels, C. W. (2012). Invertebrate lysozymes: Diversity and distribution, molecular mechanism and in vivo function. *Journal of Biosciences, 37*, 327–348.

van Niekerk, G., & Engelbrecht, A. M. (2015). On the evolutionary origin of the adaptive immune system—the adipocyte hypothesis. *Immunology Letters, 164*, 81–87.

Vasta, G. R., Ahmed, H., & Odom, E. W. (2004). Structural and functional diversity of lectin repertoires in invertebrates, protochordates and ectothermic vertebrates. *Current Opinion in Structural Biology, 14*, 617–630.

Vetvicka, V., & Sima, P. (2009). Origins and functions of annelide immune cells: The concise survey. *Invertebrate Survival Journal, 6*, 138–143.

Vilcinskas, A. (2011). Anti-infective therapeutics from the lepidopteran model host *Galleria mellonella. Current Pharmaceutical Design, 17*, 1240–1245.

Vilcinskas, A., Mukherjee, K., & Vogel, H. (2013). Expansion of the antimicrobial peptide repertoire in the invasive ladybird *Harmonia axyridis. Proceedings of the Royal Society B, 280*, 20122113.

Wang, Z., & Wang, G. (2004). APD: The antimicrobial peptide database. *Nucleic Acids Research, 32*, D590–D592.

Watson, F. L., et al. (2005). Extensive diversity of Ig-superfamily proteins in the immune system of insects. *Science, 309*, 1874–1878.

Weiss, B. L., Wu, Y., Schwank, J. J., Tolwinski, N. S., & Aksoy, S. (2008). An insect symbiosis is influenced by bacterium-specific polymorphisms in outer-membrane protein A. *Proceedings of the National Academy of Sciences of the United States of America, 105*, 15088–15093.

Weiss, B. L., Maltz, M., & Aksoy, S. (2012). Obligate symbionts activate immune system development in the tsetse fly. *Journal of Immunology, 188*, 3395–3403.

Wiesner, J., & Vilcinskas, A. (2010). Therapeutic potential of antimicrobial insect peptides. In A. Vilcinskas (Ed.), *Insect biotechnology* (pp. 29–65). Dordrech: Springer.

Williams, M. J., Rodriguez, A., Kimbrell, D. A., & Eldon, E. D. (1997). The 18-wheeler mutation reveals complex antibacterial gene regulation in *Drosophila* host defense. *EMBO Journal, 16*, 6120–6130.

Wilson, R., Chen, C. W., & Ratcliffe, N. A. (1999). Innate immunity in insects: The role of multiple, endogenous serum lectins in the recognition of foreign invaders in the cockroach, *Blaberus discoidalis*. *Journal of Immunology, 162*, 1590–1596.

Yazzie, N., Salazar, K. A., & Castillo, M. G. (2015). Identification, molecular characterization, and gene expression analysis of a CD109 molecule in the Hawaiian bobtail squid *Euprymna scolopes*. *Fish & Shellfish Immunology, 44*, 342–355.

Yi, H. Y., Chowdhury, M., Huang, Y. D., & Yu, X. Q. (2014). Insect antimicrobial peptides and their applications. *Applied Microbiology and Biotechnology, 98*, 5807e5822.

Zhang, G., & Ghosh, S. (2002). Negative regulation of Toll-like receptor-mediated signaling by Tollip. *Journal of Biological Chemistry, 277*, 7059–7065.

Zhang, Q. X., Liu, H. P., Chen, R. Y., Shen, K. L., & Wang, K. J. (2013). Identification of a serine proteinase homolog (Sp- SPH) involved in immune defense in the mud crab *Scylla paramamosain*. *PLoS ONE, 8*, e63787.

Zhang, X., et al. (2015). Phylogenetic analysis and expression profiling of the pattern recognition receptors: Insights into molecular recognition of invading pathogens in *Manduca sexta*. *Insect Biochemistry and Molecular Biology, 62*, 38–50.

Zhu, Y., Thangamani, S., Ho, B., & Ding, J. L. (2005). The ancient origin of the complement system. *EMBO Journal, 24*, 382–394.

Chapter 8
Horizontal Gene Transfer

Abstract Processes leading to the acquisition of one or more genes from a different species are reviewed for bacteria and metazoans. Evolutionary benefits related to different categories of transferred genes through acquisition and insertion mechanisms are discussed. Gene expression, prevalence and frequency of gene transfer are reviewed, together with the dimensions of insertions and their evolutionary consequences.

Keywords Conjugation • Evolution • Fitness • HGT • Gene • Metazoa • Metabolism • Operon • Parasitism • Phage • Sequence similarity • Transduction • Transformation

1 Introduction

Genomes evolution has shown a number of processes at work which remind, in some way, the four basic math operators (addition, subtraction, multiplication and division). These may be considered to correspond to the horizontal acquisition, deletion, duplication and splitting of genes. The definition, with a brief history, of horizontal (or lateral) gene transfer (HGT) has been already provided in Chap. 2. In synthesis, HGT refers to the acquisition of one or more genes not from an ancestor but through a transfer from an other organism, in most cases belonging to a different species. This process may have occurred once or more times during the evolutionary history of a species. The acquired genes are then transmitted vertically to all descendents. The recipient and donor species may be separated by a large phylogenetic distance, which in some cases may even reach the Kingdom level.

Any HGT event identified produced the stable integration of genes or operons, leading the recipient species to acquire new functional processes or other advantages. These usually involved novel enzymes or other proteins, providing a metabolic or evolutive benefit. HGT has been observed at many levels along the Tree of Life and its impact on evolution has been debated. In metazoans it involved the introgression of many genes proceeding from viruses, fungi and/or bacteria, with others acquired from plants or other metazoans (Bird et al. 2003, 2015; Zhu and Gao 2014). In bacteria, HGT yielded the acquisition or exchange of several genes

A. Ciancio, *Invertebrate Bacteriology*, DOI 10.1007/978-94-024-0884-3_8

from other donor bacterial or Archaea species (Jain et al. 1999; Koonin et al. 2001; Dunning Hotopp et al. 2007; Polz et al. 2013).

Comparative analyses of the large amounts of sequence data made available today at the issue of the more than 70,000 genomes sequenced (66,901 prokaryotic and 2982 eukaryotic, by April 2016), confirmed that HGT occurred and acted as a significant evolutionary force in many phyla (Crisp et al. 2015; Brochier-Armanet and Moreira 2015). Insect or nematode genomes show many cases of HGT, thanks to the comparative analyses allowed by a number of genome-wide sequencing projects (Bird et al. 2003; 2015). HGT has also been identified in other invertebrate phyla, including marine lineages of Cnidaria and Porifera (Degnan 2014).

The first indication of HGT is given by a high DNA or protein pair-wise sequence similarity found between genes from species placed at distant phylogenetic positions. However, the identification of HGT events is not an easy task, and possible sources of uncertainty have to be taken into account. To validate similarity-based HGT inferences, the rates of sequence evolution and conservation among genomes must be known, due to the higher rates of evolution reported among the transferred genes. Also, the representativeness of the sequenced genomes and of the sampled phyla has to be considered, including the occurrence of rare sequences in close lineages, not usually found in groups of common, core genes. Finally, phylogenetic-based approaches and comparative analyses of two or three genomes may also be required, to support HGT identification (Parkinson and Blaxter 2003; Brochier-Armanet and Moreira 2015).

In any case, the discovery and validation of several HGT cases yielded a significant progress in the knowledge about the mechanisms of evolution, opening a broader perspective and deeper view on the whole history of life. The transfer of genetic informations from even very distant organisms and their integration in germlines and eventual vertical transmission (with or without gene losses or substitutions), underpin the view that also a number of additive or reductive processes, rather than only mutations and selection, contributed to the construction and differentiation of species.

Many phylogenetic analyses carried out at the whole genome level revealed a level of complexity present in the prokaryotic evolutionary history that reflects the occurrence and effects of HGT events. Following specific analyses of COG[1] functional genes and of their topologies among Prokaryotes, it was proposed that the representation of the species radiations, as provided by a "Tree of Life", should be indeed integrated in the construction of a more effective and complex paradigm. This is rather based on a "Net of Life", a system of relationships accounting for the integration of quasi-random distributed HGT events. The complexity of these systems, however, results by other concomitant mechanisms including processes of genome reduction and parallel gene losses, which are more pronounced in Bacteria than in Archaea (Koonin and Wolf 2008; Doolittle 2009; Puigbò et al. 2010; Toft and Andersson 2010; Wolf and Koonin 2013).

---

[1] *COG* = Clusters of orthologous groups, based on genes coding proteins with the same or highly similar functions, proceeding from at least three lineages. See NCBI site http://www.ncbi.nlm.nih.gov/COG/

How much the mechanisms of foreign gene acquisition have been effective, and contributed to give shape to prokaryotic and metazoan radiations, is still object of study, and different weights have been attributed to the HGT events so far identified. Varying rates were observed in the frequency of the network evolution patterns related to HGT among prokaryotes, with higher frequencies reported for *Deinococcus* and Thermotogae, Aquificales, Cyanobacteria, Actinobacteria, Chloroflexi, Firmicutes and Fusobacteriae lineages (Puigbò et al. 2010). Similarly, due to the increase in the corresponding genome sequence data, HGT events have been identified in many invertebrates including insects and nematodes. They contributed to their evolutionary diversification and to the acquisition of novel functions and metabolic pathways, improving fitness of symbiosis or adaptation to a parasitic lifestyle.

HGT is still a growing and debated field of study, due to the deep implications it has not only in the comprehension of evolutionary trends and natural history, but also for the insurgence and diversification of many processes, spanning from symbiosis to parasitism and defense, including recent events possibly related to human health (Toft and Andersson 2010; Robinson et al. 2013).

2 Evolutionary Benefits

2.1 Transferred Gene Categories

Genome analyses showed several hundred genes that were horizontally transferred in metazoans, evidencing a number of functions positively selected throughout the subsequent evolutionary history of the recipient organisms. A large set of HGT events will remain forever undetectable because of subsequent gene losses or due to negative selection, eliminating the introgressed genes, forever. The HGT events recognized today account hence only for a fraction of successful acquisitions that contributed to improve the recipients' fitness. The success of these events mainly depended on benefits which improved existing traits, underpinning the metabolic or environmental fitness of the recipient organism, or even adding new metabolic possibilities (Table 8.1). Examples include the events leading to the acquisition of plant parasitism capabilities in four nematode lineages, or metabolic genes with low network connectivity, introgressed in marine invertebrates (Mitreva et al. 2009; Danchin et al. 2010; Haegeman et al. 2011; Degnan 2014; Bird et al. 2015). Although genes coding for metabolic functions are the most represented, a number of cases is also reported for the HGT introgression of non-metabolic genes like those involved in mismatch repair activity, biomineralization, or pore-forming toxin genes, as reported in marine invertebrates (Jackson et al. 2011; Moran et al. 2012; Degnan 2014; Ettensohn 2014).

In the free-living nematode *C. elegans*, HGT allowed the acquisition of traits like resistance to dessiccation or water stress tolerance, conferred by genes of the

Table 8.1 Examples of HGT in metazoans

Recipient[a]	Putative donors or best match	Annotations or functions	References
Drosophila ananassae (I)	*Wolbachia*	Protein coding gene GF19976	Crisp et al. (2015)
D. ananassae (I)	*Wolbachia*	(Guanine-N1)-methyltransferase; transposases; ABC transporters; NADH dehydrogenase (I, L subunits); dihydroneopterin aldolase; glutathione/dihydropteroate synthetases	Dunning Hotopp et al. (2007)
Nasonia vitripennis (I)	*Wolbachia*	DNA-directed RNA polymerase (β/β'subunits); ubiquinol-cytochrome c reductase; translation initiation factor IF-2	Dunning Hotopp et al. (2007)
Acyrthosiphon pisum (I)	Fungi	Carotenoid metabolism	Moran and Jarvik (2010)
Hypothenemus hampei (I)	*Bacillus* sp.	Mannanase (glycosil hydrolase)	Acuña et al. (2012)
Meloidogyne spp. (N)	Rhizobia	Glutamine synthetase; L-threonine aldolase; nodL	Scholl et al. (2003)
M. incognita (N)	*Arabidopsis thaliana*, *Lycopersicon esculentum*	Glyceraldehyde-3-phosphate dehydrogenase; histone H2A; citrate (SI)-synthase; LeMir (protease inhibitor)	Bellafiore et al. (2008)
M. artiellia (N)	*Mycobacterium tuberculosis*	Polyglutamate synthetase	Veronico et al. (2001)
Heterodera glycines (N)	Prokaryote	Vitamin B(6) biosynthesis pathway	Craig et al. (2008)
C. elegans (N)	Fungi, Eubacteria	Alcohol dehydrogenases	Parkinson and Blaxter (2003)
Pristionchus pacificus (N)	Leaf beetle, *Gastrophysa atrocyanea*	Diapausins	Rödelsperger and Sommer (2011)
P. pacificus (N)	Slime molds	Cellulases	Mayer et al. (2011)
Bursaphelenchus xylophilus (N)	Bacteria	Endo-β-1,3-glucanase	Kikuchi et al. (2005)
Panagrolaimus superbus (N)	*Phytophthora*	Exo-β-1,3-glucanase	Tyson et al. (2012)
L. vannamei (C)	Cloroflexi	Acetyl-coenzyme A synthetase	Yuan et al. (2013)
L. vannamei (C)	Bacteroidetes	30–50S ribosomal proteins; biopolymer transport protein; chloramphenicol acetyltransferase	Yuan et al. (2013)
L. vannamei (C)	γ-Proteobacteria	Transposase; streptomycin 3″- adenylyltransferase; repressor protein C2; dihydrofolate reductase	Yuan et al. (2013)
Nematostella vectensis (Cn)	Bacteria	Isocitrate lyase (glyoxylate cycle enzymes)	Kondrashov et al. (2006)
Hydra magnipapillata (Cn)	Bacteria	Lipopolysaccharide (LPS) biosynthetic pathway; capA homolog for poly-gammaglutamate biosynthesis	Chapman et al. (2010)

[a]*I* Insecta, *N* Nematoda, *C* Crustacea, *Cn* Cnidaria

threalose-synthesis pathway. Further benefits are the reinforcement of the innate immune defense or stress responses, through i.e. the acquisition of enzymes involved in bacterial or fungal cell wall degradation, or genes coding for heat-shock proteins. Using Gene Onthology (GO) terms, genome-wide analyses in *Caenorhabditis* showed a preponderance of foreign operational genes related to enzyme activities and membrane-bound proteins. The GO categories related to HGT and most commonly encountered in *Caenorhabditis* and *Drosophila* included "innate immune response", "lipid metabolism", "macromolecule modification", "stress response" "anti-oxidant activities" and "amino acid metabolism" (Crisp et al. 2015).

In the Pacific white shrimp *Litopenaeus vannamei* the HGT events introgressed mostly bacterial genes involved in energy metabolism and defense. They included genes coding for an acetyl-CoA synthetase, for carbohydrate metabolism, and others involved in electron transport with NAD(H) or NADP(H) oxidoreductases, energy transduction system, protein synthesis and interaction. Further acquired genes were two transferases and an SOS response gene, conferring antibiotics resistance, with an O-methyltransferase involved in growth, development and defense (Yuan et al. 2013).

HGT may occur in very distant organisms, as shown by the secretome of the phytoparasitic nematode *Meloidogyne incognita* in which five secreted proteins having homologs only in plants are present (Bird et al. 2003, 2015). They include a LeMir protease inhibitor, upregulated during plant parasitism (Bellafiore et al. 2008).

In Prokaryotes, higher HGT frequencies can be encountered for genes responsible for "ion transport", "signal transduction", "defense system components" or for proteins involved in "amino acid or carbohydrate metabolism" (Puigbò et al. 2010). Jain et al. (1999) suggested that the genes active in a relative isolation within the cell machinery, like those coding for specific enzymes or involved in simple processes, have been more easily prone to a stable integration after their horizontal transfer, rather than those involved in more complex cellular systems. This possibility was also confirmed by Crisp et al. (2015) analyses of prokayotic gene transfers to metazoans, which indicated a "one gene at a time" introgression type. The metazoan genes exchanged through HGT were often the result of a single insertion process or just part of reduced transferred clusters (Crisp et al. 2015). Net-like evolution patterns indicative of HGT, observed at the issue of whole genome analyses of prokaryotic genes, reinforced these observations. However, a different scenario may also take place, like the dense set of genes accounting for a large part of a *Wolbachia* donor genome, that has been identified in *Drosophila ananassae* (Dunning Hotopp et al. 2007).

2.2 Acquisition and Insertion Mechanisms

In bacteria, mechanisms of gene exchange and recombination, actually recognized, include: (i) transformation (with direct acquisition of short DNA fragments), (ii) conjugation (by moving long DNA fragments through plasmids or transposons) or

(iii) transduction, through the action of phages (Brochier-Armanet and Moreira 2015). HGT added further exchange or transfer mechanisms involving donors and recipients, depending on the biology and phylogenetic position of the organisms involved (i.e. exposure to viral infections, or to phagocytosis) and their relationships (i.e. endosymbionts and other similar bacteria, or parasites) (Robinson et al. 2013; McNulty et al. 2010). These mechanisms also include the nuclear introgression in Eukaryotes of genes proceeding from endosymbionts, or the effects of transposable genetic elements (Wijayawardena et al. 2013). When HGT involved both donor and recipient Prokaryotes, the known mechanisms of transformation, conjugation and transduction appear as the most likely processes involved (Brochier-Armanet and Moreira 2015).

A first requirement for HGT to occur is that both recipient and donor species occupy the same space or ecological niche. A second requirement is the occurrence of durable physical contacts or a persistent proximity of donor and recipient cells. These conditions have been met in many HGT events including, for example, the events leading to rhizobial genes acquisition by *Meloidogyne* spp., which share the same rhizosphere niche with the root symbionts and are related to the nematode phytoparasitic lifestyle (Scholl et al. 2003). Bacterial parasites or endosymbionts are considered as the most feasible candidates to acquire or exchange genes within the cellular environment represented by their common eukaryotic hosts (Robinson et al. 2013). Although the majority of HGT events consider the gene transfer from bacteria to their hosts, also a host-to-symbiont mechanism has been reported, involving the transfer of non metabolic genes of ankyrin-repeat proteins from a sponge to its bacterial symbiont, and facilitating its cell behaviour (Nguyen et al. 2014).

A possible mechanism, although not exaustive, is considered to have involved DNA acquisition through feeding, favouring a host-to-parasite gene flow, as proposed by Doolittle (1998). Also this possibility is supported by the observation of several cases of plant genes acquired by phytoparasitic bacteria (i.e. *Xylella fastidiosa*) or by nematodes, in which they integrated other genes for cell-wall degrading enzymes, transferred from fungi and bacteria (Bellafiore et al. 2008; Koonin et al. 2001; Jones et al. 2005; Mitreva et al. 2009; Bird et al. 2015). A further example of mechanism connected to feeding is given by the sea slug *Elysia chlorotica*, expressing genes of *Vaucheria litorea*, the alga on which the slug feeds on (Rumpho et al. 2008).

The transposable elements (TEs), which may move DNA fragments within and between organisms, provided further HGT mechanisms. Class I retrotrasposons are TEs active through an intermediate RNA, whereas Class II transposons directly move DNA. Their functional role in HGT could be also reinforced by the activity of viruses or by an enhanced spreading capacity, as observed for a number of parasites (Malik et al. 2000; Keeling and Palmer 2008; Wijayawardena et al. 2013). Horizonthal transfer of transposons has been documented between Isopoda and from them to insects (Dupeyron et al. 2014). A transposase identified in the shrimp *L. vannamei* was introgressed from an *E. coli* transposon together with an adiacent *kch* gene (a voltage-gated potassium channel) and a further 6913 nt long DNA fragment (Yuan et al. 2013). Transposons-mediated HGT is considered as a mechanism leading to the acquisition of further genetic material and functional genes adiacent to the transposon region, as shown in insect-associated nematodes or in shrimps (Rödelsperger and Sommer 2011; Yuan et al. 2013).

In bacteria, phage parasitism and an erroneous encapsulation of DNA fragments (other than the viral ones) can lead to the transfer of variable amounts of genetic material from a donor, parasitized cell to a recipient one. This occurs when the phage injects the non-viral fragment (instead of the viral DNA) that is stably integrated in a new host cell, at the next parasitic cycle. The size of the transferred material is considered to be limited only by the storing capacity of the phage capside (Brochier-Armanet and Moreira 2015).

Once a successful cell introgression is achieved, the foreign DNA fragment must avoid its subsequent cytoplasmic degradation by endonucleases. This process occurs in Prokaryotes by restriction-modification systems (R/M), relying on restriction enzymes, methylases and endonucleases, part of a defensive system towards phage invasions. The R/M systems recognize foreign DNA because of the differences in methylation[2] patterns, and provides to its degradation through endonuclease cleavage. Avoidance may be due to different conditions, including (i) the invasion of a single strand DNA that is synthesized as a double strand and then methylated, (ii) the introgression of an already methylated DNA fragment or (iii) its fast methylation into the recipient cell, (iv) the absence (or a low frequency) of endonuclease recognition sites present in the fragment or (v) any other interference with the R/M system (Brochier-Armanet and Moreira 2015).

Foreign DNA introgression in the bacterial chromosome is then achieved through recombinations processes which are classified as homologous (between regions of high similarity), non-homologous (also referred to as "illegitimate", a pathway for repair of double-strand DNA breaks) or homology-facilitated illegitimate (HFIR). In the latter case short regions with high homology are needed to initiate DNA recombination, which can lead to the HFIR introgression of large fragments or entire plasmids (Thomas and Nielsen 2005).

After introgression, the fate of the acquired gene varies and different outcomes have been proposed. In general, the newly acquired gene(s) can be: (i) retained after introgression for vertical transmission and be actively expressed in descendents, as shown by transcriptomic and other analyses, yielding a functional protein, (ii) retained without any further expression, (iii) exposed to mutation(s) or other duplication events, leading to the evolution in new functional gene(s), or finally (iv) lost due to negative selection deriving from deleterious introgressions, lack of vertical transmission, genetic drifts or other causes (Wijayawardena et al. 2013; Degnan 2014).

[2] *Methylation* = a process in which methyl groups (-CH_3) are linked to nucleotides in a DNA region, leading to the inactivation of its genes. Part of the epigenetic mechanisms that cells use to downregulate gene expression. See Chap. 1.

2.3 Selection and Expression

In theory, the bacterial genes introgressed by prokaryotes or metazoans represent only a small fraction of the total HGT events that occurred during their evolutionary history, in particular those that led to the conservation or reinforcement of useful traits in the recipient species. The transferred genes, detected with different statistical methods on the basis of available sequences and database comparisons, have to be fully integrated in the recipient genomes and expressed when necessary. A process of positive selection is also required for their stable introgression.

This theoretical prediction was confirmed by experimental observations. Phylogenetic analyses carried out with diplogasterid nematodes showed the acquisition and stable maintenance of functional cellulases acquired through two independent HGT events, proceeding from an amoebozoan (or a related microorganism) and a further, unknown donor. In particular, the enzymes showed longevity over evolutionary time scales, being present in the lineages of the beetle-associated nematode *Pristionchus pacificus* (congruent with a stable maintenance during its species radiation) and in another branch, with a *Koerneria* and *Aphelenchus* spp. line. When tested in controlled conditions, all the cellulases were transcribed and functional. Species-specific gene duplications and deletions were also detected in the *P. pacificus* lineage, indicative of functional selective processes. This observation was confirmed by the detection of a site-specific positive selection, active on some cellulase genes, and by the acquisition of introns, providing a clear indication of functional eukaryotic introgression (Mayer et al. 2011; Scholl and Bird 2011). In metazoans, many transferred genes acquired distinctive eukaryotic traits like the presence of introns or were subject to duplications, a further sign indicative of their functional introgression in the eukaryotic cell machinery (Veronico et al. 2001; Boschetti et al. 2012; Yuan et al. 2013; Crisp et al. 2015).

In Prokaryotes, one example of functional gene introgression is the *pho* operon, a group of around 20 genes activated by P starvation including some phospatases, a phosphate transport system (*pst*) and other metabolic enzymes, whose phylogenies do not match the corresponding evolutionary trees based on the conserved 16S rRNA genes. Two types of *pst*-operons were observed in 55 lineages, that were incongruent with the corresponding species phylogeny trees. This lack of congruent relationships among species has been considered as indicative of an early introgression of the operon, through HGT. The parallel evolution of the different operons converged towards similar 3D structures of the coded proteins, indicative of selective pressures maintaining the structural organization, function and activity of the acquired operon enzymes (Moreno-Letelier et al. 2011).

3 HGT Dimensions

3.1 Prevalence

Dagan et al. (2008) estimated that, during their evolutionary history, HGT events interested around 81 % of prokaryotic genes. A lower frequency was considered for genes introgressed in multicellular organisms. Recent analyses, however, showed that hundreds of HGT events introgressed genes active in multicellular organisms including humans, significantly contributing to their differentiation (Scholl et al. 2003; Boto 2014; Crisp et al. 2015). Depending on the stringency of the alignment matches, 68–174 HGT events were identified in *Caenorhabditis*, 4–40 in *Drosophila* with a similar level (32–109 genes) in primates, with major percent origins proceeding from bacteria and protists (Crisp et al. 2015). Around 3 % of genes in *Meloidogyne* was found to be acquired from non-animal donors (Paganini et al. 2012). HGT from *Wolbachia* was detected in approx. 33 % of the sequenced arthropod genomes (Robinson et al. 2013).

A study on HGT in the genome of the shrimp *L. vannamei* showed that 14 genes could be identified as introgressed. They were mostly bacterial in origin (from ecologically related Bacteroidetes and Proteobacteria), with two genes only assigned to Ascomycota. The genes were found at the issue of a filtering procedure based on sequence homology comparison and phylogenetic analyses (Yuan et al. 2013).

In Prokaryotes, the fraction of the genome showing a foreign origin (obtained through genome comparisons) varies from around 20 % in *Salmonella* spp. and 25 % in *E. coli* (in which 1632 additional coding genes with 131 virulence determinants were reported, mostly of viral origin) to around 50 % in *Frankia* spp. (Hayashi et al. 2001; Brochier-Armanet and Moreira 2015). Applying phylogenetic-based methods, the frequent incongruencies in the 16S rRNA ribosomal gene phylogenies suggested a high incidence of HGT in the Prokaryotes evolution and a preponderant role of this process as a leading evolutionary force (Lan and Reeves 1996).

Koonin et al. (2001) produced a detailed review of HGT in Prokaryotes, reporting also a number of genes transferred from Metazoans to bacteria ranging from 1 (found in *Neisseria meningitidis* or *Borrelia burgdorferi*) to 26 (in *Pseudomonas aeruginosa*). The genes acquired from plants ranged from 1 (found in *Borrelia burgdorferi*, *Pyrococcus abyssi* or *Ureaplasma urealyticum*) to 167 (in *Synechocystis*), whereas 1 (in *Aquifex aeolicus*, *Borrelia burgdorferi*, *Campylobacter jejuni*, *Thermotoga maritima* or *Mycoplasma pneumoniae*) to 16 genes (in *Pseudomonas aeruginosa*) were acquired from fungi. Interestingly, in a plant pathogen like *Xylella fastidiosa* around 8 % of genes were acquired from other organisms, including ten genes proceeding from plants and five with a metazoan origin (Koonin et al. 2001).

3.2 Large Genome Insertions

One of the many advantages offered by whole genome sequencing data available for several invertebrate and bacterial species is the possibility of large scale comparisons and the eventual identification and reconstruction of underlying evolutionary processes. One possibility concerns the identification of large – up to a chromosome fraction or even whole genome – insertions in the nuclear genome of a recipient organism. This process may be considered as a kind of TE mediated "genome parasitism" in which one species reproduces itself any time the recipient's genome is duplicated.

Such a process was discovered in Tsetse flies (*Glossina* spp., cyclical vector of *Trypanosoma* spp., the protozoan causal agent of human sleeping sickness and other livestock diseases). The insertions of the *Wolbachia* chromosome in the *Glossina* nuclear genome showed introgression of two large and similar fragments of 0.4 and 0.5 Mbp (corresponding to 47.5 % and 51.7 % of the bacterium genome), and a smaller (2 k bp) one, arising from at least three HGT events. The two large inserted fragments showed 159 and 197 putatively functional coding sequences with, respectively, 13 and 15 tRNAs. The insertions contained also 148 pseudogenes[3] each, plus several additional pseudogene remnants (Brelsfoard et al. 2014).

Nuclear *Wolbachia* transfers have been identified in different invertebrate hosts, including the bean beetle *Callosobruchus chinensis*, with around 30 % of a *Wolbachia* genome integrated in the X chromosome (although with a low transcription), or the longicorn beetle *Monochamus alternatus*, vector of the pine wood nematode *Bursaphelenchus xylophilus* (Nikoh et al. 2008; Aikawa et al. 2009). There is evidence that a large fraction of the species in which *Wolbachia* is present, including filarial nematodes, could show nuclear *Wolbachia* genome transfers, although of varying lengths and with different levels of transcription. A large (1.4 Mbp) nuclear transfer of a *Wolbachia* genome in the *Drosophila* fourth chromosome was studied in three *D. ananassae* populations by Klasson et al. (2014). The authors found that multiple copies of the *Wolbachia* genome have been transferred to the *Drosophila* nuclear genome, with differences observable as concerns size and sequences, among populations proceeding from different geographic regions.

References

Acuña, R., et al. (2012). Adaptive horizontal transfer of a bacterial gene to an invasive insect pest of coffee. *Proceedings of the National Academy of Science, USA, 109*, 4197–4202.

Aikawa, T., et al. (2009). Longicorn beetle that vectors pinewood nematode carries many *Wolbachia* genes on an autosome. *Proceedings of the Royal Society B, 276*, 3791–3798.

[3] *Pseudogenes* = a sequence structurally similar to a gene but not functional, having lost its expression and protein coding capacities.

Bellafiore, S., Shen, Z., Rosso, M. N., Abad, P., Shih, P., & Briggs, S. P. (2008). Direct identification of the *Meloidogyne incognita* secretome reveals proteins with host cell reprogramming potential. *PLoS Pathogens, 4*, e1000192.

Bird, D. M. K., Opperman, C. H., & Davies, K. G. (2003). Interactions between bacteria and plant-parasitic nematodes: Now and then. *International Journal for Parasitology, 33*, 1269–1276.

Bird, D. M., Jones, J. T., Opperman, C. H., Kikuchi, T., & Danchin, E. G. J. (2015). Signatures of adaptation to plant parasitism in nematode genomes. *Parasitology, 142*, S71–S84.

Boschetti, C., et al. (2012). Biochemical diversification through foreign gene expression in bdelloid rotifers. *PLoS Genetics, 8*, e1003035.

Boto, L. (2014). Horizontal gene transfer in the acquisition of novel traits by metazoans. *Proceedings of the Roal Society B, Biological Sciences, 281*, 20132450.

Brelsfoard, C., et al. (2014). Presence of extensive *Wolbachia* symbiont insertions discovered in the genome of its host *Glossina morsitans morsitans. PLoS Neglected Tropical Diseases, 8*, e2728.

Brochier-Armanet, C., & Moreira, D. (2015). Horizontal gene transfer in microbial ecosystems. In J. C. Bertrand, P. Caumette, P. Lebaron, R. Matheron, P. Normand, & T. Sime-Ngando (Eds.), *Environmental microbiology: Fundamentals and applications: Microbial ecology* (pp. 445–481). Dordrecht: Springer.

Chapman, J. A., et al. (2010). The dynamic genome of *Hydra. Nature, 464*, 592–596.

Craig, J. P., Bekal, S., Hudson, M., Domier, L., Niblack, T., & Lambert, K. N. (2008). Analysis of a horizontally transferred pathway involved in vitamin B6 biosynthesis from the soybean cyst nematode *Heterodera glycines. Molecular Biology and Evolution, 25*, 2085–2098.

Crisp, A., Boschetti, C., Perry, M., Tunnacliffe, A., & Micklem, G. (2015). Expression of multiple horizontally acquired genes is a hallmark of both vertebrate and invertebrate genomes. *Genome Biology, 16*, 50.

Dagan, T., Artzy-Randrup, Y., & Martin, W. (2008). Modular networks and cumulative impact of lateral transfer in prokaryote genome evolution. *Proceedings of the National Academy of Science, USA, 105*, 10039–10044.

Danchin, E. G. J., et al. (2010). Multiple lateral gene transfers and duplications have promoted plant parasitism ability in nematodes. *Proceedings of the National Acadeny of Science, USA, 41*, 17651–17656.

Degnan, S. M. (2014). Think laterally: Horizontal gene transfer from symbiotic microbes may extend the phenotype of marine sessile hosts. *Frontiers in Microbiology, 5*, 638.

Doolittle, W. F. (1998). You are what you eat: A gene transfer ratchet could account for bacterial genes in eukaryotic nuclear genomes. *Trends in Genetics, 14*, 307–311.

Doolittle, W. F. (2009). The practice of classification and the theory of evolution, and what the demise of Charles Darwin's tree of life hypothesis means for both of them. *Philosophical Transactions of the Royal Society of London B, Biological Sciences, 364*, 2221–2228.

Dunning Hotopp, J. C., et al. (2007). Widespread lateral gene transfer from intracellular bacteria to multicellular eukaryotes. *Science, 317*, 1753–1756.

Dupeyron, M., Leclercq, S., Cerveau, N., Bouchon, D., & Gilbert, C. (2014). Horizontal transfer of transposons between and within crustaceans and insects. *Mobile DNA, 5*, 4.

Ettensohn, C. A. (2014). Horizontal transfer of the msp130 gene supported the evolution of metazoan biomineralization. *Evolution and Development, 16*, 139–148.

Haegeman, A., Jones, J. T., & Danchin, E. G. J. (2011). Horizontal gene transfer in nematodes: A catalyst for plant parasitism? *Molecular Plant-Microbe Interactions, 24*, 879–887.

Hayashi, T., et al. (2001). Complete genome sequence of enterohe- morrhagic *Escherichia coli* O157:H7 and genomic comparison with a laboratory strain K-12. *DNA Research, 8*, 11–22.

Jackson, D. J., Macis, A., Reitner, J., & Wörheide, G. (2011). A horizontal gene transfer supported the evolution of an earlymetazoan biomineralization strategy. *BMC Evolutionary Biology, 11*, 238.

Jain, R., Rivera, M. C., & Lake, J. A. (1999). Horizontal gene transfer among genomes: The complexity hypothesis. *Proceedings of the National Acadeny of Science, USA, 96*, 3801–3806.

Jones, J. T., Furlanetto, C., & Kikuchi, T. (2005). Horizontal gene transfer from bacteria and fungi as a driving force in the evolution of plant parasitism in nematodes. *Nematology, 7*, 641–646.

Keeling, P. J., & Palmer, J. D. (2008). Horizontal gene transfer in eukaryotic evolution. *Nature Reviews Genetics, 9*, 605–618.

Kikuchi, T., Shibuya, H., & Jones, J. T. (2005). Molecular and biochemical characterization of an endo-beta-1,3-glucanase from the pinewood nematode *Bursaphelenchus xylophilus* acquired by horizontal gene transfer from bacteria. *Biochemical Journal, 389*, 117–125.

Klasson, L., et al. (2014). Extensive duplication of the *Wolbachia* DNA in chromosome four of *Drosophila ananassae. BMC Genomics, 15*, 1097.

Kondrashov, F. A., Koonin, E. V., Morgunov, I. G., Finogenova, T. V., & Kondrashova, M. N. (2006). Evolution of glyoxylate cycle enzymes in metazoa: Evidence of multiple horizontal transfer events and pseudogene formation. *Biology Direct, 1*, 31.

Koonin, E. V., & Wolf, Y. I. (2008). Genomics of bacteria and archaea: The emerging dynamic view of the prokaryotic world. *Nucleic Acids Research, 36*, 6688–6719.

Koonin, E. V., Makarova, K. S., & Aravind, L. (2001). Horizontal gene transfer in prokaryotes: Quantification and classification. *Annual Reviews in Microbiology, 55*, 709–742.

Lan, R., & Reeves, P. R. (1996). Gene transfer is a major factor in bacterial evolution. *Molecular Biology and Evolution, 13*, 47–55.

Malik, H. S., et al. (2000). Poised for contagion: Evolutionary origins of the infectious abilities of invertebrate retroviruses. *Genome Research, 10*, 1307–1318.

Mayer, W. E., Schuster, L. N., Bartelmes, G., Dieterich, C., & Sommer, R. J. (2011). Horizontal gene transfer of microbial cellulases into nematode genomes is associated with functional assimilation and gene turnover. *BMC Evolutionary Biology, 11*, 13.

McNulty, S. N., et al. (2010). Endosymbiont DNA in endobacteria-free filarial nematodes indicates ancient horizontal genetic transfer. *PLoS One, 5*, e11029.

Mitreva, M., et al. (2009). Role of horizontal gene transfer in the evolution of plant parasitism among nematodes. In M. B. Gogarten et al. (Eds.), *Horizontal gene transfer: Genomes in flux* (pp. 517–535). Dordrecht: Springer.

Moran, N. A., & Jarvik, T. (2010). Lateral transfer of genes from fungi underlies carotenoid production in aphids. *Science, 328*, 624–627.

Moran, Y., Fredman, D., Szczesny, P., Grynberg, M., & Technau, U. (2012). Recurrent horizontal transfer of bacterial toxin genes to eukaryotes. *Molecular Biology and Evolution, 29*, 2223–2230.

Moreno-Letelier, A., Olmedo, G., Eguiarte, L. E., Martinez-Castilla, L., & Souza, V. (2011). Parallel evolution and horizontal gene transfer of the *pst* operon in firmicutes from oligotrophic environments. *International Journal of Evolutionary Biology, 2011*, 781642.

Nguyen, M. T., Liu, M., & Thomas, T. (2014). Ankyrin-repeat proteins from sponge symbionts modulate amoebal phagocytosis. *Molecular Ecology, 23*, 1635–1645.

Nikoh, N., et al. (2008). *Wolbachia* genome integrated in an insect chromosome: Evolution and fate of laterally transferred endosymbiont genes. *Genome Research, 18*, 272–280.

Paganini, J., et al. (2012). Contribution of lateral gene transfers to the genome composition and parasitic ability of root-knot nematodes. *PLoS One, 7*, e50875.

Parkinson, J., & Blaxter, M. (2003). SimiTri – Visualizing similarity relationships for groups of sequences. *Bioinformatics, 19*, 390–539.

Polz, M. F., Alm, E. J., & Hanage, W. P. (2013). Horizontal gene transfer and the evolution of bacterial and archaeal population structure. *Trends in Genetics, 29*, 170–175.

Puigbò, P., Wolf, Y. I., & Koonin, E. V. (2010). The tree and net components of prokaryote evolution. *Genome Biology and Evolution, 2*, 745–756.

Robinson, K. M., Sieber, K. B., & Dunning Hotopp, J. C. (2013). A review of bacteria-animal lateral gene transfer may inform our understanding of diseases like cancer. *PLoS Genetics, 9*, e1003877.

Rödelsperger, C., & Sommer, R. J. (2011). Computational archaeology of the *Pristionchus pacificus* genome reveals evidence of horizontal gene transfers from insects. *BMC Evolutionary Biolology, 11*, 239.

Rumpho, M. E., et al. (2008). Horizontal gene transfer of the algal nuclear gene psbO to the photosynthetic sea slug *Elysia chlorotica. Proceedings of the National Acadeny of Science, USA, 105*, 17867–17871.

Scholl, E. H., & Bird, D. M. (2011). Computational and phylogenetic validation of nematode horizontal gene transfer. *BMC Biology, 9*, 9.

Scholl, E. H., Thorne, J. L., McCarter, J. P., & Bird, D. M. (2003). Horizontally transferred genes in plant parasitic nematodes: A high-throughput genomic approach. *Genome Biology, 4*, R39.

Thomas, C. M., & Nielsen, K. M. (2005). Mechanisms of, and barriers to, horizontal gene transfer between bacteria. *Nature Reviews Microbiology, 3*, 711–721.

Toft, C., & Andersson, S. G. E. (2010). Evolutionary microbial genomics: Insights into bacterial host adaptation. *Nature Reviews Genetics, 11*, 465–475.

Tyson, T., et al. (2012). A molecular analysis of desiccation tolerance mechanisms in the anhydrobiotic nematode *Panagrolaimus superbus* using expressed sequenced tags. *BMC Research Notes, 5*, 68.

Veronico, P., Jones, J., Di Vito, M., & De Giorgi, C. (2001). Horizontal transfer of a bacterial gene involved in polyglutamate biosynthesis to the plant-parasitic nematode *Meloidogyne artiellia. FEBS Letters, 508*, 470–474.

Wijayawardena, B. K., Minchella, D. J., & DeWoody, J. A. (2013). Hosts, parasites, and horizontal gene transfer. *Trends in Parasitology, 29*, 329–338.

Wolf, Y. I., & Koonin, E. V. (2013). Genome reduction as the dominant mode of evolution. *Bioassays, 35*, 829–837.

Yuan, J. B., et al. (2013). Horizontally transferred genes in the genome of Pacific white shrimp, *Litopenaeus vannamei. BMC Evolutionary Biology, 13*, 165.

Zhu, S., & Gao, B. (2014). Nematode-derived drosomycin-type antifungal peptides provide evidence for plant-to-ecdysozoan horizontal transfer of a disease resistance gene. *Nature Communications, 5*, 3154.

Chapter 9
The -Omics Race

Abstract A number of advancements concerning the genome organization and functioning of the bacterial chromosome are reviewed, including the genome GC content, gene ordering and classification, and the biological role of operons. Some bioinformatic resources available for genome analysis are briefly shown. Processes related to regulation and gene silencing are also described. Recombination and the evolution, among invertebrate-associated bacteria, of reduced genomes are also discussed. Applications of metagenomic approaches to soil and invertebrate studies are reviewed, together with some aspects of metatranscriptomics and proteomics.

Keywords Chromosome • Gene • Genome • Metabolism • Metagenomics • micro-RNA • Operon • Plasmid • RNAs • Sequencing • Transcriptomics

1 Introduction

Bacteria represent the branch of the Tree (or Net) of Life characterized by the highest biodiversity on earth (Toft and Andersson 2010). Given their importance, several sequencing projects have been carried out to date, yielding a large amount of genetic informations. The first organism ever sequenced was the ΦX174 phage, whose genome was obtained by Frederick Sanger in 1977. This achievement was followed by the first sequenced bacterial genome, belonging to the human pathogen *Haemophilus influenzae*, made available in by Craig Venter and collaborators (Fleischmann et al. 1995).

Most of the research work that is actually carried out on bacteria worldwide relies on large amounts of genomic data made available through the NCBI[1] or EMBL[2] databases for comparative and analytical purposes, produced at the issue of several ongoing or already completed sequencing projects. The knowledge available may be considered as still growing, increasing every year at high rates. Actually, as shown on the NCBI website, the number of bacterial genomes totally or partially

[1] *NCBI* = National Center for Biotechnology Information, US National Library of Medicine, MD, USA (http://www.ncbi.nlm.nih.gov/).

[2] *EMBL* = European Molecular Biology Laboratory, Hidelberg, Germany (http://www.embl.de/).

A. Ciancio, *Invertebrate Bacteriology*, DOI 10.1007/978-94-024-0884-3_9

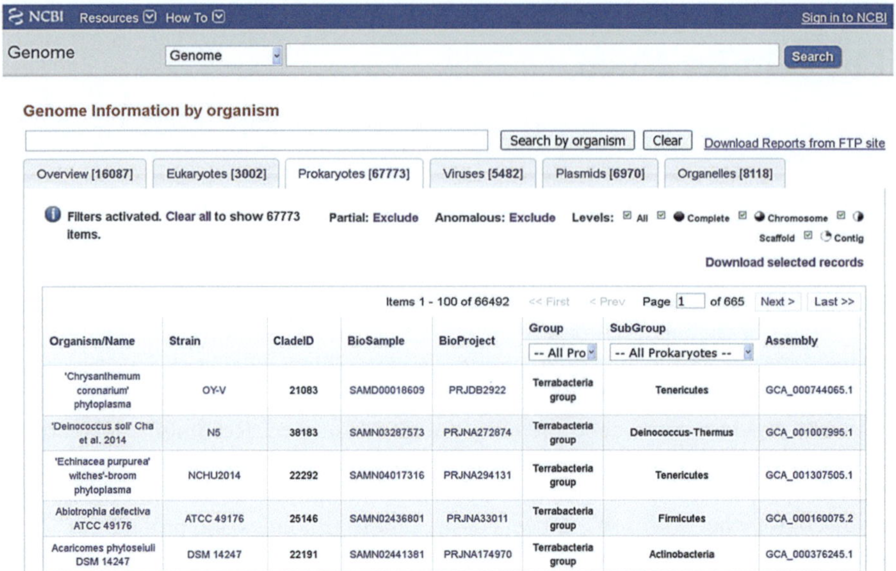

Fig. 9.1 A partial screenshot of the NCBI webpage showing database links for available genomes (http://www.ncbi.nlm.nih.gov/genome) of Prokaryotes and related informations, including project completion level

sequenced, including chromosomes and/or plasmids, is greater than 67,700 (Fig. 9.1).

At the same time, other "-omic" approaches already emerged, that were developed thanks to the progressive reduction in sequencing costs. They now represent the standard for a number of research studies and applications, including personalized medicine, based on sequencing of different kinds of RNA transcripts (transcriptomics) and small or microRNAs, as well as of metabolites (metabolomics) or proteins (proteomics) (Walling and Kaloshian 2016). In this Chapter some of these progresses will be reviewed, with additional informations on the available bioinformatic tools and online resources, also provided.

2 Genomics

Bacteria have an haploid genome and mostly one circular chromosome, which may be integrated by one or more circular plasmid(s), extrachromosomal replicons, living in alliance with their host bacteria. In some species (i.e. *Bacillus thuringiensis*) plasmids may be present in high numbers, up to 11–17 (Reyes-Ramirez and Ibarra 2008). They may vary from a few to hundred copies per cell (Nordström and Dasgupta 2006). Two chromosomes are present in some genera, like *Vibrio*, often integrated by further plasmids (Tagomori et al. 2002). The number of genes present

Table 9.1 Size and gene numbers of sequenced chromosomes from some invertebrate-associated bacteria available on BacMap, GenBank and other databases

Species	Genome id.	Associations[a]	Size (bp)	N. of genes
Anaplasma marginale str. Florida	NC_012026	Ticks (V, P)	1,202,435	980
Anaplasma marginale str. St. Maries	NC_004842	Ticks (V, P)	1,197,687	986
Bacillus thuringiensis serovar *chinensis*	CP_001907	Insects (P)	5,486,830	5596
Borrelia burgdorferi	NC_001318	Ticks (V, P)	910,724	874
Buchnera aphidicola	NC_011833	*Acyrthosiphon pisum* (E)	642,122	590
Buchnera aphidicola	NC_008513	*Cinara cedri* (E)	416,380	394
Ca. Phytoplasma mali	NC_011047	*Cacopsylla* spp. (V, P)	601,943	518
Ca. Liberibacter asiaticus str. psy62	NC_012985	Psyllidae, *Diaphorina citri*	1,226,704	1162
Photorhabdus luminescens subsp. *laumondii* TTO1	NC_005126	*Heterorhabditis bacteriophora*	5,688,987	4897
Rickettsia conorii str. Malish 7	NC_003103	Ticks (V, P)	1,268,755	1412
Rickettsia felis URRWXCal2	NC_007109	*Ixodes scapularis* (V, P)	1,485,148	1439
Wigglesworthia glossinidia	NC_004344	*Glossina brevipalpis* (E)	697,724	651
Wolbachia	NC_006833	*Brugia malayi* (E)	1,080,084	842
Wolbachia	NC_010981	*Culex quinquefasciatus* (E)	1,482,455	1312
Wolbachia	NC_002978	*Drosophila melanogaster* (E)	1,267,782	1234
Xylella fastidiosa subsp. *multiplex* M12	NC_010513	Leafhoppers (V, P)	2,475,130	2160
Xylella fastidiosa subsp. *fastidiosa* M23	NC_010577	Leafhoppers (V, P)	2,535,690	2218
Yersinia pestis biovar *Mediaevails* str. 91001	NC_005810	Fleas (V, P)	4,595,065	4097

[a]*E* endosymbiotic, *P* parasite or pathogen, *V* vectored

in the bacterial and archaeal chromosomes is in general correlated to their genome size. The species shown in Table 9.1, for example, show a significant correlation ($r=0.993$, $P<0.001$) between these two variables, a relationship that is widely recognized. The link is considered related to the fact that bacteria mostly lack introns, which instead are present in Eukaryotes. The latter show in fact a much weaker, not-significant relationship between genome size and number of genes (Gregory 2005). The absence of introns, the reduced amounts of non-coding DNA and the lower number of codons[3] per gene allow a greater concentration of the information content in bacterial chromosomes, which show a gene density ca ten-fold greater than that of Eukaryotes.

2.1 *Organization*

2.1.1 GC Genome Content

The nucleotide gene composition of bacterial genomes has been known since many years as being not uniform. It shows a remarkably wide range of variation among phyla. The GC (guanine-cytosine) genome contents vary largely, within a range spanning from 13.5 % (as shown by the endosymbiont *Ca*. Zinderia insecticola) to 74.9 % (in *Anaeromyxobacter dehalogenans*) (Zhou et al. 2014). This unexplained heterogeneity, and the GC content distribution among taxa, have been possibly considered as the footprint of mutations or, in alternative, as the result of selective evolutionary processes. In particular, the GC content measured at each of the three codon positions appears related to the whole GC content of the coding genes. The widest variability has been reported for the third codon position, which is often a synonymous[4] nucleotide or determines coding of similar amino acids (Bernardi 1985; Brocchieri 2014).

The mutation rates observed in pathogenic species or in other phylogenetically diverse bacteria appear universally biased towards AT (adenine-thiamine). These nucleotides are linked by two hydrogen bonds, whereas GC show three bonds. Modeling the changes in AT-rich genes, however, confirmed that the AT → GC nucleotidic switch is unfavoured, either in coding and non coding, intergenic DNA regions. Mutational bias towards an excess of ATs have also been modeled as favoured in bacteria. Data showed that the high levels of GC contents instead observable in some bacterial lineages had hence to be maintained by the action of one or more unknown selective forces, possibly not related to the codon usage (Hershberg and Petrov 2010; Hildebrand et al. 2010; Brocchieri 2014).

The GC bias has been reported either in bacteria and Eukaryotes. A mechanism called GC-Biased Gene Conversion (gBGC) has been considered as a source of interference with natural selection, affecting the evolution of genomes in a non-adaptive way, since it also interests non-coding intergenic DNA regions. The GC bias was proposed to depend on gene expression, chromosome organization and/or gene recombination, consolidating higher frequencies of GCs instead of the more

[3] *Codon* = a sequence of three nucleotides (or triplet) corresponding in translation to a given amino acid. In Prokaryotes the start codon ATG initiates translation, coding for a modified methionin. It is preceded by a 5′ untranslated region with the ribosome binding site. Other codons correspond to a stop signal, terminating DNA translation to messenger mRNAs (termination codons are UAG, UAA, UGA in RNA and TGA, TAA and TAG in DNA).

[4] A synonymous substitution (or nucleotide) that does not affect the amino acid encoded by the triplet.

favoured AT ending codons. The identification of the molecular or evolutionary factors behind gBGC represents hence an important, although controversial, research field. It appears necessary in fact to identify the gBGC underlying mechanisms and origins, either in case it acts as a selective force or, in case of a non-adaptive neutral processes, to discriminate its effects from those which are effectively active in selection (Hildebrand et al. 2010; Wu et al. 2012; Lassalle et al. 2015).

The three classes of the catalytic α subunits of the DNA polymerase III were proposed as a possible cause affecting the GC content of bacterial genomes. This was indicated by the GC-enriching polymerase E1/E2 subunits, associated to aerobic bacteria with higher GC levels (Wu et al. 2012). These authors also reported a putative relationship between optimal growth temperature and genomic GC contents, as shown in groups of bacteria characterized by different polymerase types (Wu et al. 2012). However, the factors underpinning the evolutionary changes affecting the nucleotide composition of genomes are not yet completely understood, and the evidence proposed as supportive of environmental or other factors was not yet considered as sufficient to explain the heterogeneities of the observed genome composition.

The evolutionary mechanisms determining genome nucleotide composition represent indeed a debated field of study. The availability of many genomes opens the way to more complex and detailed analyses, spanning over large phylogenetic distances. By studying 2670 prokaryotic (Bacteria and Archaea) genomes, Zhou et al. (2014) showed that the nucleotide, codon and amino acid frequency-based distances, calculated for genome pairs, increased with the genomes GC content and were positively correlated (in particular the nucleotide usage) with their GC differences. Data also showed that the A at the first codon position, as the triplet AAA of the Lys (lysine) amino acid codon, decreased almost linearly with increasing genomic GC content. Furthermore, genomes with a highest or lowest GC contents adopted similar patterns. The authors also reported that the GC content has an impact on base, codon and aminoacid usage, a factor resulting stronger than the effects of the lineage phylogenetic position (Zhou et al. 2014).

A bias towards higher AT genome contents was reported in a number of insect endosymbionts, in particular in those species that evolved towards a reduced genome size like *Buchnera aphidicola* (Fig. 9.2, see par. 2.6) (Moran 2002; McCutcheon and Moran 2010).

2.1.2 Gene Ordering and Classification

As soon as sequenced genomes have been made available, the order of genes and its conservation along the bacterial chromosomes has emerged as an informative parameter, providing an indication of evolutionary distances and allowing a further insight on the functioning of genes (Tamames 2001). Using data from several genomes, gene ordering has been exploited for comparison of Gram positive taxa and to measure their distance and relatedness (House et al. 2015). The order of

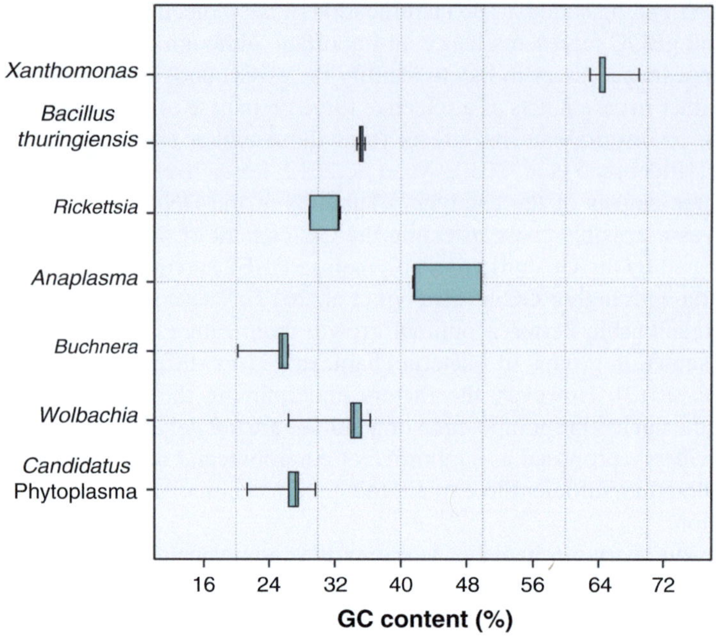

Fig. 9.2 Variation in genome *GC* content among some invertebrate-associated bacteria. *Left* and *right box* sides in the whisker plot show first and third quartiles, respectively. *Vertical bar* in the *box* shows the median. *Left* and *right lateral bars* show minimum and maximum values, respectively (Data source: NCBI)

genes, and the disposition of orthologs[5] along the chromosome, reflect the evolutionary distances among species. They account for the effects of events like inversions, duplications, insertions and transpositions, that occurred after first divergence and during the eventual phylogenetic radiations, leading to the different speciation paths that we observe today.

The presence of common genes or gene families, together with their distances, may be used to construct phylogenetic trees and to estimate evolutionary relationships among genomes, on a basis wider than that provided by one or a few genes, or by the presence/absence of given gene pairs (Tamames 2001; Wolf et al. 2001; House et al. 2015). Gene order conservation is called *synteny* and is highest among phylogenically close lineages. This relative position conservation decreases as evolutionary distances among phyla increase, with changes that may even affect the integrity and occurrence of operons (Tamames 2001; Junier and Rivoire 2013). This approach revealed higher order conservation among Firmicutes and Actinobacteria (two Gram positive lineages) for which a common, rather than multiple, origin was

---

[5] *Ortholog* = genes present in different species but with the same ancestral origin. They evolved from a common ancestor sequence and have very similar functions. Orthologs can be identified through reciprocal BLAST best hits and can be used for functional predictions in newly sequenced genomes.

proposed, and a similar trend also in Proteobacteria. A frequent occurrence of inversions was observed in other lineages, even interesting large genome fractions, hampering by this way an assignment of more precise phylogenetic reconstructions (Darling et al. 2008; House et al. 2015).

High conservation levels of some genes and operons have been, however, also observed. Using the *E. coli* genome as a reference, a number of operons and genes like the ribosomal and the *dcw* clusters showed conserved orders between genomes from even distant phyla, covering genomic regions not limited to single operons. This conservation level was possibly due to factors like a divergence between the examined species that occurred in recent times, HGT of entire gene clusters, or positive selection, being the ordering required for proper gene functioning. The functional need for ordering was proposed to depend on processes favouring the interactions of close, encoded proteins and/or mRNAs, or reflecting the occurrence of HGT (Tamames et al. 2001).

Gene products are usually classified depending on the functional or biological processes in which they are involved. The COG[6] database (Tatusov et al. 1997; Galperin et al. 2015) found at http://www.ncbi.nlm.nih.gov/COG/, is a fundamental, useful tool for identification of proteins. It lists the classification of proteins from conserved homologous genes, whose coded products share common similarities and are classified according to the processes and functions in which they take part (Fig. 9.2). The COG functional annotations are needed to integrate the informations produced experimentally on a number of proteins, assigning tentative functions to newly sequenced gene products, through an orthology-based comparison with the best known functional homologs found in the database (Galperin et al. 2015).

A further NCBI on-line tool for protein function identification is the Conserved Domain Database (Marchler-Bauer et al. 2015). It can be interrogated by uploading protein or gene sequences at http://www.ncbi.nlm.nih.gov/Structure/cdd/wrpsb.cgi.

2.1.3 Operons and Gene Expression

Gene expression and control are achieved in bacteria through the operons. These are clusters of genes localized in close proximity to each other in some regions along the bacterial circular chromosome, and activated by a single, upstream promoter[7] and regulated by a common operator.[8] In operons all genes are transcribed in the same direction and all together in a single, polycistronic[9] mRNA. Their conserved

[6] *COG* = clusters of orthologous groups, based on patterns of sequence similarities.

[7] A site for attachment of the RNA polymerase, initiating transcription.

[8] A gene upstream region at which a repressor factor binds, down regulating transcription until its removal by an inducer molecule.

[9] A polycistronic mRNA includes all the operon mRNAs in a single filament produced for transcription. It is then split in each cistron, each one coding for a given product.

spatial arrangements and order are functional to expression and regulation (Tamames 2001; Yang and Sze 2008).

The *lac* operon was the first one to be identified, by François Jacob and Jaques Monod (1961). The operon is involved in the production of the three enzymes required for the metabolism of lactose. The enzymes are produced just when needed, since the operon is activated when lactose is present in the cell environment. This was one of the first examples of metabolic efficiency described in cell biology. The lactose molecule is the inducer that binds to and removes the repressor from the operator DNA region. The latter inhibits gene transcription and thus down regulates the operon. The operon transcription by the RNA polymerase of the three genes *lacZ* (coding a β-galactosidase cleaving lactose in its monosaccharid constituents, glucose and galactose), *lacY* (a membrane permease favouring lactose cell assumption) and *lacA* (coding for an acetyl-transferase), and the eventual splitting of the polycistronic mRNAs in each cistron, produce the final polypeptides required to catabolize lactose (Fig. 9.3).

A number of algorithms has been applied to the study of bacterial operons, including their organization and evolution across multiple genomes. By using sequences and gene arrangements, Yang and Sze (2008) were able to set up a genome querying strategy based on a defined list of clusters across hundreds of genomes. Through this approach, and on the basis of the coded protein-protein

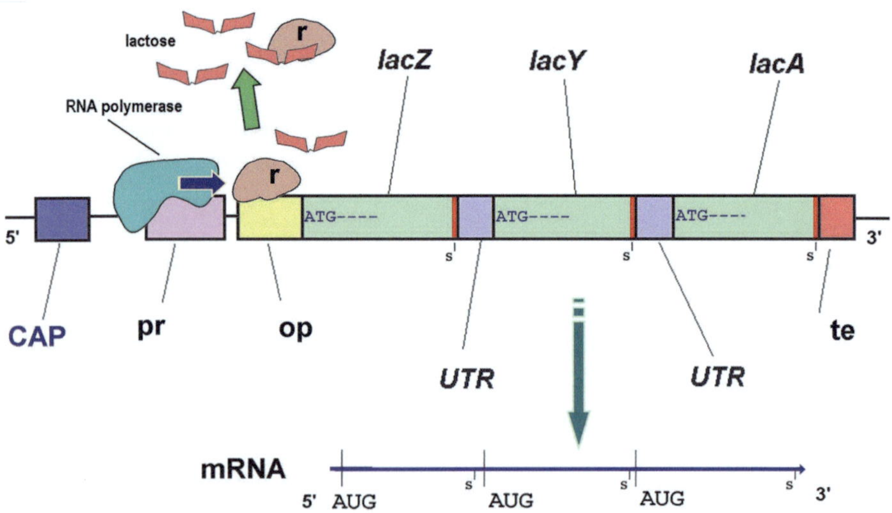

Fig. 9.3 A schematic drawing of the 5′–3′ polarity of the polycistronic lac operon. The structure and expression mechanism are shown, with the promoter (*pr*), operator (*op*), repressor (*r*) and terminator (*te*) sequences. The Catabolite Activator Protein (CAP, also known as CAMP) opens the DNA double strand to allow sense transcription *by the RNA* polymerase (*blue arrow*). The repressor protein inhibits the operon transcription to the polycistronic messenger RNA (mRNA, *dark green arrow*), until its removal by an inducer molecule (in this case this is the disaccharide lactose, *light green arrow*). Also shown are the stop (*S*, *red bars*) and the start codons (ATG/AUG), and untranscribed regions (*UTR*)

BLAST identities, it was possible to identify the occurrence of several operons, their gene orientations and rearrangements (Yang and Sze 2008).

Operons played important roles in bacterial evolution, and significantly contributed to the success of endosymbiotic adaptations to the host cell environment, through processes leading to drastic reshaping and resizing of genomes. Some operons (including a non-coding intergenic sequence neutral to selection) like *groE* (encoding the heat shock proteins HSP10 and HSP60), appeared involved in mutualistic associations, and have been used to reconstruct unresolved phylogenetic relationships among *Wolbachia* endosymbionts (Masui et al. 1997).

Operons may be acquired through HGT events that contributed largely to the evolutionary history of endosymbionts. A specific, additional gene cluster encoding for the biotin (vitamin B7) complete synthetic pathway (5.4 kb, six genes) was found in a *Wolbachia* endosymbiont of the bedbug *Cimex lectularius* (Nicoh et al. 2014). The capacity to synthesize biotin (vitamin B7), adds to other pathways for riboflavin (vitamin B2), pyridoxine (vitamin B6, partial) and folate (vitamin B9) that are present in most *Wolbachia* lineages. However, the biotin operon that was found only in the *Wolbachia* from *C. lectularius*, was also reported from the genome of *Cardinium hertigii* (a facultative endosymbiont inducing cytoplasmic incompatibility in *Encarsia pergandiella*), in the insect pathogen *Lawsonia intracellularis* and in the plasmid of a *Rickettsia* sp. from *Ixodes scapularis* (Gillespie et al. 2012; Penz et al. 2012; Sait et al. 2013).

Phylogenetic analyses suggested an origin of the *Wolbachia* biotin operon through HGT that involved an ancestor, possibly an unrelated endosymbiont. Similarly, also a partial operon with the thiamine synthesis genes *tenA1*, *thiD* and *ψthiM* (in common with other *Wolbachia*) was found in the *C. lectularius* endosymbiont. The operons conferred the endosymbiont an adaptive advantage for its stable recruitment by *C. lectularius*, providing the insect host with benefits concerning some key dietary supplements, important because of the bedbug biotin-deficient diet. This process appears essential for the survival of both organisms and represents a basic, metabolic building block of their obligate nutritional mutualism (Nicoh et al. 2014).

Two other operons were identified in the genomes of two *Wolbachia* lines, from Hymenoptera and from *Drosophila simulans*. The first operon showed five *vir* genes (*virB8*, *virB9*, *virB10*, *virB11* and *virD4*) with a downstream *wspB* locus. The second showed three genes (*virB3*, *virB4*, and *virB6*) with four additional open reading frames[10] (*orf1* to *orf4*), in the same direction. Both operons appeared involved in the production of transport membrane proteins of the Type IV SS, indicating that the system is functional and active in the symbiosis (Rancès et al. 2008).

In bacteria that evolved towards a drastically reduced genome, like the aphid endosymbiont *Buchnera aphidicola*, operons and genome functionning have been challenged by the loss of a number of functional genes and products, as well as by the constraints imposed by the metabolic adaptation to the biochemical interchange

[10] *Open reading frame* = a region of a gene sequence that is transcribed from the start codon (that marks the beginning of the transcribed region), until a stop codon is found.

with the host. The *B. aphidicola* transcription units (TUs)[11] evolved rapidly at the time of the first host-symbiont association (150–200 Myrs ago). This shows a drastic genome reduction due to gene losses and a high AT content bias (up to 70% of the genome) (Fig. 9.2). Further changes include, among others, the lack of recombination, the loss of two-components regulatory systems and of the operons coding for essential amino acid pathways (Wernegreen 2002). The subsequent evolutive pressures acted on the gene ordering and operon map through mechanisms like inversions, transpositions, deletions, insertions and duplications. These processes acted on a progressively reduced genome mostly at the operon level, re-shaping chromosomal maps and organization, including the assemblage of unrelated genes under the same TUs (Briza et al. 2010). By classifying the TUs in the genome of *B. aphidicola* a comparative analysis was carried out using the polycistronic orthologs of *E. coli*, a species close to the aphid endosymbiont, using dedicated databases like BioCyc (http://biocyc.org/), DOOR (http://csbl.bmb.uga.edu/DOOR/) and MicrobesOnline (http://www.microbesonline.org/). On average, the *B. aphidicola* TUs showed denser gene contents than *E. coli*, with 133 predicted polycistronic and 155 monocistronic TUs.

The operon *luxICDABEG* of *Vibrio* (=*Aliivibrio*) *fischeri* is central to the production of bioluminescence, a process activated as the result of a multiple bacterial behaviour in the symbiosis with the bacterium squid host. The bacterial response is governed by a quorum-sensing signal, active at high cell densities. The operon is activated for protection of the host, which exploits the light emission to mimik the nocturnal incident light, in an attempt to evade predators. Other bioluminescent strains of *V. fisheri* are also present in light organs of fishes. All of the eight genes required for bioluminescence are clustered in a unique operon in the second chromosome. The process relies on the dimeric enzyme luciferase, whose components are encoded by the genes *luxA* and *luxB*. Luciferase catalyzes oxidation of an aldehyde and a reduced flavin mononucleotide. Other genes of the operon are involved in the recruitment of substrate fatty acids, and in other functional reactions.

The quorum sensing behaviour acts through a positive feedback mechanism, related to the accumulation of an autoinducer molecule (N-3-oxohexanoyl-homoserine lactone) and a transcription factor encoded by two other genes, *luxI* and *luxR*, respectively (Schaefer et al. 1996; Ruby et al. 2005; Miyashiro and Ruby 2012; Septer and Stabb 2012). These genes represent a signaling module widespread among bacteria and involved in a number of reactions including pathogenesis and antibiotic production. *LuxR* is also an up regulator of 24 other genes within 13 operons (Miyashiro and Ruby 2012).

Other operons, important for practical exploitation of invertebrate pathogenic bacteria, are those clustering the many hundred *cry* genes present in *B. thuringiensis* and other *Bacillus* spp., used for the genetic engineering of plants for expression of the insecticidal crystals, active against herbivorous pests (Kota et al. 1999).

---

[11] *Transcription unit* (*TU*) = the ensemble of genes under control of a single promoter. A single operon represents a single TU. Complex operons with several promoters have many TUs.

2.1.4 Plasmids

Expression vectors naturally occur in many bacteria. They are plasmids including an enhancer and promoter regions, bound to a specific gene. They are often used in biotechnology to introduce a gene in a recipient cell, and to achieve by its expression the encoded protein production, through eventual mass culturing. Plasmids are also involved as self-transferable elements in processes like transformation (the introgression of free DNA) or in transfer through conjugation (the direct introgression of a whole plasmid from another cell). The two processes take place with highest frequencies in conditions in which high densities of metabolically active bacteria are present, like in the rhizosphere or in special trophic niches with high densities of bacteria, like the gut of earthworms or termites (Tebbe et al. 2006).

The mechanism of bacterial conjugation includes also the possible presence of a second type of mobilizing plasmids required in some cases, that can be provided either by the donor or by a third cell, in a sort of tri-parental mating. This mechanism was experimentally observed in the gut of Collembola. However, higher frequencies were observed for self-transferable elements, as mobilizing plasmids were transferred when the mobilizing genes were present in the donor cells (Tebbe et al. 2006).

Plasmids also represent a detection target to determine at which extent gene recombination and conjugation occur among bacterial populations in natural environments, like soil or rhizosphere. They can also be exploited to identify the role played by species or groups associated to invertebrates. This is particularly important when monitoring genetically modified bacteria used for pest management, or when tracking the spread of invasive pathogenic species or strains, by detecting the plasmids that they carry.

Many biotechnological applications rely on the use of plasmids to transfer a number of genes to recipient bacteria, for applied purposes of genetic engineering, including further transfer to plants or production of non-bacterial products (i.e. vaccines or insulin). The ingestion of genetically modified bacteria by invertebrates may yield different outcomes, ranging from inactivation and lysis through digestion to an increase of the bacterial density in faeces, due to enhanced multiplication rates. In the latter case the higher numbers of a given species or strain may also induce higher rates of genetic recombination through gene exchange or transfer.

To study this process, the millipede *Pachyiulus flavipes*, the woodlouse *Armadillidium vulgare* and the earthworms *Aporrectodea caliginosa* and *Eisenia fetida* have been tested in laboratory microcosms to determine their effect on the transfer of plasmids from *Pseudomonas* spp. and other bacteria, to other bacteria present in soil, litter or vermicompost. Data showed that a recombinant conjugative plasmid (pLV1017) carrying a gene conferring resistance to antibiotics was transferred to two facultatively anaerobic gram-negative bacteria found in the *A. vulgare* faeces. Also a naturally occurring plasmid (pAMb1) encoding resistance to erythromycin was transferred from a *Streptococcus faecalis* strain to *B. thuringiensis* var. *israelensis*, isolated from the vermicompost (Byzov et al. 1999).

Self-replicating large plasmids are present in almost all strains of *B. thuringiensis* (Schnepf et al. 1998). They carry several *cry* and β-exotoxin genes and were also used for identification and characterization of strains at the sub-serotype level (Reyes-Ramírez and Ibarra 2008). Plasmids have been applied as expression vectors of the *cry* genes of *B. thuringiensis* in *E. coli*, or to study other sporulation genes in *Bacillus* spp. (Klier et al. 1982).

2.1.5 Regulation

Gene silencing (GS) is a process allowing control of gene expression through an induced down regulation. It differs from gene knockout because GS acts at the transcription or translation levels, whereas in the latter the gene is eliminated. GS has been widely applied in many research fields, being very informative about the role of genes and is considered as a source of possible, therapeutic applications. In bacteria, GS is applied through antisense RNAs, complementary to the mRNAs or tRNAs to down regulate, delivered through phage or plasmid vectors, and other mechanisms.

The bacterial chromosome is formed by two complementary DNA filaments, linked in a double helix strand. When a gene or operon is expressed, transcription by the RNA polymerase produces a DNA-RNA hybrid molecule which is then released as a mRNA. The RNA polymerase measures ca 400 kd, and is composed by four subunits, one of which (the σ subunit) detects a promoter DNA site to trigger transcription, that starts after the separation of the σ subunit from the rest of the molecule. The enzyme opens and unwinds the double DNA helix for ca 17 nucleotides, and runs in one direction, also referred to as "sense" and conventionally defined as 5′–3′ direction, to produce a DNA-mRNA hybrid. The promoters are specific sequences located 35 or 10 nucleotides upstream the transcription initiation site, and characterized by two universal motifs (Fig. 9.4). Transcription ends when a GC-AT rich stem and loop structure, the terminator, is encountered by the enzyme,

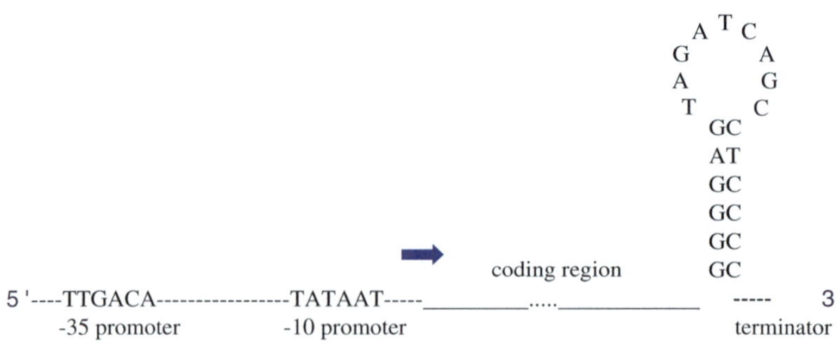

Fig. 9.4 A schematic drawing showing the location of promoters and of terminator loop sequence, in relation to the gene coding region (*arrow* shows the RNA polymerase 5′–3′ direction of transcription)

releasing the newly produced mRNA and refolding the template DNA. This mRNA is the complement of the DNA template, corresponding to the same sequence (except of U instead of T) of the coding DNA (Berg et al. 2002).

Many gene silencing mechanisms are present in bacteria result from the evolution of adaptive immune defense systems (see Wiedenheft et al. (2012), for a review).

Antisense ("as") oligonucleotides are small fragments of DNA with a complementary sequence which, once expressed in the bacterial cell, binds to its target in mRNA. By this way, the mRNA is no more traduced to an aminoacidic sequence, and the protein production is stopped. Antisense strands have very wide dimensional ranges (from ten to thousand nucleotides) and act as post-transcriptional regulators, orchestrating the expression of many genes in the bacterial cell. This process has been applied to study DNA expression in bacteria and to identify the role of specifically silenced genes. The mechanism is based on antisense RNA targeting pre-mRNAs and/or mRNAs, then inducing their eventual degradation by the cell enzymatic machinery. Further antisense products blocking access of the cellular machinery to the RNA, not leading to its degradation, appear promising for the developed of one or more "therapeutic" strategies and may also be proposed as a mechanism for noxious bacteria control (Kole et al. 2012; Nakashima et al. 2012).

Among the regulatory elements present in bacteria, the small RNAs (sRNA) are classified in two types, *cis* and *trans*. The *cis*-encoded (or antisense) RNAs correspond to the opposite coding DNA gene strand and have perfect complementarity with their targets. *Trans*-encoded RNAs are instead produced on distant genome sites, with a lower complementarity and multiple targets. Silencing occurs through mRNA folding, hiding the binding site to the ribosome and reversibly preventing initiation of the protein synthesis, by degradation of the mRNA::as RNA complex, or by a combination of both (Thomason and Storz 2010; Good and Stach 2011).

The clustered, regularly inter-spaced short palindromic repeats (CRISPR, see Chap. 1) are part of a bacterial defensive system towards bacteriophages, relying on a sort of molecular "memory" stored as small DNA sequences interleaved with unique spacers. This system provides a molecular archive whose expression yields CRISPR-derived RNAs, corresponding to the foreign sequences previously encountered (Wiedenheft et al. 2012). The CRISPR DNA interleaved repeats are flanked by a set of CRISPR-associated (Cas) nuclease/helicase coding genes. A duplex formed by a CRISPR and a transactivating RNA directs, as a guide RNA, a Cas endonuclease to a complementary target region, inducing a double strand break. The recognition results in an imperfect repair with insertions-deletions and eventual inactivation of the DNA of the recurrent invader (like viruses or plasmids) or silencing its RNA, in a specific and precise, sequence-depending mechanism (Horvath and Barrangou 2010; Jinek et al. 2012). Three main CRISPR types have been recognized, together with a wide range of *cas* genes (Wiedenheft et al. 2012). CRISPR-Cas can be used as a fast gene editing technique, relying on the CAS9 enzyme and an expression vector.

This system is promising as a potential mechanism for regulating or even editing the genome of bacteria including symbionts like *Wolbachia*, present in nematodes or in some insects which are vectors of important human diseases. It has been also identified in some invertebrate-associated bacteria of medical importance like *Francisella tularensis*, the causal agent of tularemia (Schunder et al. 2013). CRISPR-Cas has also been used for genome editing through targeted mutagenesis, via injected duplexes, in a number of invertebrates including *C. elegans* and *Daphnia magna* (Friedland et al. 2013; Nakanishi et al. 2014; Zhao et al. 2014).

Further gene silencing mechanisms present in some bacteria include the group II and the RNase P ribozymes.[12]These are self-slicing catalytic RNA molecules cleaving other RNA molecules. Group II introns have been found in Rickettsiales and, in particular, in high numbers in genomes of arthropods-associated *Wolbachia*. In these genomes, 36 elements from 5 distinct families were found, together with insertion sequence (IS) transposable elements, introgressed as the result of frequent HGT events (Leclercq et al. 2011).

In eukaryotes, RNA interference (RNAi) is an epigenetic[13] mechanism in which a dsRNA can interfere with and shut down the expression of a specific gene target. It is a sequence-specific and post-transcriptional gene silencing mechanism, largely and successfully applied in basic studies of *C. elegans* and *D. melanogaster*, to investigate gene expression and function. The invertebrates were targeted through a method based on ingestion, by simply soaking the animals in a dsRNA solution. In *C. elegans* the RNAi was shown to be inherited and may diffuse from cell to cell. It is considered as a evolutionarily conserved, defensive system (Wilkins et al. 2005; Kole et al. 2012).

MicroRNAs (miRNAs) are endogenous and non-coding sRNAs (20–22 nt long) which are involved, through gene regulation, in many eukaryotic cellular processes, including development and immunity. Their synthesis occurs through expression of specific encoding genes, in the cell nucleus (Fig. 9.5). They act post-transcriptionally, binding to a complementary mRNA, affecting translation and leading to its cellular degradation. MiRNAs are often involved in defensive reactions to silence genes of bacterial or virus invaders, and may also act as inducers of gene expression. They are transcribed by RNA Polymerase II as a long primary transcript (pri-miRNAs), which may contain more than one miRNA. In the nucleus, pre-miRNAs are processed to hairpin-like elements by the RNAse III-like enzyme Drosha. Pre-miRNAs are then exported to the cytosol by Exportin 5. In the cytosol RNAse III-like Dicer processes the precursors to yield mature miRNAs which are incorporated in RISC (miRNA-induced silencing complex). The miRNAs with high homology to the target mRNA lead to mRNA cleavage, whereas those with imperfect base pairing to the target mRNA lead to translational repression and/or mRNA degradation.

MiRNAs are active in many invertebrate-bacteria interactions, and their study represents a fertile and growing research field. Among the most studied endosymbionts, *Wolbachia* was shown to be capable of interfering with the host cell biochemistry and miRNA pathways.

---

[12] *Ribozyme* = a RNA whose sequence and folding allow its functionning as an enzyme.

[13] *Epigenetic* = the vertical transmission to descendents of a trait or condition based on a mechanism different from gene inheritance, usually via a DNA methylation or other sRNA-based reactions.

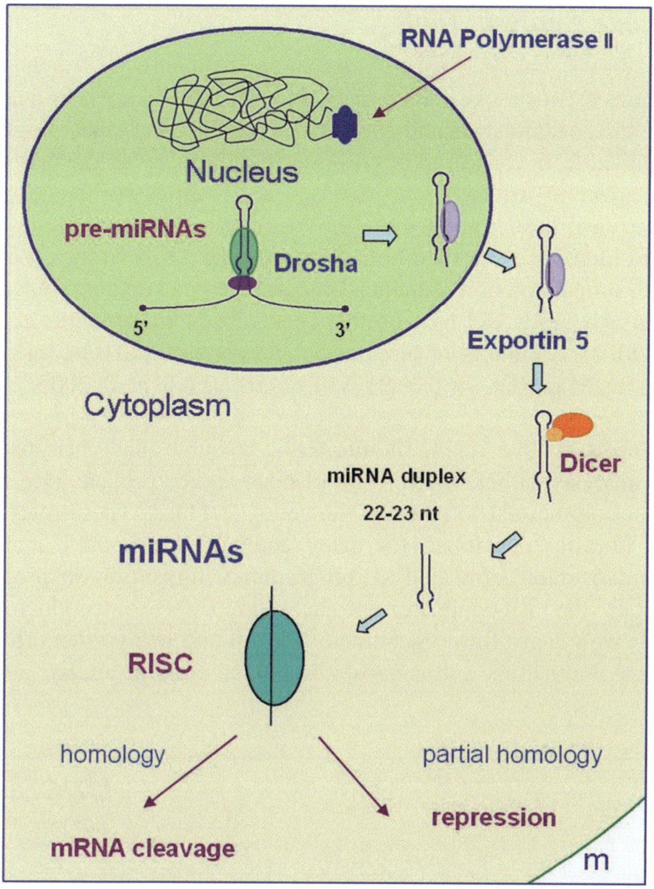

Fig. 9.5 A schematic drawing of miRNA expression and activity (m = cell membrane, for details see text)

Activity concerned the induction of different intracellular distributions of Argonaute proteins. In the dengue mosquito vector *Aedes aegypti*, *Wolbachia* was shown to manipulate the miRNAs profile of several host genes to facilitate its self-maintenance (Hussain et al. 2011, 2013). Analysis of miRNAs in the two-spotted spider mite *Tetranychus urticae*, with or without *Wolbachia*, showed that the endosymbiont differentially expressed miRNAs that regulated 90 mRNAs in females and 9 in males. These were involved in sphingolipid metabolism, lysosome function, apoptosis and lipid transport in both sexes and, in females only, in reproduction (Rong et al. 2014). For databases of deposited miRNA sequences see miRBase (ftp://mirbase.org/pub/mirbase/).

2.2 Genome Analysis Tools

Representations by linear or circular maps (genome atlases) are useful drafting tools allowing the fast visualization and comparison, on different lanes, of genetic informations obtained from a plasmid or chromosome. In this way the genetic information is summarized in order to show properties and features like the genes and ORFs positions, the local base composition, the location of annotated genes on the two strands, or to identify other specific regions of interest. A further use of genome atlases is the comparison of a sequenced genome mapped on a reference one, or the identification of best BLAST hits (Stothard et al. 2005; Wassenaar et al. 2010; Ozen and Ussery 2014). A number of bioinformatic tools are available, for either online analysis like BacMap (Fig. 9.6), or BLAST queries (Field et al. 2005; Alikhan et al. 2011).

Programmes and other online bioinformatic resources have ben made available for genome analysis, either as the result of open source projects (see for example Fig. 9.6) or as commercial software (Zhang et al. 2011). A. The former include, for example, R libraries developed for many application purposes, including gene expression and/or microarray analysis programmes (https://www.r-project.org/), or software from the Bioconductor project (http://www.bioconductor.org/), the Genometools web pages (http://genometools.org/) and many others (http://grouth-bio.com/Home_Page.html, https://www.broadinstitute.org/cancer/cga/, J. Craig

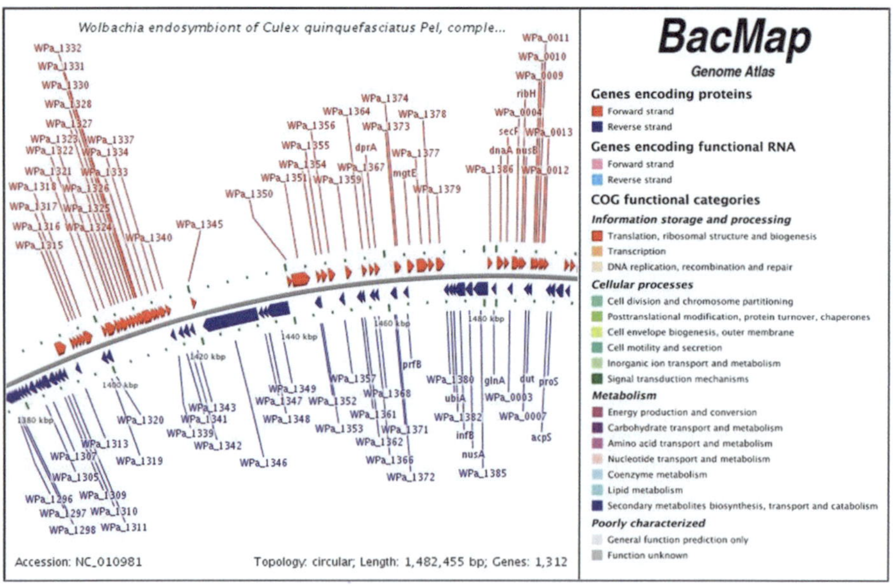

Fig. 9.6 A screenshot of the BacMap output showing the location of forward and reverse protein encoding genes on an enlarged region of the *Wolbachia* NC_010981 chromosome, and other COG mapping markers (For further details see http://wishart.biology.ualberta.ca/BacMap/)

Venter's Institute pages at http://www.jcvi.org/cms/research/software/#c614, or other listing webpages like http://pages.stat.wisc.edu/~yandell/statgen/). Commercial software include several packages like CLCbio® (www.clcbio.com), the Illumina GenomeStudio® (http://www.illumina.com/) or the pyrosequencing DNASTAR-454® software (http://www.454.com/products/analysis-software/).

2.3 Recombination

Genetic recombination (GR) is a process in which parts of a gene are substituted by homologous fragments proceeding from another donor and eventually vertically transmitted by the recipient organism, originating descendent lines carrying the recombined gene. This kind of genetic remixing with a foreign DNA (proceeding or not by the same species) may occur more than once, and may be subsequently masked by other evolutionary processes. When possible, GR may be detected through a number of bioinformatic analyses and resources. Programmes like MaxChi, Chimera, RDP and Geneconv may be applied to the analysis of gene alignments, and are considered as informative about the evolutionary history of the microbial (viral, bacterial or fungal) strains or populations tested (Martin et al. 2005).

GR has been considered a mechanism of evolution and genetic adaptation frequent among bacteria, in relation to the high degree of homology of the DNA fragments introgressed. It was detected among a number of pathogens (*Anaplasma*, *Neisseria*, *Leptospira*, *Borrelia*), in particular as concerns genes for surface proteins involved in pathogenicity and recognition by the immune system, thus contributing to the rapid insurgence of new virulence and pathogenicity traits (Howell-Adams and Seifert 2000; Meeus et al. 2003; Haake et al. 2004; Rich et al. 2001).

Among many intracellular endosymbionts, GR was considered a rare event, given the protected and stable cell environment in which they live and the reduced exposure to selective pressures, the lower number of mobile gene elements and the reduced genome. However, a pervasive GR was reported within and across *Wolbachia* genomes, whose chimeric nature appears related to peculiar traits of the endosymbiont lineages, accounting for its complex phylogeny. These traits include a wide host range, the capabilities for either horizontal and vertical transmission (with several cases of manipulation of the host reproductive behaviour), high rates of mobile elements and recombinations (Baldo et al. 2006).

2.4 Genome Reductionism

Genome sequencing studies showed a number of phylogenetically distinct, invertebrate-associated bacteria characterized by a remarkable reduction of their genome size, with five insect symbionts providing the smallest genomes (less than 300 kb, the size of a large virus) so far sequenced (McCutcheon and Moran 2012).

The symbionts (*Ca*. "Sulcia muelleri", *Ca*. "Zinderia insecticola", *Ca*. "Carsonella ruddii", *Ca*. "Hodgkinia cicadicola" and *Ca*. "Tremblaya princeps") represent the final outcome of an irreversible evolutive path, leading to highest specialized lifestyles, reminiscent in their evolution of other specialized cell organelles like mitochondria and chloroplasts, but lacking any functional relationship with recruitment of nucleus encoded proteins.

Although similar in their final outcome, the evolutionary events leading to minimal genomes appeared in different phyla (α-, β-, γ-Proteobacteria and Bacteroidetes) independently, with variable traits and patterns of gene losses and erosion typical for each lineage, including the loss of mobile elements and very high levels of AT content. These extreme bacteria show a higher level of genome reductionism, when compared to other obligate endosymbionts or pathogens known for their reduced genome sizes, like *Ca*. "Baumannia cicadellinicola", *B. aphidicola*, *Wigglesworthia* and *Blattabacterium* spp., whose genomes range in the order of 350–600 kb (Van Ham et al. 2003).

The cellular host environments represent a "space" with reduced selective pressures, as asexuality, minimal drifts and low recombination rates contributed to stabilize the tiny genomes. The trends towards size reduction have been flanked by other evolutionary outcomes like fast sequence evolution rates, codon reassignments and extreme biases in gene composition, with higher rates of protein misfolding. The common trait of these tiny, obligate bacteria is that they adapted to endosymbiosis through a mechanism based on a severe gene loss, including the loss of cell envelope genes involved in fatty acid, phospholipid and peptidoglycan metabolism, and of genes for expression regulation, DNA repair and recombination. The tiny bacteria retain mainly genes involved in basic cellular metabolism (replication, transcription and translation) and those involved in their mutualistic contributions, providing their hosts essential nutritional factors (amino acid biosynthesis), which represent the conserved core of their specialized metabolism (McCutcheon and Moran 2012).

A particular case is shown by *Ca*. Tremblaya princeps, an endosymbiont of the mealybug *Planococcus citri*. The bacterium shows a tiny genome and a rare example of symbiosis, since it hosts a further endosymbiont, *Ca*. Moranella endobia. *Ca*. Tremblaya princeps was shown to have a sparse and degenerated genome, lacking tRNA synthetase and other functional genes, and is integrated for its nutritional requirements by the products supplied by its endosymbiont, *Ca*. Moranella endobia (McCutcheon and von Dohlen 2011; McCutcheon and Moran 2012).

3 Other -Omics

3.1 Metagenomics

Metagenomic studies allow the identification of almost all the bacterial species present in a given environment, by sequencing the whole range of 16S rRNA ribosomal or other genes available for taxonomic purposes. The RNA transcripts are previously extracted by the samples examined (whose weights are usually in the order of 1–2 g), and processed with the so-called Next Generation Sequencing (NGS) approaches. These are provided by devices like the 454 Pyrosequencing™, Illumina™ or Ion Torrent™, capable to yield 150–300 bases-long sequence reads, in the order of several million per sample.

A further approach recently made available is the nanopore sequencing technology including devices like, among others, the MinION™ (Oxford Nanopore Tech., UK), a small apparatus with higher levels of miniaturization, portability and ease of use (Benítez-Páez et al. 2016). One innovation of the latter technology is given by the long length of the produced reads, which may be several kb long. The gene mostly used in metagenomics is the 16S ribosomal rRNA, in particular its most diverse regions V3 and V4 (Handelsman 2004; Caporaso et al. 2011). Also the gene encoding for the chaperonin-60 (cpn60) is a reliable universal target for bacterial barcoding. Both provide extensive metagenomic informations, as shown by a large number of applications (Hill et al. 2002, 2004; Links et al. 2012; Miller at el. 2013; Allard et al. 2015).

The operational taxonomic units (OTUs) identified through bioinformatic analyses from the sequenced contigs[14] are then used to assess the biodiversity levels present in a sample. Biodiversity is usually defined as the richness of species that may be found *i)* in a given site or sample (α-diversity, Hα) or *ii)* along a gradient or among samples, as an indication of species turnover (β-diversity, Hβ). A more comprehensive view is given by the γ-diversity ($H\gamma = H\alpha + H\beta$), which accounts for the global diversity of all samples (Veech and Crist 2010; Whittaker et al. 2001; Anderson et al. 2011). Several indices can be used to represent and compare the species diversity in a sample or population. As shown by Jost (2006), it is worth to recall that diversity corresponds to the effective number of species present in a given space, volume or population, and that an index is just its representation. The most useful and applied index is the Shannon biodiversity or entropy index (Shannon-Wiener index) (Jost 2006; Normand et al. 2015).

Being culture-independent and by-passing the isolation and culturing steps once required for identification, metagenomic approaches represent an actual, advanced standard for the study of microbial communities and their ecology, in many conditions (Patrick and Handelsman 2005; Susannah and Edward 2005; Sogin et al. 2006).

Several factors, including the biogeochemistry of an area sampled, affect the composition and stability of the bacterial communities within the environmental niches they occupy. These considerations mostly apply to many interactions with invertebrates. The levels of biodiversity encountered in metagenomic analyses may hence vary, depending on the occurrence of ecological or environmental peculiarities, or in function of the different treatments that have been applied to the samples. These may be usually performed to identify the effects, on microbial communities, of a number of variables accounting for physical or ecological conditions (i.e. dry-

[14] *Contigs* = sequences of different lengths produced by joining two or more partially overlapping reads, produced by a sequencing device.

ing, temperatures, storage, climate), of specific compounds (i.e. pesticides, fertilizers, pollutants or other products) or of further technical factors (i.e. effects due to cropping, fishing, diet or other environmental exploitation process).

Metagenomic analyses have been applied to soil and water samples (ranging from oceans to freshwater or extreme environments), including niches particularly rich in invertebrate species. As shown by various reports, metagenomic approaches may allow the study of single genes and pathways (e.g. antibiotic synthesis), as well as the identification and characterization of group of organisms and/or of whole microbial communities (Healy et al. 1995; Rondon et al. 2000; Tyson et al. 2004; Vogel et al. 2009).

Many other applications have been also performed including, among others, the metagenomics of food, animals or man, of some specific niches or organs in different organisms (i.e. gut, intestine, lungs), supporting a prevalent role of metagenomics in applied experimental sciences (Caporaso et al. 2011). These approaches are extremely useful in the study of bacterial ecology, because it is now recognized that only a small number of OTUs (0.1–1 %) can be identified through traditional methods (cultivation and isolation). The majority of bacterial species in fact are in large part undetected and their effects only partially determined (Rosello-Mora et al. 2001; Roesch et al. 2007).

3.1.1 Soil Metagenomics

Soil bacteria are closely associated to solid particles and organic fractions, and their metabolism reflects the properties of the local microhabitat conditions, which may vary over very small scales. The structural, physiochemical heterogeneity of the soil microhabitat is also affected by soil microfauna, or by additional anthropic factors, including aboveground plant communities, forests or deforestation, agricultural practices applied to cropping, use of fertilizers and organic matter, or the chemicals applied for pest and weed control (Souza et al. 2013).

A number of evolutive pressures is also active at the microhabitat scale, making metagenomic analyses a valuable tool for identification of unsuspected bacteria-invertebrate associations. The growing interest in soil metagenomics is in fact related to the deep insight made available by these techniques on bacterial species or groups. They account for a number of specialized functions and show a high biotechnological potential, related to exploitation of secondary metabolites or useful processes like antibiosis or biological control (Schloss and Handelsman 2003; Riesenfeld et al. 2004; Van Elsas et al. 2008).

Possible fallouts expected by soil metagenomics are in fact the identification of unculturable bacteria (including new taxa with industrial potential) or of species from which new possible applications could emerge. Microbial diversity in soil may range from a few thousand OTUs $\cdot$ g^{-1} of soil to very large densities up to 10^7 bacterial cells $\cdot$ g^{-1}, of which only 0.1 % appears culturable (Torsvik et al. 1996; Kellenberger 2001).

3.1.2 Invertebrate Metagenomics

Sequencing the ribosomal 16S rRNA has been applied to explore the diversity of bacterial clades associated to many soil invertebrates, including nematodes (Tian et al. 2015). In free living or plant parasitic nematodes the number of species associated, endosymbiotic or regulating their density is progressively increasing. New taxa have been discovered or isolated thanks to advanced molecular techniques, like i.e. specialized parasitic endosymbionts of the genus *Pasteuria* (Zou et al. 2015). Rhizosphere bacteria have practical utility including pests regulation or growth promotion, and the pool of soil microorganisms associated to nematodes and yet undescribed or unrecovered represents an unexplored biological reservoir, with a high application potential, not yet fully exploited.

Invertebrates harbor many unculturable bacteria, which in some samples may account for more than 90 % of total species. This largely unexplored "universe" has a high potential value, not only for basic research. It is generally accepted that the large number of microorganisms include species useful because of their biology (i.e. as pest regulators) or coding for a number of potentially useful products. These include for example enzymes for cellulose or lignin degradation, antibiotics and/or other metabolites not yet characterized, and worth efforts for identification and exploitation (Krishnan et al. 2014).

The insects gut is considered as a hotspot for bacterial biodiversity. However, rather than numerical diversification, it is characterized by a functional diversity, given the high density of metabolically active cells, and their high conjugation rates. In Collembola, a specific microflora rich in fungi and bacteria is associated to a diet rich in biopolymers, that they contribute to degrade. In particular, many members of the gut microbiota digest chitin, which is the major constituent of fungal cell walls and is also present in the insect after molt as esuvial remnant. Among chitin digestors a number of *Bacillus* spp. and other bacteria like *Stenotrophomonas maltophilia* and *Curtobacterium* sp. were found. The bacterial flora of Collembola also includes several intracellular *Wolbachia* sp. and other endosymbionts, present in fat bodies, ovaries and other tissues (Tebbe et al. 2006).

Gut bacteria have a role critical to insect nutrition, physiology, immune responses and pathogen resistance (Engel and Moran 2013a). A particular attention has been paid to the microorganisms present in the gut of termites, in view of the possible exploitation of their enzymes in the production of biofuel (ethanol), starting from wood (Sethi et al. 2013). Termites host a complex microbial community in their hindgut, integrating bacteria, protozoa and archaea in the digestion of lignocellulose, the main insect diet component, and release of acetate as insect basic nourishment (Boucias et al. 2013; Scharf 2015).

A metagenomic study of the hindgut of *Reticulitermes flavipes* showed that more than 580 bacterial species (mostly from the phyla *Spirochaetes*, *Elusimicrobia*, *Bacteroidetes*, *Firmicutes* and *Proteobacteria*) were present. A reduced number of OTUs with 69 core taxa, conserved among more than 95 % of samples, accounted for the majority (67 %) of the 16S rRNA sequences (V4 region). The most represented bacterium of *R. flavipes* belonged to the genus *Treponema*, which was present in the common set of core sequences, in all samples. Differences were, however, observed among colonies as well as between the alate and worker termite castes, as shown by β-diversity analyses, possibly due to their different dietary requirements (Benjamino and Graf 2016).

The debated "core microbiota" concept concerns a sort of functional fingerprint of the host, since the species commonly shared in different samples and situations account for a number of required processes, that allow host stability and resilience. The microbiota composition is based on the view that a common set of functions and metabolic activities determine the taxonomic composition of an organ or tissue, and is conserved along a range of environmental changes. Similar studies in insects and other invertebrate hosts revealed varying but low numbers of gut core microbiota OTUs, often restricted to a reduced set of phyla (Table 9.2). The gut microbiome of the honeybee *Apis mellifera*, for example, shows a reduced core microbiome set (five to nine species) which are common among social bees, playing symbiotic roles critical to the host health and survival. In the tsetse flies *Glossinia* spp., the most dominant and common core gut microbial group was represented only by *Wigglesworthia* spp. (Sabree et al. 2012; Engel and Moran 2013b; Aksoy et al. 2014).

3.2 Metatranscriptomics

The ensemble of genes expressed in a given environment and/or time, by all the organisms therein present, that hence includes all transcripts expressed in the active metabolic pathways, is referred to as the **metatranscriptome**. Its analysis is now technically affordable thanks to reduced sequencing costs. Metatranscriptomic data may result very informative about the processes and species active in a determined condition, as well as about their relationships and role played in a given environment or community, or on the way they react to changes (i.e. pollution or anthropic activities) occurring in their local environments.

This global transcriptomic approach is useful when the samples or organisms to study are difficult to reach or collect, or live in extreme conditions which do not permit fast and easy samplings, thus hindering the reproducibility of the studies. Further circumstances justifying such a deep sequencing approach concern the need for a fast sample processing given the effects of transcript changes related to storage, or acting on a very short time scale.

Metatranscriptomic analyses were used, for example, in the study of gill symbionts of *Alviniconcha*, a snail living in proximity of deep-sea hydrothermal vents. The snails are associated to two γ-proteobacteria and an ε-proteobacterium. The symbionts metabolism (and indirectly the ecosystem and species in which they live) are independent from sunlight, gaining energy from the oxidation of reduced compounds of geothermal origin (Sanders et al. 2013). The physiology and the spatial patterns of the snail associated bacterial species was found to depend on the local geochemistry conditions. Since transcriptomic studies suffered from an effect of

Table 9.2 Examples of invertebrate core gut microbiota and most represented phyla

Hosts	Core OTUs	Most represented phyla	Reference
Termites, fungus-growing termites	2	*Firmicutes, Bacteroidetes, Spirochaetes, Proteobacteria, Synergistes*	Benjamino and Graf (2016) and Otani et al. (2014)
Honeybees	5	*Firmicutes, Proteobacteria, Actinobacteria,*	Sabree et al. (2012)
Cimex lectularius	2	*Proteobacteria*	Meriweather et al. (2013)
Anopheles gambiae	2	*Cyanobacteria, Proteobacteria, Bacteroidetes, Actinobacteria, Firmicutes*	Wang et al. (2011)
Glossina spp.	1	*γ-Proteobacteria*	Aksoy et al. (2014)
Penaeus monodon	14–88	*Actinobacteria, Fusobacteria, Proteobacteria, Firmicutes, Bacteroidetes*	Rungrassamee et al. (2013, 2014)
Crassostrea virginica	5–44	*Firmicutes, Planctomycetes, γ-Proteobacteria*	King et al. (2012)
Hirudo verbana	2	*γ-Proteobacteria, Bacteroidetes*	Maltz et al. (2014)

sampling and storage (changing gene expression among samples), a device was applied for sampling and processing, extracting the RNAs *in situ*, halting gene expression and stabilizing the transcripts for later processing, pyrosequencing and metatranscriptome analysis. Data showed differences in the expression of genes for hydrogen and sulfur oxidation among different symbionts and samples associated to distinct geochemical environments, as well as common patterns for usage of genes involved in nitrogen metabolism (Sanders et al. 2013). A similar approach was used to study bacterioplankton taxa and the succession of metabolically distinct groups, involved in the aerobic degradation of methylphosphonate, a process contributing to the methane supersaturation of oceanic waters (Martinez et al. 2014).

3.3 Metabolomics and Proteomics

One of the main advantages and benefits expected by the progressively reduced costs of the genomic and other molecular biology techniques is the possibility to develop more feasible and low cost treatments for diseases and parasites that affect million people worldwide, and in particular in poorest regions. Progress also concerns the basic understanding of many processes that underpin disease transmission and vector associations, as well as the biology of the pathogens involved and possible therapeutic targets (Slatcko et al. 2014).

Metabolomic approaches aim at the identification of all the small metabolic products and intermediate metabolites, present in a given organism or tissue, and interacting with (or resulting by) external factors. These may include not only environmental conditions but also anthropic drivers like nutritional status and diet, as well as social environment or income. Marine bacteria and symbiont are considered as a fundamental biological resource for mining novel secondary metabolites with pharmacological properties, and their metabolome is actively investigated. By applying liquid chromatography-high resolution mass spectrometry and nuclear magnetic resonance spectroscopy to marine sponge symbionts, Macintyre et al. (2014) were able to identify and select three bacterial strains (*Bacillus* sp. 4117, *Rhodococcus* sp. Zs402 and *Vibrio splendidus* lgp32) based on their metabolomic profile. A further basic resource for applied studies is the human metabolome database (HMDB) (www.hmdb.ca) listing more than 40,000 annotated metabolite entries, including the expected secondary products of several biochemical pathways (Wishart et al. 2013). Coupled to metagenomics, the investigation on novel metabolic products and/or species, including the unculturable ones, may disclose new perspectives for the practical the exploitation of minor, or even neglected, bacterial species (Wilson and Piel 2013).

Similarly, **proteomics**, a concept developed by Wilkins et al. (1996) aims at identifying, through mass spectroscopy, all the proteins present in a determined condition and, when a genome is available, their corresponding encoding genes. This approach is very dynamic, since dramatic changes in protein encoding may occur even on a short time scale. It is applied to reconstruct the functionning and response of the cellular biochemical machinery, in conditions varying from normal metabolism to pathogenesis. The proteome and transcriptome may in fact differ, due to post-transcriptional silencing or degradation of mRNAs, gene expression occurring at a low levels or changes like acylation, methylation, glycosylation, phosphorylation or proteolysis, that may affect function and structure of proteins, in the cell environment (Finetti-Sialer and Manzanilla-López 2011).

A large-scale proteomic characterization analysis applied to the human filarial parasite *Brugia malayi* yielded also a significant part of its *Wolbachia* predicted proteome. Nematode proteins identified were related to the *B. malayi* differentiation stages. Proteome data also showed stage-specific protein expression by its endosymbiont (Bennuru et al. 2011). Apart of altering the reproductive behaviour of its hosts, *Wolbachia* may also protect them from pathogenic species or interfere with the transmission of diseases, when associated to arthropod vectors. It also appears as a feasible therapeutic target. Tetracycline antibiotics are prescribed to control lymphatic filariasis in humans exposed to river blindness (caused by *Onchocerca volvulus*) due to the bacterium sensitivity to the antibiotic. A proteomic study carried out *in vitro* on *D. melanogaster* cells infected with a *Wolbachia* endosymbiont, showed that the bacterium responds to the antibiotic deoxycyline by modulating translation, upregulating nucleotide synthesis and energy metabolism, increasing translocase and ABC transporters activity, and downregulating outer membrane proteins. All these data indicate a *Wolbachia* capacity to adapt to stress by a responsive behaviour. They appear hence important in order to improve the antibiotic efficacy in chemotherapy and understand the bacterium capacity to adapt to the chemical stress induced by exposure to antibiotics (Darby et al. 2014).

Quantitative proteomic analyses of cytoplasmic incompatibility induced in *Drosophila melanogaster* by *Wolbachia* showed molecular mechanisms based on a set of 83 proteins differentially expressed when compared to the uninfected control. They were related to metabolism, immunity and reproduction. The endosymbiont was capable to change proteins abundance in the insect spermathecae and seminal receptacles by affecting the proteolysis mediated by the ubiquitin–proteasome complex (Yuan et al. 2015).

Proteomic analyses also result informative about the invertebrate and pathogens interaction in productive ecosystems. Hou et al. (2016) applied proteomic analyses to the study of the immune response of hemocytes of the freshwater prawn *Macrobrachium rosenbergii* during infection by pathogenic *Spiroplasma eriocheiris*. The disease has a dramatic impact on the aquaculture industries in China and South East Asia, and is particularly severe in high-density farming systems. The prawn hemocytes are involved in the defense response through processes like phagocytosis, non-self recognition, encapsulation and elimination of reactive oxygen intermediates. Proteomic analyses through a stable isotopes and mass spectrometry method, performed before and after infection, showed 49 up-regulated and 20 down-regulated proteins, indicative of a host reaction to infection. Up regulated proteins included immunologic (like vitellogenin or a Gram-negative binding proteins involved in LPS recognition), as well as physiologic and intracellular proteins. Down-regulated proteins were involved in hemocytes immunity reactions, with two common pattern recognition receptors (Hou et al. 2016).

Due to the ease of *in vitro* cultivation and the large number of pathogenic bacteria which may be assumed by feeding, *C. elegans* represents a suitable model to study the proteome after exposure to a bacterial infection, with potential fallouts in the study also of vertebrate and human pathogens. In this nematode, 65 known proteins were differentially synthesised after infection with the gram-negative pathogen *Aeromonas hydrophila*, when compared to control. The proteins were involved in immunity, and included galectins, C-type lectins and lipid binding proteins (Bogaerts et al. 2010). A quantitative proteome analysis performed through isotopic labeling and mass spectrometry, upon exposure of *C. elegans* to Bt also showed 288 proteins differentially abundant, including lectins, lysozymes, and transthyretin-like proteins involved in the innate immune defense, together with other proteases and apoptosis related proteins (Treitz et al. 2015). Transcriptomic-proteomic analyses also confirmed differential changes of C-type lectin containing proteins, indicative of post transcriptional protein changes, influenced by AMP-activated protein kinases (AMPKs), acting as regulators in the nematode immune response (Yang et al. 2015).

References

Aksoy, E., et al. (2014). Analysis of multiple tsetse fly populations in Uganda reveals limited diversity and species-specific gut microbiota. *Applied and Environmental Microbiology, 80*, 4301–4312.

Alikhan, N. F., Petty, N. K., Ben Zakour, N. L., & Beatson, S. A. (2011). BLAST Ring Image Generator (BRIG): Simple prokaryote genome comparisons. *BMC Genomics, 12*, 402.

Allard, G., Ryan, F. J., Jeffery, I. B., & Claesson, M. J. (2015). SPINGO: A rapid species-classifier for microbial amplicon sequences. *BMC Bioinformatics, 16*, 324.

Anderson, M. J., et al. (2011). Navigating the multiple meanings of β diversity: A roadmap for the practicing ecologist. *Ecology Letters, 14*, 19–28.

Baldo, L., Bordenstein, S., Wernegreen, J. J., & Werren, J. H. (2006). Widespread recombination throughout *Wolbachia* genomes. *Molecular Biology and Evolution, 23*, 437–449.

Benítez-Páez, A., Portune, K. J., & Sanz, Y. (2016). Species-level resolution of 16S rRNA gene amplicons sequenced through the MinION™ portable nanopore sequencer. *GigaScience, 5*, 4.

Benjamino, J., & Graf, J. (2016). Characterization of the core and caste-specific microbiota in the termite, *Reticulitermes flavipes. Frontiers in Microbiology, 7*, 171.

Bennuru, S., et al. (2011). Stage-specific proteomic expression patterns of the human filarial parasite *Brugia malayi* and its endosymbiont *Wolbachia. Proceedings of the National Academy of Science USA, 108*, 9649–9654.

Berg, J. M., Tymoczko, J. L., & Stryer, L. (2002). *Biochemistry* (5th ed., 1100 pp.). New York: W H Freeman.

Bernardi, G. (1985). Codon usage and genome composition. *Journal of Molecular Evolution, 22*, 363–365.

Bogaerts, A., et al. (2010). A differential proteomics study of *Caenorhabditis elegans* infected with *Aeromonas hydrophila. Developmental and Comparative Immunology, 34*, 690–698.

Boucias, D. G., et al. (2013). The hindgut lumen prokaryotic microbiota of the termite *Reticulitermes flavipes* and its responses to dietary lignocellulose composition. *Molecular Ecology, 22*, 1836–1853.

Brinza, L., et al. (2010). Structure and dynamics of the operon map of *Buchnera aphidicola* sp. strain APS. *BMC Genomics, 11*, 666.

Brocchieri, L. (2014). The GC content of bacterial genomes. *Journal of Phylogenetic & Evolutionary Biology, 2*, 1000e108.

Byzov, B. A., et al. (1999). Plasmid transfer between introduced and indigenous bacteria in leaf litter, soil and vermicompost as affected by soil invertebrates. *Biology and Fertility of Soils, 28*, 169–176.

Caporaso, J. G., et al. (2011). Global patterns of 16S rRNA diversity at a depth of millions of sequences per sample. *Proceedings of the National Academy of Science USA, 108*, 4516–4522.

Darby, A. C., et al. (2014). Integrated transcriptomic and proteomic analysis of the global response of *Wolbachia* to doxycycline-induced stress. *ISME Journal, 8*, 925–937.

Darling, A. E., Miklós, I., & Ragan, M. A. (2008). Dynamics of genome rearrangement in bacterial populations. *PLoS Genetics, 4*, e1000128.

Engel, P., & Moran, N. A. (2013a). The gut microbiota of insects – diversity in structure and function. *FEMS Microbiology Reviews, 37*, 699–735.

Engel, P., & Moran, N. A. (2013b). Functional and evolutionary insights into the simple yet specific gut microbiota of the honey bee from metagenomic analysis. *Gut Microbes, 4*, 60–65.

Field, D., Feil, E. J., & Wilson, G. A. (2005). Databases and software for the comparison of prokaryotic genomes. *Microbiology, 151*, 2125–2132.

Finetti-Sialer, M. M., & Manzanilla-López, R. H. (2011). Exploiting "omics" and molecular approaches in plant nematology research. In R. Rodríguez Herrera, C. N. Aguilar, J. Kilpatrick Simpson-Williamson, & G. Gutierrez Sanchez (Eds.), *Phytopathology in the Omics era* (pp. 39–68). Trivandrum: Research Signpost.

Fleischmann, R. D., et al. (1995). Whole-genome random sequencing and assembly of *Haemophilus influenzae* Rd. *Science, 269*(5223), 496–512.

Friedland, A. E., et al. (2013). Heritable genome editing in *C. elegans* via a CRISPR-Cas9 system. *Nature Methods, 10*, 741–743.

Galperin, M. Y., Makarova, K. S., Wolf, Y. I., & Koonin, E. V. (2015). Expanded microbial genome coverage and improved protein family annotation in the COG database. *Nucleic Acids Research, 43*, D261–D269.

Gillespie, J. J., et al. (2012). A *Rickettsia* genome overrun by mobile genetic elements provides insight into the acquisition of genes characteristic of an obligate intracellular lifestyle. *Journal of Bacteriology, 194*, 376–394.

Good, L., & Stach, J. E. M. (2011). Synthetic RNA silencing in bacteria – Antimicrobial discovery and resistance breaking. *Frontiers in Microbiology, 2*, 185.

Gregory, T. R. (2005). Synergy between sequence and size in large-scale genomics. *Nature Reviews Genetics, 6*, 699–708.

Haake, D. A., et al. (2004). Molecular evolution and mosaicism of leptospiral outer membrane proteins involves horizontal DNA transfer. *Journal of Bacteriology, 186*, 2818–2828.

Handelsman, J. (2004). Metagenomics: Application of genomics to uncultured microorganisms. *Microbiology and Molecular Biology Reviews, 68*, 669–685.

Healy, F. G., et al. (1995). Direct isolation of functional genes encoding cellulases from the microbial consortia in a thermophilic, anaerobic digester maintained on lignocellulose. *Applied Microbiology and Biotechnology, 43*, 667–674.

Hershberg, R., & Petrov, D. A. (2010). Evidence that mutation is universally biased towards AT in bacteria. *PLoS Genetics, 6*, e1001115.

Hildebrand, F., Meyer, A., & Eyre-Walker, A. (2010). Evidence of selection upon genomic GC-content in bacteria. *PLoS Genetics, 6*, e1001107.

Hill, J. E., et al. (2002). Extensive profiling of a complex microbial community by high-throughput sequencing. *Applied and Environmental Microbiology, 68*, 3055–3066.

Hill, J. E., Penny, S. L., Crowell, K. G., Goh, S. H., & Hemmingsen, S. M. (2004). cpnDB: A chaperonin sequence database. *Genome Research, 14*, 1669–1675.

Horvath, P., & Barrangou, R. (2010). CRISPR/Cas, the immune system of bacteria and archaea. *Science, 327*, 167–170.

Hou, L., et al. (2016). iTRAQ-based quantitative proteomic analysis of *Macrobrachium rosenbergii* hemocytes during *Spiroplasma eriocheiris* infection. *Journal of Proteomics, 136*, 112–122.

House, C. H., Pellegrini, M., & Fitz-Gibbon, S. T. (2015). Genome-wide gene order distances support clustering the gram-positive bacteria. *Frontiers in Microbiology, 5*, 785.

Howell-Adams, B., & Seifert, H. S. (2000). Molecular models accounting for the gene conversion reactions mediating gonococcal pilin antigenic variation. *Molecular Microbiology, 37*, 1146–1158.

Hussain, M., Frentiu, F. D., Moreira, L. A., O'Neill, S. L., & Asgari, S. (2011). *Wolbachia* uses host microRNAs to manipulate host gene expression and facilitate colonization of the dengue vector *Aedes aegypti*. *Proceedings of the National Academy of Science USA, 108*, 9250–9255.

Hussain, M., O'Neill, S. L., & Asgari, S. (2013). *Wolbachia* interferes with the intracellular distribution of Argonaute 1 in the dengue vector *Aedes aegypti* by manipulating the host microRNAs. *RNA Biology, 10*, 1868–1875.

Jacob, F., & Monod, J. (1961). Genetic regulatory mechanisms in the synthesis of proteins. *Journal of Molecular Biology, 3*, 318–356.

Jinek, M., et al. (2012). A programmable dual-RNA-guided DNA endonuclease in adaptive bacterial immunity. *Science, 337*, 816–821.

Jost, L. (2006). Entropy and diversity. *Oikos, 113*, 363–375.

Junier, I., & Rivoire, O. (2013). Synteny in bacterial genomes: Inference, organization and evolution. *arXiv*,1307.4291.

Kellenberger, E. (2001). Exploring the unknown: The silent revolution of microbiology. *EMBO Reports, 2*, 5–7.
King, G. M., Judd, C., Kuske, C. R., & Smith, C. (2012). Analysis of stomach and gut microbiomes of the eastern oyster (*Crassostrea virginica*) from coastal Louisiana, USA. *PLoS ONE, 7*, e51475.
Klier, A., Fargette, F., Ribier, J., & Rapoport, G. (1982). Cloning and expression of the crystal protein genes from *Bacillus thuringiensis* strain berliner 1715. *EMBO Journal, 1*, 791–799.
Kole, R., Krainer, A. R., & Altman, S. (2012). RNA therapeutics: Beyond RNA interference and antisense oligonucleotides. *Nature Reviews in Drug Discovery, 11*, 125–140.
Kota, M., Daniell, H., Varma, S., Garczynski, S. F., Gould, F., & William, M. J. (1999). Overexpression of the *Bacillus thuringiensis* (Bt) Cry2Aa2 protein in chloroplasts confers resistance to plants against susceptible and Bt-resistant insects. *Proceedings of the National Academy of Science USA, 96*, 1840–1845.
Krishnan, M., et al. (2014). Insect gut microbiome – An unexploited reserve for biotechnological application. *Asian Pacific Journal of Tropical Biomedicine, 4*(Suppl. 1), S16–S21.
Lassalle, F., et al. (2015). GC-content evolution in bacterial genomes: The biased gene conversion hypothesis expands. *PLoS Genetics, 11*, e1004941.
Leclercq, S., Giraud, I., & Cordaux, R. (2011). Remarkable abundance and evolution of mobile Group II introns in *Wolbachia* bacterial endosymbionts. *Molecular Biology and Evolution, 28*, 685–697.
Links, M. G., Dumonceaux, T. J., Hemmingsen, S. M., & Hill, J. E. (2012). The Chaperonin-60 universal target is a barcode for bacteria that enables de novo assembly of metagenomic sequence data. *PLoS ONE, 7*, e49755.
Macintyre, L., et al. (2014). Metabolomic tools for secondary metabolite discovery from marine microbial symbionts. *Marine Drugs, 12*, 3416–3448.
Maltz, M. A., et al. (2014). Metagenomic analysis of the medicinal leech gut microbiota. *Frontiers in Microbiology, 5*, 151.
Marchler-Bauer, A., et al. (2015). CDD: NCBI's conserved domain database. *Nucleic Acids Research, 43*(D), 222–226.
Martin, D., Williamson, C., & Posada, D. (2005). RDP2: Recombination detection and analysis from sequence alignments. *Bioinformatics, 21*, 260–262.
Martínez, A., Ventouras, L. A., Wilson, S. T., Karl, D. M., & DeLong, E. F. (2014). Metatranscriptomic and functional metagenomic analysis of methylphosphonate utilization by marine bacteria. *Frontiers in Microbiology, 4*, 340.
Masui, S., Sasaki, T., & Ishikawa, H. (1997). *groE*-homologous operon of *wolbachia*, an intracellular symbiont of arthropods: A new approach for their phylogeny. *Zoological Science, 14*, 701–706.
McCutcheon, J. P., & Moran, N. A. (2010). Functional convergence in reduced genomes of bacterial symbionts spanning 200 My of evolution. *Genome Biology & Evolution, 2*, 708–718.
McCutcheon, J. P., & Moran, N. A. (2012). Extreme genome reduction in symbiotic bacteria. *Nature Reviews Microbiology, 10*, 13–26.
McCutcheon, J. P., & von Dohlen, C. D. (2011). An interdependent metabolic patchwork in the nested symbiosis of mealybugs. *Current Biology, 21*, 1366–1372.
Meeus, P. F. M., Brayton, K. A., Palmer, G. H., & Barbet, A. F. (2003). Conservation of a gene conversion mechanism in two distantly related paralogues of *Anaplasma marginale. Molecular Microbiology, 47*, 633–643.
Meriweather, M., Matthews, S., Rio, R., & Baucom, R. S. (2013). A 454 survey reveals the community composition and core microbiome of the common bed bug (*Cimex lectularius*) across an urban landscape. *PLoS ONE, 8*, e61465.
Miller, R. R., Montoya, V., Gardy, J. L., Patrick, D. M., & Tang, P. (2013). Metagenomics for pathogen detection in public health. *Genome Medicine, 5*, 81.
Miyashiro, T., & Ruby, E. G. (2012). Shedding light on bioluminescence regulation in *Vibrio fischeri. Molecular Microbiology, 84*, 795–806.

Moran, N. A. (2002). Microbial minimalism: Genome reduction in bacterial pathogens. *Cell, 108*, 583–586.

Nakanishi, T., Kato, Y., Matsuura, T., & Watanabe, H. (2014). CRISPR/Cas-mediated targeted mutagenesis in *Daphnia magna. PLoS ONE, 9*, e98363.

Nakashima, N., Goh, S., Good, L., & Tamura, T. (2012). Multiple-gene silencing using antisense RNAs in *Escherichia coli*. In M. Kaufmann & C. Klinger (eds.), *Functional genomics: Methods and protocols* (Methods in molecular biology, Vol. 815, pp. 307–319). New York: Humana Press.

Nikoh, N., et al. (2014). Evolutionary origin of insect–*Wolbachia* nutritional mutualism. *Proceedings of the National Academy of Science USA, 111*, 10257–10262.

Nordström, K., & Dasgupta, S. (2006). Copy-number control of the *Escherichia coli* chromosome: A plasmidologist's view. *EMBO Reports, 7*, 484–489.

Normand, P., et al. (2015). Biodiversity and microbial ecosystems functioning. In J.-C. Bertrand et al. (Eds.), *Environmental microbiology: Fundamentals and applications: Microbial ecology* (pp. 261–291). Dordrecht: Springer.

Otani, S., et al. (2014). Identifying the core microbial community in the gut of fungus-growing termites. *Molecular Ecology, 23*, 4631–4644.

Ozen, A. I., & Ussery, D. W. (2014). Genome atlases, potential applications in study of metagenomes. In *Encyclopedia of metagenomics* (pp. 1–4). New York: Springer.

Patrick, D. S., & Handelsman, J. (2005). Metagenomics for studying unculturable microorganisms: Cutting the Gordian knot. *Genome Biology, 6*, 229.

Penz, T., et al. (2012). Comparative genomics suggests an independent origin of cytoplasmic incompatibility in *Cardinium hertigii. PLoS Genetics, 8*, e1003012.

Rancès, E., Voronin, D., Tran-Van, V., & Mavingui, P. (2008). Genetic and functional characterization of the type IV secretion system in *Wolbachia. Journal of Bacteriology, 190*, 5020–5030.

Reyes-Ramírez, A., & Ibarra, J. E. (2008). Plasmid patterns of *Bacillus thuringiensis* type strains. *Applied and Environmental Microbiology, 74*, 125–129.

Rich, S. M., Sawyer, S. A., & Barbour, A. G. (2001). Antigen polymorphism in *Borrelia hermsii*, a clonal pathogenic bacterium. *Proceedings of the National Academy of Science USA, 98*, 15038–15043.

Riesenfeld, C. S., Goodman, R. M., & Handelsman, J. (2004). Uncultured soil bacteria are a reservoir of new antibiotic resistance genes. *Environmental Microbiology, 6*, 981–989.

Roesch, L. F. W., et al. (2007). Pyrosequencing enumerates and contrasts soil microbial diversity. *ISME Journal, 1*, 283–290.

Rondon, M. R., et al. (2000). Cloning the soil metagenome: A strategy for accessing the genetic and functional diversity of uncultured microorganisms. *Applied and Environmental Microbiology, 66*, 2541–2547.

Rong, X., Zhang, Y. K., Zhang, K. J., & Hong, X. Y. (2014). Identification of *Wolbachia*-responsive microRNAs in the two-spotted spider mite, *Tetranychus urticae. BMC Genomics, 15*, 1122.

Rosello-Mora, R., et al. (2001). The species concept for prokaryotes. *FEMS Microbiology Reviews, 25*, 39–67.

Ruby, E. G., et al. (2005). Complete genome sequence of *Vibrio fischeri*: A symbiotic bacterium with pathogenic congeners. *Proceedings of the National Academy of Science USA, 102*, 3004–3009.

Rungrassamee, W., et al. (2013). Bacterial population in intestines of the black tiger shrimp (*Penaeus monodon*) under different growth stages. *PLoS One, 8*, e60802.

Rungrassamee, W., et al. (2014). Characterization of intestinal bacteria in wild and domesticated adult black tiger shrimp (*Penaeus monodon*). *PLoS One, 9*, e91853.

Sabree, Z. L., Hansen, A. K., & Moran, N. A. (2012). Independent studies using deep sequencing resolve the same set of core bacterial species dominating gut communities of honey bees. *PLoS ONE, 7*, e41250.

Sait, M., et al. (2013). Genome sequence of *Lawsonia intracellularis* strain N343, isolated from a sow with hemorrhagic proliferative enteropathy. *Genome Announcements, 1*, e00027–e13.

Sanders, J. G., Beinart, R. A., Stewart, F. J., Delong, E. F., & Girguis, P. R. (2013). Metatranscriptomics reveal differences in *in situ* energy and nitrogen metabolism among hydrothermal vent snail symbionts. *The ISME Journal, 7*, 1556–1567.

Schaefer, A. L., Val, D. L., Hanzelka, B. L., Cronan, J. E., Jr., & Greenberg, E. P. (1996). Generation of cell-to-cell signals in quorum sensing: Acyl homoserine lactone synthase activity of a purified *Vibrio fischeri* LuxI protein. *Proceedings of the National Academy of Science USA, 93*, 9505–9509.

Scharf, M. E. (2015). Omic research in termites: An overview and a roadmap. *Frontiers in Genetics, 6*, 76.

Schloss, P. D., & Handelsman, J. (2003). Biotechnological prospects from metagenomics. *Current Opinions in Biotechnology, 14*, 303–310.

Schnnepf, E., et al. (1998). *Bacillus thuringiensis* and its pesticidal crystal proteins. *Microbiology and Molecular Biology Reviews, 62*, 775–806.

Schunder, E., Rydzewski, K., Grunow, R., & Heune, K. (2013). First indication for a functional CRISPR/Cas system in *Francisella tularensis. International Journal of Medical Microbiology, 303*, 51–60.

Septer, A. N., & Stabb, E. V. (2012). Coordination of the arc regulatory system and pheromone-mediated positive feedback in controlling the *Vibrio fischeri* lux operon. *PLoS ONE, 7*, e49590.

Sethi, A., Slack, J. M., Kovaleva, E. S., Buchman, G. W., & Scharf, M. E. (2013). Lignin-associated metagene expression in a lignocellulose-digesting termite. *Insect Biochemistry and Molecular Biology, 43*, 91–101.

Slatko, B. E., Luck, A. N., Dobson, S. L., & Foster, J. M. (2014). *Wolbachia* endosymbionts and human disease control. *Molecular & Biochemical Parasitology, 195*, 88–95.

Sogin, M. L., et al. (2006). Microbial diversity in the deep sea and the underexplored "rare biosphere". *Proceedings of the National Academy of Science USA, 103*, 12115–12120.

Souza, R. C., Cantão, M. E., Ribeiro Vasconcelos, A. T., Nogueira, M. A., & Hungriaa, M. (2013). Soil metagenomics reveals differences under conventional and no-tillage with crop rotation or succession. *Applied Soil Ecology, 72*, 49–61.

Stothard, P., et al. (2005). BacMap: An interactive picture atlas of annotated bacterial genomes. *Nucleic Acids Research, 33*, D317–D320.

Susannah, G. T., & Edward, M. R. (2005). Metagenomics: DNA sequencing of environmental samples. *Nature Reviews Genetics, 6*, 805–814.

Tagomori, K., Iida, T., & Honda, T. (2002). Comparison of genome structures of Vibrios, bacteria possessing two chromosomes. *Journal of Bacteriology, 184*, 4351–4358.

Tamames, J. (2001). Evolution of gene order conservation in prokaryotes. *Genome Biology, 2*, 1–11.

Tamames, J., Gonzalez-Moreno, M., Mingorance, J., Valencia, A., & Vicente, M. (2001). Bringing gene order into bacterial shape. *Trends in Genetics, 17*, 124–126.

Tatusov, R. L., Koonin, E. V., & Lipman, D. J. (1997). A genomic perspective on protein families. *Science, 278*, 631–637.

Tebbe, C. C., Czarnetzki, A. B., & Thimm, T. (2006). Collembola as a habitat for microorganisms. In H. König & A. Varma (Eds.), *Intestinal microorganisms of termites and other invertebrates* (pp. 133–153). Heidelberg: Springer Science & Business Media.

Thomason, M. K., & Storz, G. (2010). Bacterial antisense RNAs: How many are there and what are they doing? *Annual Reviews Genetics, 44*, 167–188.

Tian, B. Y., Cao, Y., & Zhang, K. Q. (2015). Metagenomic insights into communities, functions of endophytes, and their associates with infection by root-knot nematode, *Meloidogyne incognita*, in tomato roots. *Scientific Reports, 5*, 17087.

Toft, C., & Andersson, S. G. E. (2010). Evolutionary microbial genomics: Insights into bacterial host adaptation. *Nature Reviews Genetics, 11*, 465–475.

Torsvik, V., Sorheim, R., & Goksøyr, J. (1996). Total bacterial diversity in soil and sediment communities – A review. *Journal of Industrial Microbiology, 17*, 170–178.

Treitz, C., et al. (2015). Quantitative proteome analysis of *Caenorhabditis elegans* upon exposure to nematicidal *Bacillus thuringiensis. Journal of Proteomics, 113*, 337–350.

Tyson, G. W., et al. (2004). Community structure and metabolism through reconstruction of microbial genomes from the environment. *Nature, 428*, 37–43.

Van Elsas, J. D., et al. (2008). The metagenomics of disease suppressive soils – Experiences from the METACONTROL project. *Trends in Biotechnology, 26*, 591–601.

Van Ham, R. C., et al. (2003). Reductive genome evolution in *Buchnera aphidicola. Proceedings of the National Academy of Science USA, 100*, 581–586.

Veech, J. A., & Crist, T. O. (2010). Diversity partitioning without statistical independence of alpha and beta. *Ecology, 91*, 1964–1969.

Vogel, T. M., et al. (2009). TerraGenome: A consortium for the sequencing of a soil metagenome. *Nature Reviews Microbiology, 7*, 252.

Walling, L. L., & Kaloshian, I. (2016). Plant-herbivore interactions in the era of big data. In H. Czosnek & M. Ghanim (Eds.), *Management of insect pests to agriculture* (pp. 3–48). Cham: Springer International Publishing.

Wang, Y., Gilbreath, T. M., Kukutla, P., Yan, G., & Xu, J. (2011). Dynamic gut microbiome across life history of the malaria mosquito *Anopheles gambiae* in Kenya. *PLoS ONE, 6*, e24767.

Wassenaar, T. M., Binnewies, T. T., Hallin, P. F., & Ussery, D. W. (2010). Tools for comparison of bacterial genomes. In K. N. Timmis (Ed.), *Handbook of hydrocarbon and lipid microbiology* (pp. 4314–4327). Berlin: Springer.

Wernegreen, J. J. (2002). Genome evolution in bacterial endosymbionts of insects. *Nature Review Genetics, 3*, 850–861.

Whittaker, R. J., et al. (2001). Scale and species richness: Towards a general, hierarchical theory of species diversity. *Journal of Biogeography, 28*, 453–470.

Wiedenheft, B., Sternberg, S. H., & Doudna, J. A. (2012). RNA-guided genetic silencing systems in bacteria and archaea. *Nature, 482*, 331–338.

Wilkins, M. R., et al. (1996). From proteins to proteomes: Large scale protein identification by two-dimensional electrophoresis and amino acid analysis. *Biotechnology, 14*, 61–65.

Wilkins, C., et al. (2005). RNA interference is an antiviral defence mechanism in *Caenorhabditis elegans. Nature, 436*, 1044–1047.

Wilson, M. C., & Piel, J. (2013). Metagenomic approaches for exploiting uncultivated bacteria as a resource for novel biosynthetic enzymology. *Chemistry & Biology, 20*, 636–647.

Wishart, D. S., et al. (2013). HMDB 3.0 – The human metabolome database in 2013. *Nucleic Acids Research, 41*, D801–D807.

Wolf, Y. I., Rogozin, I. B., Grishin, N. V., Tatusov, R. L., & Koonin, E. V. (2001). Genome trees constructed using five different approaches suggest new major bacterial clades. *BMC Evolutionary Biology, 1*, 8.

Wu, H., Zhang, Z., Hu, S., & Yu, J. (2012). On the molecular mechanism of GC content variation among eubacterial genomes. *Biology Direct, 7*, 2.

Yang, Q., & Sze, S. H. (2008). Large-scale analysis of gene clustering in bacteria. *Genome Research, 18*, 949–956.

Yang, W., et al. (2015). Overlapping and unique signatures in the proteomic and transcriptomic responses of the nematode *Caenorhabditis elegans* toward pathogenic *Bacillus thuringiensis. Developmental and Comparative Immunology, 51*, 1–9.

Yuan, L. L., et al. (2015). Quantitative proteomic analyses of molecular mechanisms associated with cytoplasmic incompatibility in *Drosophila melanogaster* induced by *Wolbachia. Journal of Proteome Research, 14*, 3835–3847.

Zhang, W., et al. (2011). A practical comparison of *de novo* genome assembly software tools for next-generation sequencing technologies. *PLoS ONE, 6*, e17915.

Zhao, P., Zhang, Z., Ke, H., Yue, Y., & Xue, D. (2014). Oligonucleotide-based targeted gene editing in *C. elegans* via the CRISPR/Cas9 system. *Cell Research, 24*, 247–250.

Zhou, H. Q., Ning, L. W., Zhang, H. X., & Guo, F. B. (2014). Analysis of the relationship between genomic GC content and patterns of base usage, codon usage and amino acid usage in Prokaryotes: Similar GC content adopts similar compositional frequencies regardless of the phylogenetic lineages. *PLoS ONE, 9*, e107319.

Zou, X. X., et al. (2015). Population diversity of *Pasteuria penetrans* from pepper fields and its genetic variation from single root-knot nematodes. *Nematology, 17*, 865–876.

Part III
Applied Approaches

Chapter 10
Applications in Farming

Abstract A number of processes concerning the role of invertebrates in farming are described, including mechanisms of density regulation and the effects of associated bacteria. Some aspects related to productions by useful species, including pollinators, parasitoids or aquaculture, and the effects of endosymbionts or disease causal agents, are reviewed. Mechanisms of natural regulation in crops, including density dependence and biological antagonisms, are examined. Some issues concerning the potential of genetically modified crops and the bacterial toxins involved are briefly reviewed.

Keywords *Bacillus* • Biological control • Crop production • Food • Modeling • *Pasteuria* • Regulation • Nematode • Parasitism • Pathogen • Pests • Population density

1 Introduction

Natural regulation is common in many ecosystems, resulting by the balancing effects exerted by one or more organisms within a food web. Ideally, any "wild" environment should display a sort of natural regulation or balance, achieved at the end of million years of evolution. However, this situation may be found only in a limited number of ecosystems. It is indeed rarely observed, being human activities highly invasive and persistent. Many invertebrate populations studied thus far in natural conditions cannot be considered at equilibrium any more, even if they occur at low levels and persist in their niches or habitats, under minimal disturbance. Several factors (too often of anthropic origin, including global temperature increase), interfere with the regulation factors of insects or other invertebrates, and often become evident in the form of outbreaks difficult to control.

When the impact of the changes become evident, usually the situation is close to be irreversible. Some examples are provided by the epidemics caused by invasive pests, and the outbreaks of the pathogens that they eventually transmit. Further examples concern the insurgence and spread of bacterial diseases in marine populations like bivalves or shrimps raised at very high population densities for human

A. Ciancio, *Invertebrate Bacteriology*, DOI 10.1007/978-94-024-0884-3_10

consumption, or the effects of contamination by pollutants dispersed in the environment as a consequence of human activities.

In a general perspective, we may consider "nature" simply as a (*1*) huge, complex and long-lasting system (on a scale of billion years) producing and processing information (of any kind, including geochemical, mineral or genetic information), which is (*2*) dependent on one or more energy flows (solar, but also geothermal) and (*3*) evolving, under the pressure of many factors, including its own complexity and chaoticity. This system has, in any case, its roots controlled by the law(s) of physics. When considering an invertebrate population in a given ecosystem, the different density changes encountered for the organisms present may simply be regarded as the results of the system initial conditions and of its own components' behaviour and peculiarities, either genetic and physical, in function of its energy flow. Invariably, the final result is not sum of any single, individual component effect.

Although the rationale for the economics and the technical use of bacteria for invertebrate integrated or organic management are out of the scopes of this volume, many processes linking bacteria to farming and production, and the knowledge produced in basic research, may result useful in applied science. In this Chapter we will hence examine some of these systems, together with the biological and ecological processes underpinning their behaviour and stability.

2 Applied Aspects

2.1 Crops

Given the economic importance that invertebrates have in many productions, there is a rich literature describing the mechanisms, outcomes and impairs related to their natural regulation. Many products from crop and water food chains that feed our urban and rural societies are the results of a range of invertebrate-bacteria interactions, which ultimately exert strong social and economic impacts. Even more ones may be expected in the years to come, due to the consumption of many natural resources and of induced climate changes. In fact, many invertebrates enter our alimentary food chain directly as i.e. pollination or products like honey, insects, molluscs, shrimps and other aquaculture food. Indirectly, invertebrates are involved in many food production systems as pests. They also interact with their complex microbial consortia including transmission of plant, human or livestock pathogens. Insects that are indirectly introduced in the human food chain as livestock or aquaculture feed, may also be added to the list.

2.1.1 Pollinators

Pollinators like honey bees, bumblebees or wasps play a fundamental and essential role in plants ecology and in crop production as well, with a contribution to food estimated around 7 % of fruit and seed yields (Schulp et al. 2014). Economic value may be estimated in the order of 1–2 hundred billion €, worldwide (Williams 1994; Gallai et al. 2009; Lautenbach et al. 2012; Schulp et al. 2014). Enormous ecosystem services are provided by pollinators, of which insects are a significant component, but often they are not considered when natural ecosystems are destroyed for other anthropic uses (see the FAO *International Initiative for the Conservation and Sustainable Use of Pollinators*, http://www.fao.org/pollination/en/). Pollinators also represent fundamental indicators of an ecosystem's health status, and are indirect targets of pesticides like neonicotinoid and fipronil insecticides, or of other environmental contaminants (Abrol 2012; Chagnon et al. 2015).

Given their importance, monitoring the microorganisms present in pollinators may help in the evaluation of their effects on pollination and food production, and of the status of wild populations. The microbial consortia associated to pollinators affect their performance in a variable number of ways, including reproduction or as pathogens responsible of population declines. A recent survey carried out in the UK on declining populations of bees and wasps showed high prevalence by parasites, including *Wolbachia* and other viral parasites like the deformed wing virus, found in *Bombus terrestris*, in around 30 % of *Vespula vulgaris* and almost 100 % of sampled honey bees (Evison et al. 2012).

Honeybees provide a paramount example of evolutionary adaptations to pathogens, since they deploy a number of defense systems. These include the production of propolis whose flavonoid and polyphenolic constituents have antibacterial effects (Mihai et al. 2012). Furthermore, they developed a sort of social immunity, being capable to modulate the composition of their nectar food, including floral preference for presence of antimicrobial plant compounds. These are preferred depending on the occurrence of pathogens like *Paenibacillus larvae*, the causal agent of American foulbrood, or other bacteria associated to European foulbrood. Self-medication efficacy appears related to selectivity in plant preference that may be very high, including preference for plants, yielding bacterial strain-specific monofloral honey (Erler et al. 2014).

Honeybees also developed a capacity to vertically transfer an immunity towards pathogens to their larvae. This process involves the egg-yolk protein vitellogenin as a carrier of immune elicitors like PAMS, LPS, peptidoglycan and zymosan, characterizing pathogenic bacteria like *P. larvae* or other gram negative species. These were experimentally assumed by workers through bacteria-contaminated pollen or nectar and then transferred to the royal jelly, which is fed exclusively to the queen. The queen can digest the pathogens, storing fragments of immune system elicitors in her fat body and eventually transferring them to the developing eggs, in which the larvae become immunized (Salmela et al. 2015).

The associations of pollinators with bacterial are not only limited to pathogens, as endosymbionts also played a fundamental role in their evolutionary history. In fig wasps pollinators, for example, *Wolbachia* showed a strong correspondence with the evolutionary divergence of distincts clades, with the appearance of cryptic species in a strict fig and pollinating wasps associations (Sun et al. 2011). Butterflies provide further examples of endosymbionts affecting speciation paths and phenotypes. The maternal inheritance of *Wolbachia* in five distinct lepidopteran lineages led to changes in the sex ratio and to male killing or sperm-egg incompatibility, resulting in reproductive interference (Salunkhe et al. 2014).

2.1.2 Beneficial Insects

Parasitoids

Parasitoids are insects laying eggs in the body or eggs of other insects, and represent a particular feeding behaviour. When the affected victim is a parasitoid, the attacker is identified as a hyperparasitoid. Many parasitoids can kill or severely impair the reproductive capacity of their hosts, diverting significant amounts of resources to their own reproduction. Depending on their targets, parasitoids may result beneficial for plants, or may act as natural regulator of other insects, playing different roles in their ecosystems.

Several parasitoid-associated bacteria have been reported and usually confer distinct traits to their hosts, enhancing host selection behaviour and transmission (Zchori-Fein et al. 2001) or changing their sex ratio. Systemic and chronic bacterial infections in adults of the parasitoid wasp *Nasonia vitripennis* were reported to induce a higher mortality of males. The female-biased ratio was considered to reduce longevity and fecundity of adult females (Huger et al. 1985). Parasitoids may also be carriers of *Wolbachia* when probing infected or healthy whitefly hosts, leading to phoretic relationships and horizontal transmission (Ahmed et al. 2015). Bacteria have also been used, as biological markers, to follow the parasitoid oviposition (Jackson et al. 2004).

Endosymbiotic bacteria may also act on the host's side. Protective effects have been demonstrated for invasive pests attacked by parasitoids, whose biocontrol efficacy was limited as hosts received a benefit by endosymbiotic associations, involving *Wolbachia* and other bacteria (White et al. 2015). An enhancement in host survival was also experimentally demonstrated in *Drosophila hydei* associated to *Spiroplasma* sp. (Xie et al. 2010).

Silkworm

Silk is a polymer based on several protein types produced by rearing the larvae of the domesticated lepidopteran *Bombyx mori*. The moth is central to sericulture since its origins around 2700 BC in China. The silk industry warrants an income to low revenue farmers in many agricultural systems in East Asia. Bacterial diseases of silkworm are a major threat and have many aetiologic agents. Losses may reach 30–40 % of yields, and are mainly managed through the use of antibiotics (Sheebha et al. 2008; Mohanta et al. 2015).

HGT events were observed in the *B. mori* genome, of which nine were characterized by having a bacterial origin, conferring useful traits including resistance to pathogens (Zhu et al. 2011). For the causal agents of mulberry silkworm diseases see Chap. 4.

2.2 *Aquatic Industries*

Shellfish and molluscs are important components of our dietary protein provision (Roberts et al. 2005). Together with shrimp industry they generate high income in many countries, based either on coastal sea or freshwater infrastructures. Bacterial contamination of waters and insurgence of cell lines resistant to the antibiotics that are largely applied to control shrimp diseases represent a severe threat for this industry. *Vibrio* spp. are among the most ubiquitous marine bacteria found in these waters. Species include *Vibrio alginolyticus*, *V. parahaemolyticus*, *Shigella flexneri* and *E. coli*. High bacterial loads of these and other species (i.e. *Salmonella* spp.) detected in waistewater effluents from tiger shrimp farms, represent also a serious threat to human health. Water quality assessment is considered as an indispensable requisite to prevent emergencies and control the bacterial populations eventually present (Wahid et al. 2014).

Vibrio parahaemolyticus is the causal agent of a severe disease of shrimps called Early Mortality Syndrome (EMS), also known as Acute Hepatopancreatic Necrosis Disease. Susceptible species include *Penaeus vannamei*, *P. monodon* and *P. chinensis*, in which the most visible sign of disease is an abnormally reduced or atrophied hepatopancreas, leading to death in 3–4 weeks. The disease causes mass mortalities of larvae in East Asia and Central America farms, with losses that may reach 80 % of total products. Yearly economic losses account for hundred million USD, worldwide. Conditions conducive to outbreaks include high stocking densities and sensitivity, high waters salinity, poor hygiene in water treatment, management and feed conditions, stress exposure of shrimps during farming, and anoxic waters. The bacterium is sensitive to heating (55 °C for 30 min or 80 °C for 1 min can reduce significantly the cell loads), to refrigeration and freezing (−18 °C to −24 °C). Shrimps stored frozen for months showed undetectable pathogen levels (Zorriehzahra and Banaederakhshan 2015).

A further consequence of intensive shrimp farming is given by algal blooms due to water eutrophization effects on phyto- and zooplankton. In particular, the damages are related to dinoflagellates, which induce high shrimp mortalities as well as damage the ecosystem. Algicidal bacteria like *Marinobacteria salsuginis* have been identified and investigated at this regard. In particular applications were considered for mitigation of algal blooms caused by *Noctiluca* sp., in ponds used for shrimp farming (Keawtaweem et al. 2011).

Bacteria like *V. splendidus*, *V. aestuarianus* and *V. harveyi* have been associated to high rates of oyster (*Crassostrea gigas*) summer mortality in French shellfish plants, together with *Shewanella colwelliana*. A further species, *V. tubiashii*, is responsible of high mortality rates on *C. virginica* in North America (Saulnier et al. 2010).

3 Regulation

3.1 Biological Control and Management

The interest of producers and consumers for organic food productions increased sharply in recent years. Organic productions are characterized by the exclusive use of natural resources or of compounds already present in nature. Organic horticultural and other crops represent still a reduced fraction of the global agricultural land, reflecting a market and consumers demand for organic food which is mainly localized in developed countries. The expansion of these productions, furthermore, requires the development of new tools, among which new products and procedures based on biological control agents, as practical alternatives to pesticides. At this regard the study of microbial consortia involved in natural management of pests appears indispensable, as pesticides need to be substituted or at least integrated, in conventional agriculture, with more environment friendly organisms and procedures, at a fast growing pace.

Many parasitic bacteria have been tested and used in biological control assays for a variety of noxious invertebrate pests. One of the most studied and applied since the 1970s is undoubtly Bt. The benefits obtained through the application of Bt strains include the selective efficacy shown in the development of Integrated Pest management (IPM) and/or biological control procedures. It is worth to recall, for clarity, that many (if not even all) human economic activities benefit from the impact that bacteria have on the crop production cycles. Some of these depend on the peculiar traits of the bacteria utilized. The relative success of Bt as a biopesticide, for example—it has been estimated that 2 % of total insecticides used worldwide are Bt-based (Bravo et al. 2011)—has been underpinned by the consideration that treatments are safe for humans, do not have adverse effects on other vertebrates and are safe for the environment (Ansari et al. 2012). The bacterium is widely applied on very large surface, like in North America where treatments against the spruce budworm *Choristoneura fumiferana* cover more than 8 M ha of forests.

Negative feedbacks observed concern the side effects on other, non-target insects, and the insurgence of resistance in pest populations, leading to reduced efficacy of treatments (Huang et al. 2001). The latter effect is due to the selective pressures (of the Red Queen type) introduced in insect populations when exposed to an artificial increase of the spore density of a naturally occurring regulator. Other limiting factors for insect biological control include the bacterium efficacy, mainly on larval stages of leaf feeder insects, and the spore sensitivity to sunlight (Bravo et al. 2011).

3.2 Regulation Mechanisms

Many theoretical works applied to the study of population density regulation yielded several specific models, applied to describe the population dynamics and density changes of invertebrates, mainly insects, observed in real environments. They showed different levels of correspondence to the observed data, with possible outcomes related to the possible use either as predictors of biocontrol efficacy or as descriptors of the underpinning regulation mechanisms (Anderson and May 1981; Barlow 2004). Simulations may describe indeed the evolution in time of a natural system but, given the external perturbations and the presence of own chaotic components, even most accurate predictions rapidly lose their adherence to real data. In most cases, however, simulations allow understanding the mechanisms of natural regulation or pest suppressivity.

The behaviour and dynamics of even a simple two-species system (like those formed by single host and bacterial parasite) is not only affected by the components' own biological properties, but also by the densities at which both organisms occur. Changes during the system evolution may even yield a cycle path leading to a local extinction of one or both species. Applied models may hence permit estimating the doses and the time required to reach an equilibrium between host and parasite, or to induce a local extinction. This may be achieved when routine treatments with introduction of one or more biocontrol agents are possible. Understanding the nature of the regulation mechanisms active in a given environment is hence a useful and informative approach, which may yield many practical outcomes and benefits (Gutierrez et al. 1985; Jaffee 1993).

3.2.1 Density Dependence

A mechanism linking two or more populations in which their densities act as main drivers affecting one or more ecosystem variables is called "density-dependence". Density-dependent population changes may be also affected by several, additional external factors, including the presence of one or more biocontrol agents. Induced changes usually appear as time and space fluctuations of densities. Populations are affected at a given time delay by their own density and/or by the numbers of other co-occurring organisms, either parasitic or predatory. These effects are more visible as densities increase, and are in general non linear. Populations are sensitive to their own numbers through a self-regulation component that depends on the progressive reduction of the resources available for growth. Many biological control applications rely on data and models concerning density changes in time and space, and the underlying bio-physical processes and forces responsible for the observed variations (Juliano 2007).

Density-dependence is referred to as "negative" when a progressive increase in population size determines a reduction in either its growth rate or transmission (either concerning the same population or an antagonistic one). In "positive" density-dependence instead a size increase enhances transmission or other processes (i.e. survival). Density-dependent self-regulatory factors are usually observed in population ecology studies. Dynamical links accounting for this kind of relationships have been also observed for many herbivores and either their parasites or predators or for the plants they feed on (Jaffee et al. 1992; Weed and Schwarzländer 2014).

Density-dependent processes regulating growth and mortality have been described in many invertebrate populations, as a component of complex regulatory networks, including nematode, insect or shrimp populations, interacting with additional environmental factors (Pérez-Castañeda and Defeo 2005). In general, knowledge about the natural regulation mechanisms is required for exploitation of natural stock populations by controlled catches in fisheries, requiring data on the age composition, growth and mortality of juveniles (Gazey et al. 2008). Similarly, monitoring is required when managing farming infrastructures, in which density levels have to be monitored and maintained around their optima, in order to sustain productions by reducing natural or density-induced mortalities. Applications concern many productive ecosystems, including aquaculture productions or fisheries (Bell et al. 2005; Forrest et al. 2013).

Invertebrate-associated microbial and antagonists communities have been the object of increasing research investments, aiming at identifying ecological factors and processes leading to methods for control or management of plant or human pests. Improved capacities to manage insect populations may have important consequences, as in the case of insect vectors of human diseases. Studies on their population size showed that many density-dependent factors interact with insects population dynamics and regulation. Two lines of research work are required at this regard: the first concerning the collection of quantitative data (i.e. population density in time and space, changes observed at regular time intervals, mortalities and other descriptive variables, including the presence and densities of predators or parasites). The second research line aims at the identification of the biological processes and forces, including the biology of predation/parasitism and any other qualitative information (including genetic data) that may reveal any vulnerability in the population studied and/or forecast the effects of man-driven interventions or decisions. Modeling and data analysis play a key role at this regard, in many applications (Juliano 2007).

Several microorganisms, including fungi and bacteria (i.e. *Pasteuria* or *Bacillus* spp., and others), have been identified as potential candidates for management of nematode pests, worldwide, allowing the development of industrial biological pesticides (Kerry 2000). However, to exploit bacteria as biocontrol agents, data on their ecology as well as low-cost production in artificial conditions, are needed. Depending on pests and crops, evaluation studies have been performed in naturally infested fields or by applying endospores or other propagules produced by artificial culturing. Production of some biocontrol bacteria in controlled conditions may be challenging, in particular for species having very narrow requirements for culturing.

Herbivores of economic importance mainly include insects and nematodes which represent, with mites, the most important invertebrate pests of crops. Biocontrol strategies of insects show many applicative examples yielding benefits related to one or more *Bacillus* spp., in particular for Bt that accounts for more than 50 % of bioinsecticidal market share (Lacey et al. 2015). They have been applied since many decades for management of several insect crop pests, as well as for biocontrol of human or animal disease vectors (Table 10.1) (Federici 2005; Berry 2012; Guidi et al. 2013; see also Chap. 4 for more details). Other species showed

Table 10.1 Some examples of effects of bacteria-based biopesticides applied for biocontrol, and target pests

Bacterial species	Host	Efficacy[a]	Reference
Bt (subsp.)	*Helicoverpa armigera, Exelastia atomosa*	23–88	Puntambekar et al. (1997)
Bt mM2	Coleoptera	4–30	Kati et al. (2007)
Bt Berliner	*Aethina tumida*	5	Buchholz et al. (2006)
Bt Berliner	*Tuta absoluta*	2–3	González-Cabrera et al. (2011)
Bt Berliner	*Chilo infuscatellus*	14–83	Kesavan et al. (2003)
Bt *thuringiensis*, Bt *israelensis*	*Meloidogyne incognita*	53–66	Sharma (1994)
Bt subsp. *aegypti*	*Meloidogyne arenaria*	18	Mokbel (2013)
Bt CR371	*Meloidogyne hapla*	50	Chen et al. (2000)
Lysinibacillus sphaericus	*Culex quinquefasciatus*	58	Lacey et al. (1987)
P. penetrans	*Meloidogyne incognita*	10	Stirling (1984)
P. penetrans	*Meloidogyne* spp.	0–49	Tzortzakakis and Gowen (1994)
P. penetrans	*Meloidogyne incognita*	25–31	Jonathan et al. (2000)

[a]Expressed as % of untreated control

variable results, as reported for formulations of *B. popilliae* (Redmond and Potter 1995).

Bacillus spp. have been also investigated in relation to their use against plant parasitic nematodes, either as direct antagonists or for activity of some toxins, with variable results (Table 10.1) (Sharma 1994; Chen et al. 2000). Further bacteria investigated for their beneficial effects in controlling root-knot and cyst nematodes were *Paenibacillus polymyxa*, *P. lentimorbus*, *Rhizobium etli* and *Streptomyces costaricanus* (Esnard et al. 1995; Hallmann et al. 2001; Son et al. 2009). However, due to their widespread occurrence and effect, most bacteria studied for nematode biocontrol include members of the genus *Pasteuria*. Applications assays with durable endospore of these obligate parasites have been carried out in pots, microplots or semicontrolled conditions with artificially infected hosts (Stirling and Wachtel 1980; Davies et al. 1988; Oostendorp et al. 1991). Further studies consistently showed that some species, i.e. *P. penetrans*, can effectively control root-knot nematodes densities with a high biocontrol potential, providing quantitative data on biological and ecological factors influencing the levels of parasitism that are observed in the field (Table 10.1) (Verdejo Lucas 1992; Dabiré et al. 2007; Mateille et al. 2009, 2010).

Progress in the cultivation of *Pasteuria* spp. in fermentation industrial units encourages studies on the ecology of these antagonists and on the effects that other soil microbial consortia have on the biocontrol efficacy achievable with introduction of endospores produced *in vitro* (Kokalis-Burelle 2015).

Root-knot nematodes have been the most common pests targeted in biological control assays. Field studies on *Pasteuria* spp. and associated hosts showed that both organisms persist in time (Verdejo Lucas 1992; Ciancio and Quénéhervé 2000; Sorribas et al. 2000; Mateille et al. 2009, 2010). A dose-response or density-dependent relationships were observed in greenhouse and field conditions for *P. penetrans* (Oostendorp et al. 1991; Verdejo Lucas 1992; Kasumimoto et al. 1993; Ciancio and Quénéhervé 2000). Further data have been made available on biological and ecological processes affecting prevalence levels in natural and artificial environments, as well as on the ecological factors sustaining host density regulation and biocontrol efficacy (Atibalentja et al. 1998; Mateille et al. 2009, 2010; Luc et al. 2010).

Population dynamics and spatial sampling data are needed to provide experimental evidence of time-delayed, density dependent relationships. In nematodes, these have been observed among different populations of plant parasitic species and associated *Pasteuria* spp. (Fig. 10.1). Changes in bacterial prevalence usually follow the nematode density increase after a time lapse. In a classic density dependent relationship the increase of the parasite/predator density is directly related to the higher host/prey abundance previously encountered (Christiansen and Fenchel 1977; Hassel 1978). This situation originates a feedback mechanism which mutually links host density and prevalence by the biological antagonist.

Many informations may be gained by a model construction and eventual application. However, it should be always considered that the constants estimated and used in models, like Model-G (Anderson and May 1981), represent a simplification of the mechanisms at work in the real environment. Caution should be applied in modeling, since non linear and complex models strongly depend on their initial conditions and on the variations and meaning of the constants used.

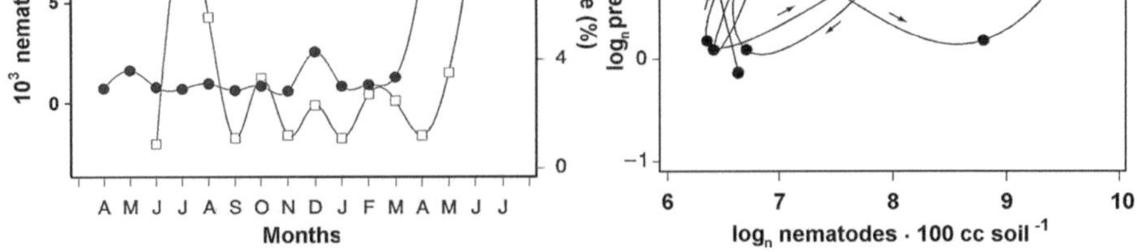

Fig. 10.1 Density-dependent relationship observed for monthly changes of Pasteuria sp. prevalence (**a**, *square*) and motile stages of the citrus nematode Tylenchulus semipenetrans in a citrus grove soil in Southern Italy. The same data on the corresponding phases space plot show a varying cycle amplitude (**b**, *arrows* show time direction) (Data from Ciancio et al. 2016)

As shown by plotting a time series data on its phase space, a nematode and *Pasteuria* relationship may show substantial equilibrium during a large fraction of the period of observation (Fig. 10.1b). Spatial samplings performed only once may also be used to detect this kind of relationship, in which the observations appear distributed around an equilibrium point. This may be identified in a region in the plot in which the host-prevalence cycle fluctuates. Some models like Model-G may calculate the coordinates of this theoretical point. For the *Pasteuria* and *T. semipenetrans* populations, plotting the regular fluctuations in time showed a time delay 3 months out of phase between host and prevalence, suggesting that the entire cyclic path was completed within this period (Fig. 10.1a). The cycle amplitude, however, depends on the initial densities of both organisms. This suggests that, when measuring the host density changes induced by inoculations or, if endospores are available in mass, by inundative treatments, a few months would be required to have a *Pasteuria* spp. biocontrol effect, depending on the initial nematodes numbers and prevalence.

Small changes concerning these factors may yield significant changes in the cycles amplitudes and in the numbers observed after a given time lapse (Gressel 2005). In nature, the host growth rate is not a constant but a function of other variables like food abundance and other regulating factors. Similarly, parasite transmission and mortality rates are affected by physical variables like temperature, soil microcosm volume and texture, or water content. Introducing temperature effects may increase the adherence of a model to real data, however validation assays by comparing calculated and observed values should be applied. Since temperature affects either the host and parasite's metabolism at the same time, it is reasonable to check if either the host and parasite share common optimal temperatures for development. Experimental data are needed for measuring this factor, since temperatures may have different effects on the functions describing parasitism and on those accounting for the development of different host stages (i.e. eggs hatching, juvenile stages, motility, etc.).

Models suggests that stochastic fluctuations of host's cycles may affect the levels of endospore loads and parasite dispersal achieved at a given time. Furthermore, at a critical distance from the equilibrium zone, a temporary host local extinction may occur. This could happen when the cycle satellite line reaches the prevalence axe. In these situations extremely low densities of the host may be achieved, a level at which: (*i*) the host population can be considered locally extinct, and (*ii*) *Pasteuria* as other persistent parasites can remain in the microcosm as a durable propagule, a trait that probably evolved in many parasites for this purpose.

The *Pasteuria* prevalence levels observed in the field and the mutual host-parasite interactions are controlled by a few basic biological characters of the organisms involved. In particular, host specificity appears as a key driver at this regard, since it may affect biocontrol efficacy. Furthermore, the parasite has been observed to persist for decades in natural conditions. The *Pasteuria* sp. associated to the citrus nematode found in specimens of *T. semipenetrans* was still detectable more than 12 years after the last samplings. Similarly, an undescribed *Pasteuria* sp. was still found attacking a population of the virus vector nematode *Xiphinema diversicaudatum*, more than a decade after the last sampling.

3.2.2 Density Independence

A number of factors not related to population size have been considered to act invariably when small or large densities occur, exerting their weights by interacting with other density-dependent forces. These factors included processes like intra- and interspecific competition, weather (thermal thresholds, rainfall) or genetic changes, allowing for resistance or improved fitness towards parasitism or predation (Young 2012).

4 GM Crops

Main GM crops modified with Bt toxin genes include tobacco, potato, maize, cotton and rice (Sanchis and Bourguet 2008; Li et al. 2003; Meiyalaghan et al. 2006). However, the selection of individual pests and populations resistant to the Bt toxins represents a major concern. Some measures to counteract this effect may be adopted by i.e. sowing susceptible (non-GM) plants as field "islands" or at the edges of the GM plantation. Non-GM plants allow natural (Bt susceptible) pest individuals to develop and mate with the resistant ones, thus originating heterozygotic progenies. Since the insect resistance genes are recessive, the new generations lose the resistant character and can be controlled by the GM crop (Ostlie et al. 1997).

A further approach consists in the use of GM crops expressing two or more Bt toxin genes, for which the insurgence of combined resistance traits is unlikely (Roush 1998). Further negative effects deriving by the use of GM plants include the possibility of a pest shift, allowing the substitution of a GM-controlled pest by a different one, previously neglected, for which no plant protection mechanism has been made yet available in the field (Sanchis and Bourguet 2008). The interest in GM based pest control, however, is still very high, as also shown by the large number of patents which have been filed on new *cry* sequences or on methods for toxins exploitation in pest management (Crickmore et al. 1998).

References

Abrol, D. P. (2012). Pollinators as bioindicators of ecosystem functioning. In *Pollination biology: Biodiversity conservation and agricultural production* (pp. 509–544). Dordrecht: Springer.

Ahmed, M. Z., et al. (2015). The intracellular bacterium *Wolbachia* uses parasitoid wasps as phoretic vectors for efficient horizontal transmission. *PLoS Pathogens, 11*, e1004672.

Anderson, R. M., & May, R. M. (1981). The population dynamics of microparasites and their invertebrate hosts. *Philosophical Transactions of the Royal Society of London, 291*, 451–524.

Ansari, N. S., Ahmad, N., & Hasan, F. (2012). Potential of biopesticides in sustainable agriculture. In A. Malik & E. Grohmann (Eds.), *Environmental protection strategies for sustainable development, strategies for sustainability* (pp. 529–595). Dordrecht: Springer.

Atibalentja, N., Noel, G. R., Liao, T. F., & Gertner, G. Z. (1998). Population changes in *Heterodera glycines* and its bacterial parasite *Pasteuria* sp. in naturally infested soil. *Journal of Nematology, 30*, 81–92.

Barlow, N. (2004). Models in biological control: A field guide. In B. A. Hawkins & H. V. Cornell (Eds.), *Theoretical approaches to biological control* (pp. 43–70). Cambridge: Cambridge University Press.

Bell, J. D., et al. (2005). *Advances in marine biology. Restocking and stock enhancement of marine invertebrate fisheries* (374 pp.). San Diego: Elsevier Academic Press.

Berry, C. (2012). The bacterium, *Lysinibacillus sphaericus*, as an insect pathogen. *Journal of Invertebrate Pathology, 109*, 1–10.

Bravo, A., Likitvivatanavong, S., Gill, S. S., & Soberón, M. (2011). *Bacillus thuringiensis*: A story of a successful bioinsecticide. *Insect Biochemistry and Molecular Biology, 41*, 423–431.

Buchholz, S., Neumann, P., Merkel, K., & Hepburn, H. R. (2006). Evaluation of *Bacillus thuringiensis* Berliner as an alternative control of small hive beetles, *Aethina tumida* murray (Coleoptera: Nitidulidae). *Journal of Pest Science, 79*, 251–254.

Chagnon, M., et al. (2015). Risks of large-scale use of systemic insecticides to ecosystem functioning and services. *Environmental Science and Pollution Research, 22*, 119–134.

Chen, J., Abawi, G. S., & Zuckerman, B. M. (2000). Efficacy of *Bacillus thuringiensis, Paecilomyces marquandii*, and *Streptomyces costaricanus* with and without organic amendments against *Meloidogyne hapla* infecting lettuce. *Journal of Nematology, 32*, 70–77.

Christiansen, F. B., & Fenchel, T. M. (1977). *Theories of populations in biological communities* (Ecological studies series, 20, 144 pp.). Berlin: Springler-Verlag Ed.

Ciancio, A., & Quénéhervé, P. (2000). Population dynamics of *Meloidogyne incognita* and infestation levels by *Pasteuria penetrans* in a naturally infested field in Martinique. *Nematropica, 30*, 77–86.

Ciancio, A., Roccuzzo, G., & Ornat, C. (2016). Modeling regulation of the citrus nematode *Tylenchulus semipenetrans* by a *Pasteuria* sp. bacterial microparasite. *BioControl* 61, in print.

Crickmore, N., et al. (1998). Revision of the nomenclature for the *Bacillus thuringiensis* pesticidal crystal proteins. *Microbiology and Molecular Biology Reviews, 62*, 807–813.

Dabiré, R. K., Ndiaye, S., Mounport, D., & Mateille, T. (2007). Relationships between abiotic soil factors and epidemiology of the biocontrol bacterium *Pasteuria penetrans* in a root-knot nematode *Meloidogyne javanica*-infested field. *Biological Control, 40*, 22–29.

Davies, K. G., Kerry, B. R., & Flynn, C. A. (1988). Observations on the pathogenicity of *Pasteuria penetrans*, a parasite of root-knot nematodes. *Annals of Applied Biology, 112*, 491–501.

Erler, S., Denner, A., Bobiş, O., Forsgren, E., & Moritz, R. F. (2014). Diversity of honey stores and their impact on pathogenic bacteria of the honeybee, *Apis mellifera. Ecology and Evolution, 4*, 3960–3967.

Esnard, J., Potter, T. L., & Zuckerman, B. M. (1995). *Streptomyces costaricanus* sp. nov., isolated from nematode-suppressive soil. *International Journal of Systematic Bacteriology, 45*, 775–779.

Evison, S. E. F., et al. (2012). Pervasiveness of parasites in pollinators. *PLoS ONE, 7*, e30641.

Federici, B. A. (2005). Insecticidal bacteria: An overwhelming success for invertebrate pathology. *Journal of Invertebrate Pathology, 89*, 30–38.

Forrest, R. E., McAllister, M. K., Martell, S. J. D., & Walters, C. J. (2013). Modelling the effects of density-dependent mortality in juvenile red snapper caught as bycatch in Gulf of Mexico shrimp fisheries: Implications for management. *Fisheries Research, 146*, 102–120.

Gallai, N., Salles, J.-M., Settele, J., & Vaissière, B. E. (2009). Economic valuation of the vulnerability of world agriculture confronted with pollinator decline. *Ecological Economics, 68*, 810–821.

Gazey, W. J., Gallaway, B. J., Cole, J. G., & Fournier, D. A. (2008). Age composition, growth, and density-dependent mortality in juvenile red snapper estimated from observer data from the Gulf of Mexico penaeid shrimp fishery. *North American Journal of Fisheries Management, 28*, 1828–1842.

González-Cabrera, J., Mollá, O., Montón, H., & Urbaneja, A. (2011). Efficacy of *Bacillus thuringiensis* (Berliner) in controlling the tomato borer, *Tuta absoluta* (Meyrick) (Lepidoptera: Gelechiidae). *BioControl, 56*, 71–80.

Gressel, J. (2005). Problems in qualifying and quantifying assumptions in plant protection models: Resultant simulations can be mistaken by a factor of million. *Crop Protection, 24*, 1007–1015.

Guidi, V., Lehner, A., Lüthy, P., & Tonolla, M. (2013). Dynamics of *Bacillus thuringiensis* var. israelensis and *Lysinibacillus sphaericus* spores in urban catch basins after simultaneous application against mosquito larvae. *PLoS One, 8*, e55658.

Gutierrez, A. P., Messenger, P. S., & van den Bosch, R. (1985). *An introduction to biological control* (246 pp.). New York: Springer.

Hallmann, J., Quadt-Hallmann, A., Miller, W. G., Sikora, R. A., & Lindow, S. E. (2001). Endophytic colonization of plants by the biocontrol agent *Rhizobium etli* G12 in relation to *Meloidogyne incognita* infection. *Phytpathology, 91*, 415–422.

Hassel, M. P. (1978). *The dynamics of arthropod predator-prey systems* (237 pp.). Princeton: Princeton University Press.

Huang, F., Buschman, L. L., & Higgins, R. A. (2001). Larval feeding behavior of Dipel-resistant and susceptible *Ostrinia nubilalis* on diet containing *Bacillus thuringiensis* (Dipel ES™). *Entomologia Experimentalis et Applicata, 98*, 141–148.

Huger, A. M., Skinner, S. W., & Werren, J. H. (1985). Bacterial infections associated with the son-killer trait in the parasitoid wasp *Nasonia* (= *Mormoniella*) *vitripennis* (Hymenoptera: Pteromalidae). *Journal of Invertebrate Pathology, 46*, 272–280.

Jackson, T., McNeill, M., & Madurappulige, D. (2004). Use of *Serratia* spp. bacteria for monitoring behaviour of parasitoids. *International Journal of Pest Management, 50*, 173–176.

Jaffee, B. (1993). Density-dependent parasitism in biological control of soil-borne insects, nematodes, fungi and bacteria. *Biocontrol Science and Technology, 3*, 235–246.

Jaffee, B., Phillips, R., Muldoon, A., & Mangel, M. (1992). Density-dependence host-pathogen dynamics in soil microcosms. *Ecology, 73*, 495–506.

Jonathan, E. I., Barker, K. R., Abdel-Alim, F. F., Vrain, T. C., & Dickson, D. W. (2000). Biological control of *Meloidogyne incognita* on tomato and banana with rhizobacteria, actinomycetes, and *Pasteuria penetrans*. *Nematropica, 30*, 231–240.

Juliano, S. A. (2007). Population dynamics. *Journal of the American Mosquito Control Association, 23*(2 Suppl), 265–275.

Kasumimoto, T., Ikeda, R., & Kawada, H. (1993). Dose response of *Meloidogyne incognita* infecting cherry tomatoes to application of *Pasteuria penetrans*. *Japanese Journal of Nematology, 23*, 10–18.

Kati, H., Sezen, K., & Demirbag, Z. (2007). Characterization of a highly pathogenic *Bacillus thuringiensis* strain isolated from commoncockchafer, *Melolontha melolontha*. *Folia Microbiologica, 52*, 146–152.

Keawtaweem, T., Fukami, K., Songsangjinda, P., & Muangyao, P. (2011). Isolation and characterization of *Noctiluca*-killing bacteria from a shrimp aquaculture pond in Thailand. *Fish Science, 77*, 657–664.

Kerry, B. R. (2000). Rhizosphere interactions and the exploitation of microbial agents for the biological control of plant-parasitic nematodes. *Annual Review of Phytopathology, 38*, 423–424.

Kesavan, S. R., Easwaramoorthy, S., & Santhalakshmi, G. (2003). Evaluation of different formulations of *Bacillus thuringiensis* against sugarcane early shoot borer *Chilo infuscatellus*. *Sugar Tech, 5*, 51–55.

Kokalis-Burelle, N. (2015). *Pasteuria penetrans* for control of *Meloidogyne incognita* on tomato and cucumber, and *M. arenaria* on snapdragon. *Journal of Nematology, 47*, 207–213.

Lacey, L. A., Day, J., & Heitzman, C. M. (1987). Long-term effects of *Bacillus sphaericus* on *Culex quinquefasciatus*. *Journal of Invertebrate Pathology, 49*, 116–123.

Lacey, L. A., et al. (2015). Insect pathogens as biological control agents: Back to the future. *Journal of Invertebrate Pathology, 132*, 1–41.

Lautenbach, S., Seppelt, R., Liebscher, J., & Dormann, C. F. (2012). Spatial and temporal trends of global pollination benefit. *PLoS ONE, 7*, e35954.

Li, R., et al. (2003). Transgenic plants expressing *Bacillus thuringiensis* delta-endotoxins. *Entomologia Sinica, 10*, 155–166.

Luc, J. E., Pang, W., Crow, W. T., & Giblin-Davis, R. M. (2010). Effects of formulation and host nematode density on the ability of in vitro-produced *Pasteuria* endospores to control its host *Belonolaimus longicaudatus. Journal of Nematology, 42*, 87–90.

Mateille, T., Fould, S., Dabiré, K. R., Diop, M. T., & Ndiaye, S. (2009). Spatial distribution of the nematode biocontrol agent *Pasteuria penetrans* as influenced by its soil habitat. *Soil Biology & Biochemistry, 41*, 303–308.

Mateille, T., Dabiré, K. R., Fould, S., & Diop, M. T. (2010). Host-parasite soil communities and environmental constraints: Modelling of soil functions involved in interactions between plant-parasitic nematodes and *Pasteuria penetrans. Soil Biology & Biochemistry, 42*, 1193–1199.

Meiyalaghan, S., Jacobs, J. M. E., Butler, R. C., Wratten, S. D., & Conner, A. J. (2006). Transgenic potato lines expressing *cry1Ba1* or *cry1Ca5* genes are resistant to potato tuber moth. *Potato Research, 49*, 203–216.

Mihai, C. M., et al. (2012). Interactions among flavonoids of propolis affect antibacterial activity against the honeybee pathogen *Paenibacillus larvae. Journal of Invertebrate Pathology, 110*, 68–72.

Mohanta, M. K., et al. (2015). Characterization of *Klebsiella granulomatis* pathogenic to silkworm, *Bombyx mori* L. *3 Biotech, 5*, 577–583.

Mokbel, A. A. (2013). Impact of some antagonistic organisms in controlling *Meloidogyne arenaria* infecting tomato plants. *Journal of Life Sciences and Technologies, 1*, 69–74.

Oostendorp, M., Dickson, D. W., & Mitchell, D. J. (1991). Population development of *Pasteuria penetrans* on *Meloidogyne arenaria. Journal of Nematology, 23*, 58–64.

Ostlie, K. R., Hutchison, W. D., & Hellmich, R. L. (Eds.). (1997). *Bt corn and european corn borer. Long-term success through resistance management* (North Central Region Extension Publication NCR 602). St. Paul: University of Minnesota.

Pérez-Castañeda, R., & Defeo, O. (2005). Growth and mortality of transient shrimp populations (*Farfantepenaeus* spp.) in a coastal lagoon of Mexico: Role of the environment and density-dependence. *ICES Journal of Marine Science, 62*, 14e24.

Puntambekar, U. S., Mukherjee, S. N., & Ranjekar, P. K. (1997). Laboratory screening of different *Bacillus thuringiensis* strains against certain lepidopteran pests and subsequent field evaluation on the pod boring pest complex of pigeonpea (*Cajanus cajan*). *Antonie van Leeuwenhoek, 71*, 319–323.

Redmond, C. T., & Potter, D. A. (1995). Lack of efficacy of *in-vivo* and putatively *in-vitro* produced *Bacillus popilliae* against field populations of Japanese beetle (Coleoptera: Scarabaeidae) grubs in Kentucky. *Journal of Economic Entomology, 88*, 846–854.

Roberts, T. A., et al. (2005). Micro-organisms in foods 6. In T. A. Roberts et al. (Eds.), *Microbial ecology of food commodities* (pp. 174–249). New York: Springer.

Roush, T. R. (1998). Two–toxin strategies for management of insecticidal transgenic crops: Can pyramiding succeed where pesticide mixtures have not? *Philosophical Transactions of the Royal Society of London B, 353*, 1777–1786.

Salmela, H., Amdam, G. V., & Freitak, D. (2015). Transfer of immunity from mother to offspring is mediated via egg-yolk protein vitellogenin. *PLoS Pathogens, 11*, e1005015.

Salunkhe, R. C., Narkhede, K. P., & Shouche, Y. S. (2014). Distribution and evolutionary impact of *Wolbachia* on butterfly hosts. *Indian Journal of Microbiology, 54*, 249–254.

Sanchis, V., & Bourguet, D. (2008). *Bacillus thuringiensis*: Applications in agriculture and insect resistance management. A review. *Agronomy for Sustainable Development, 28*, 11–20.

Saulnier, D., et al. (2010). A large-scale epidemiological study to identify bacteria pathogenic to pacific oyster *Crassostrea gigas* and correlation between virulence and metalloprotease-like activity. *Microbial Ecology, 59*, 787–798.

Schulp, C. J. E., Lautenbach, S., & Verburg, P. H. (2014). Quantifying and mapping ecosystem services: Demand and supply of pollination in the European Union. *Ecological Indicators, 36*, 131–141.

Sharma, R. D. (1994). *Bacillus thuringiensis*: A biocontrol agent of *Meloidogyne incognita* on barley. *Nematologia brasileira, 18*, 79–84.

Sheebha, A., Quraiza, F., Mdfhilsath, Manohar, D., Sam, S., & Bai, R. (2008). Effect of prophylactic antibiotic treatment on the growth and cocoon characteristics of *Bombyx mori* L. *Journal of Basic Applied Biology, 2*, 19–22.

Son, S. H., Khan, Z., Kim, S. G., & Kim, Y. H. (2009). Plant growth-promoting rhizobacteria, *Paenibacillus polymyxa* and *Paenibacillus lentimorbus* suppress disease complex caused by root-knot nematode and fusarium wilt fungus. *Journal of Applied Microbiology, 107*, 524–532.

Sorribas, F. J., Verdejo-Lucas, S., Forner, J. B., Alcaidel, A., Pons, J., & Ornat, C. (2000). Seasonality of Tylenchulus semipenetrans Cobb and Pasteuria sp. in citrus orchards in Spain. *Journal of Nematology, 32*, 622–632.

Sorribas, F. J., Verdejo-Lucas, S., Forner, J. B., Alcaidel, A., Pons, J., Ornat, C. (2000). Seasonality of *Tylenchulus semipenetrans* Cobb and *Pasteuria* sp. in citrus orchards in Spain. *Journal of Nematology*, 32, 622–632. Stirling, G. (1984). Biological control of *Meloidogyne javanica* with *Bacillus penetrans. Phytopathology, 74*, 55–60.

Stirling, G. R., & Wachtel, M. F. (1980). Mass production of Bacillus penetrans for the biological control of root-knot nematodes. *Nematologica, 26*, 308–312.

Sun, X. J., Xiao, J. H., Cook, J. M., Feng, G., & Huang, D. W. (2011). Comparisons of host mitochondrial, nuclear and endosymbiont bacterial genes reveal cryptic fig wasp species and the effects of *Wolbachia* on host mtDNA evolution and diversity. *BMC Evolutionary Biology, 11*, 86.

Tzortzakakis, E. A., & Gowen, S. (1994). Evaluation of *Pasteuria penetrans* alone and in combination with oxamyl, plant-resistance and solarization for control of *Meloidogyne* spp on vegetables grown in greenhouses in Crete. *Crop Protection, 13*, 455–462.

Verdejo Lucas, S. (1992). Seasonal population fluctuations of *Meloidogyne* spp. and the *Pasteuria penetrans* group in kiwi orchards. *Plant Disease, 76*, 1275–1279.

Wahid, M. A., et al. (2014). Antibiotic resistance bacteria in coastal shrimp pond water and effluent. In R. Hassan et al. (eds.), *InCIEC 2014* (pp. 1011–1018).

Weed, A. S., & Schwarzländer, M. (2014). Density dependence, precipitation and biological control agent herbivory influence landscape-scale dynamics of the invasive Eurasian plant *Linaria dalmatica. Journal of Applied Ecology, 51*, 825–834.

White, J. A., et al. (2015). Endosymbiotic candidates for parasitoid defense in exotic and native New Zealand weevils. *Microbial Ecology, 70*, 274–286.

Williams, I. H. (1994). The dependences of crop production within the European Union on pollination by honey bees. *Agricultural Zoology Review, 6*, 229–257.

Xie, J., Vilchez, I., & Mateos, M. (2010). *Spiroplasma* bacteria enhance survival of *Drosophila hydei* attacked by the parasitic wasp *Leptopilina heterotoma. PLoS One, 5*, e12149.

Young, A. M. (2012). *Population biology of tropical insects* (524 pp.). Springer.

Zchori-Fein, E., et al. (2001). A newly discovered bacterium associated with parthenogenesis and a change in host selection behavior in parasitoid wasps. *Proceedings of the National Academy of Science USA, 98*, 12555–12560.

Zhu, B., et al. (2011). Horizontal gene transfer in silkworm, *Bombyx mori. BMC Genomics, 12*, 248.

Zorriehzahra, M. J., & Banaederakhshan, R. (2015). Early Mortality Syndrome (EMS) as new emerging threat in shrimp industry. *Advances in Animal and Veterinary Sciences, 3*, 64–72.

Chapter 11
Environmental Interactions

Abstract Some aspects concerning invertebrates, bacteria and the environment are examined, including the response to man-induced changes. Some services deployed by invertebrates are briefly reviewed. The effects of different contamination types on species involved in farming are described, including the effects of heavy metals, pesticides, and oil spills. The link of climate changes with bacterial diseases is discussed, with a brief review on the role of invasive species in agroecosystems and marine habitats.

Keywords Bioindicators • Heavy metals • Invasive species • Pesticides • Pollution • Sponges

1 Introduction

Many services fundamental for the biosphere have their roots in invertebrate and bacteria interactions and are fundamental in primary productions, in a disparate range of terrestrial and aquatic environments. Invertebrates cover around 80 % of all eukaryotic species described, and are ubiquitous in earth ecosystems (Prather et al. 2013). Apart of direct or indirect involvements in biomass or food productions (Zulka and Götzl 2015), services deployed include many elemental cycles, of which C storage and recycling is one of the most important. Many examples may be found among populations of herbivores or biomass decomposers and in communities of organisms accumulating or recycling a wide range of organic compounds and natural polymers like chitin.

Invertebrate and bacteria communities are also subject to ongoing changes that affect their populations stability, in particular to those with an anthropic origin, related to the human activities and population growth. Among disturbance factors, soil or water pollutants and thermal shifts appear of a paramount importance because of their unpredictable and/or health or nutritional consequences. The list of endangered habitats at risk, however, is large and threats are global. Soil and groundwater contamination add to increasing mean temperatures due to the accumulation of greenhouse gases (methane, CO_2) in the atmosphere. Many marine environments

A. Ciancio, *Invertebrate Bacteriology*, DOI 10.1007/978-94-024-0884-3_11

are also at risk as shown, for example, by the losses of coral reef habitats exposed to water acidification.

Diffuse knowledge about the way these ecosystems are affected by global changes is often lacking, and data are urgently needed by research work focusing on resistant or resilient responses (Prather et al. 2013). At the ecosystem level, **resistance** may be considered as the capacity to retain functioning under an external disturbance, without changes or with minimal effects. The concept of **resilience** involves instead a more dynamic and elastic reaction, in which a capacity is present to recover, after a given time delay, the ecosystem function and integrity, following an external disturbance affecting its performance (Tugel et al. 2005; Herrick and Wander 1998; Seybold et al. 1999) (Fig. 11.1).

All ecosystems, including those far from the urban concentrations, contribute to the global balance by services whose effects may be observed even at very large distances, either in space and time. They range from agricultural land to forests and wildlife, including food webs established in extreme situations like deep sea and benthic environments. Examples of the latter include hydrothermal vent invertebrates like anellids or methanotrophs-associated sponges, contributing to elemental recycling through the food webs they sustain. Fixation of atmospheric N_2 or O_2 released by Cyanobacteria significantly occur in oceanic environments, sustaining marine productions through zooplankton food webs. Even simple sponges may significantly contribute to nitrification and recycling, in their water columns (Bell 2008).

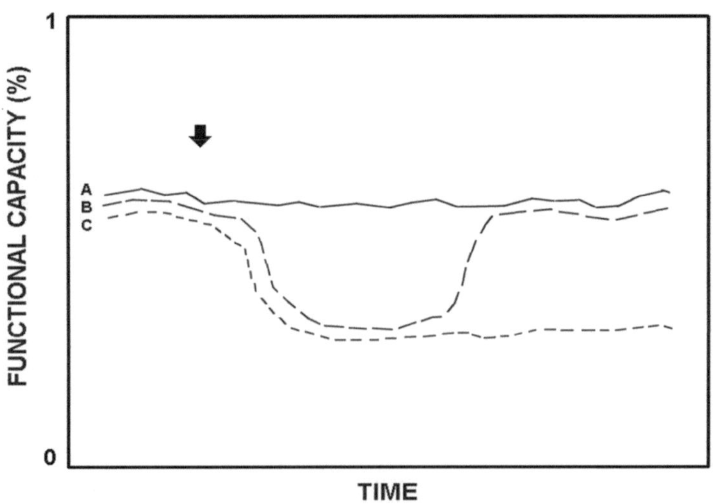

Fig. 11.1 A scheme showing the effect in time of resistance and resilience on a generic functional capacity in disturbed ecosystems. *Arrow* shows the time of disturbance. *Lines* refer to the reactions of highly resistant (**A**) or poorly resistant ecosystems, with high (**B**) and low (**C**) levels of resilience

Many human health issues are also connected to invertebrates as already described in other Chapters. Although a complete description and generalization of all the environmental interactions taking place on earth and directly or indirectly linked with human society represent an unaffordable task, some aspects concerning invertebrates, bacteria and the environment are worth a more detailed examination, provided in this Chapter.

2 Effects on Ecosystem Services

2.1 Invertebrate Services

Soil is one of the most important environments on earth, and our knowledge about the diversity of its biological constituents, as well as role and conservation, is growing rapidly (see Chap. 9). The conservation of soil productive potentialities is central to any sustainable agroecosystem management. Soil biological quality can be evaluated by monitoring the changes in the community structure of its microfauna and of the associated microorganisms. Biological soil components originate a complex system of food webs (Kampichler 1999; Brussaard et al. 2007; Ferris 2010), responsible for services like nutrient recycling (decomposition) and crop production. Invertebrates also contribute by their ecosystem engineering activities, with provision of services like water infiltration and storage, porosity, C storage in biogenic aggregates and conservation of microbial diversity (Lavelle et al. 2006).

The fundamental components of the soil microfauna include nematodes and higher invertebrates, which play a significant ecological role, not only as pests but also as regulators or nutrient recyclers. They are present in trophic webs with different roles, feeding on algae, bacteria, fungal hyphae, plant roots or other small animals. Given their abundance, they increase the turnover of nutrients and indirectly influence decomposition by feeding on soil microbial decomposers. Moreover, some species may be very sensitive to disturbances of various types. In the last decades many studies analyzed nematode community structures in relation to environmental changes or disturbance. Data showed that nematodes are better indicators of changes in soil quality than other organisms sensitive to disturbance, since they are widespread and may be easily isolated, counted and identified (Bongers et al. 2001; Neher 2010; Ferris et al. 2012).

Primary producers affect the complexity of soil microfauna, as shown by food webs in primary successions of remediated soils (Hohberg 2003). In connection with plants, soil macroinvertebrate communities are active in the maintenance of fertility and other services provided i.e. in tropical soils, through the intense remix that they operate and the interactions evolved with other soil organisms (Lavelle et al. 2006). In natural savannah soils, for example, earthworms may ingest, mix and redistribute 800–1200 tons of dry soil · ha^{-1} · $year^{-1}$. Termites and ants are also active ecosystem engineers that contribute to some general effects of great importance for soils, including biodiversity, through the selection and redistribution of microbial communities in the upper 20 cm of soil, and the production of energetic substrates analogous to root exudates (Lavelle et al. 2004). Other effects concern fertility, through plant growth stimulation and accelerated nutrients release, protection from parasites (especially nematodes), stimulation of root symbionts, physical conditioning of soil and production of plant growth promoting compounds via selective stimulation of microorganisms (Brown et al. 1999; Scheu 2003). Additional services are the hydraulic effects through the maintenance of porosity, active macroaggregation of soil in the upper 20 cm with positive effects on water infiltration and storage, and C sequestration in stable biogenic aggregates.

Similarly, in sea and freshwater habitats, invertebrates are central to many services produced. Marine species contribute to the biosphere balance with a wide range of ecosystem services including provision of food and other metabolites and products, sustaining and regulating effects on marine biodiversity, participation as nourishment in food webs, recycling and C sequestration, shoreline erosion control, habitat creation, water filtration and purification, and climate regulation. Many of these services involve bacteria, as direct energy sources in many water food webs, or as decomposers and recyclers, or through the regulation of invertebrate populations. Indirectly, bacteria are involved in the release of a wide range of bioproducts they synthesize, or in biodegradation and bioremediation services (Sharp et al. 2007; Levin and Sibuet 2012; Blackburn et al. 2014; Katsanevakis et al. 2014).

2.2 Heavy Metals Pollution

Several gasteropods, anellids or molluscs are capable to accumulate or metabolize heavy metals and other pollutants, thanks to speciliked tissues and organs or to a number of metabolic and physiological adaptive processes. The latter include the synthesis of metallothioneins, low molecular weight and cystein-rich stress-induced proteins involved in protection of cells from the effects of toxic metals, with high affinities for elements like Cd, Cu and Zn (Dallinger 1994). Resistance to high concentrations of heavy metals in bacteria is associated to their capacity to produce biopolymers based on exopolysaccharides, involved in biosorption processes, and to chelate heavy metal ions, a process that may play an important role in resilience and natural bioremediation (Iyer et al. 2005).

Heavy metal contaminations of industrial origin, however, are responsible of significant environmental changes. In seawater and benthic communities they include the loss of entire invertebrate populations. A paramount example concerns the disappearance of the pearl-bearing oyster banks along the Indian coasts (Selvin et al. 2009). Other invertebrates, however, may activate specific biochemical pathways related to stress response and express indicative genes like heat shock proteins, that can be used as stress and environmental bioindicators (Zhang et al. 2010). Benthic sedentary organisms like sponges can accumulate heavy metals when filtering contaminated sea water for food assumption (Bell 2008). Given their filtering capacity, sponges are indicators of environmental health or may even be considered as living databases, storing traces of the metals and concentrations they have been exposed to, at different times (Rao et al. 2006).

Endosymbiotic bacteria are a significant component of the sponge tissues, and are receiving increasing interest due to their capacity to produce a wide range of secondary metabolites of biotechnological interest (Prem Anand et al. 2006; see Chap. 3). Exposure to heavy metals may induce significant biochemical and metabolic adaptations. A number of bacteria from the marine sponge *Fasciospongia cavernosa* collected along the southern coast of India showed, in experimental conditions, the occurrence of isolates characterized by resistance to one or more metal ions. The bacteria, accounting only for a limited fraction (around 1 %) of culturable species among those present in the sponges, included members of the genera *Alteromonas*, *Micromonospora*, *Pseudomonas*, *Roseobacter*, *Saccharomonospora*, *Salinobacter*, *Streptomyces* and *Vibrio*. Resistance to a wide range of antibiotics and heavy metals was observed among isolates, mostly capable to grow at increased concentrations of Cd, Hg and Cu as the result of environmental adaptations, with a reduced capability to grow in presence of Pb, Co and Ni (Selvin et al. 2009).

By manipulating the Pb content in the diet, decreased resistance and longevity, with lower pupae and adults were observed in *Drosophila melanogaster* infected by a *Wolbachia* strain, in laboratory conditions. The endosymbiont appeared to limit the production of peroxides by the host and inactivated its protective response, by altering the expression of the immune related *relish* and *CecA2* genes (Wang et al. 2012). As suggested by the authors, these finding should raise caution in the use of *Drosophila* populations in toxicology studies, suggesting that a previous evaluation of their endosymbionts activity should be performed, to avoid bias in the analysis of data and derived effects.

2.3 Pesticides

As many farmers know by their own experience in pests control, the way invertebrates respond to pesticides may vary largely, depending on the active ingredients applied and on peculiar adaptive traits of the target organisms. Insurgence of resistance is commonly encountered among insect pest populations, worldwide, and is one of the main factors responsible for the loss of efficacy in their management. This is due to the appearance of specific mechanisms and improved metabolic capabilities, including changes in insecticide target cells or biochemical pathways, as well as increased degradation and excretion capabilities (Dennehy et al. 2005; Kikuchi et al. 2012). Similarly, resistance has been observed toward Bt toxins, and in genetically transformed plants carrying Bt toxin encoding genes (Tabashnik et al. 2013).

Symbionts have been shown to confer several resistance traits to their hosts, including resistance to diseases or parasitoids (see Chap. 4). They may also induce resistance to pesticides. Infection by a fenitrothion-degrading symbiont of the genus

Burkholderia was shown to confer immediate resistance capability to the beanbug *Riptortus pedestris*. The bacterium is horizontally transmitted and can be acquired by the insect nymphal stages from soil. Treatments of soil with the insecticide rapidly selected bacterial populations capable to degrade fenitrothion, conferring this trait to their hosts after eventual acquisition. This experimental mechanisms was validated by field observations carried out in sugarcane fields in Japan, showing a population in which around 8 % of insects already developed resistance to fenitrothion (Kikuchi et al. 2012). This discovery is worth attention not only as demonstrating a previously unsuspected adaptive mechanism in a pest, but also for the implications that it has on the insurgence of resistance, in particular when insecticides represent the main protection strategy from insects vectors of human diseases.

Insecticide resistance has also been exploited to suppress mosquito-induced diseases, by coupling introduction of resistant traits together with the addition of a *Wolbachia* line. In this case the symbiont carries a useful trait, reducing the transmission capability of its host. Once introduced in vectors like *Aedes aegypti* and *A. albopictus*, *Wolbachia* was observed to reduce transmission of human diseases like dengue, chikungunya, yellow fever or malaria (Brelsfoard and Dobson 2009). However, to spread "beneficial" vectors and/or displace wild types, *Wolbachia*-infected individuals have to invade and replace populations already present and lacking infection. In two North Australia release sites, successful introductions showed deleterious effects like a reduced viability or fecundity of released insects, inhibiting further spatial spread. An approach was developed by coupling *Wolbachia* introduction with insecticide resistance, relying on maternal transmission, male-female incompatibility and use of chemical control of susceptible wild types, to favour the spread of modified vectors and integrate them into existing control strategies (Hoffmann and Turelli 2013).

Entomopathogenic nematodes contribute to natural regulation of many insects. Their populations can be artificially increased and in most cases they are a valid alternative to chemical pest management, in many agroecosystems around the world (Campos-Herrera et al. 2015; De Luca et al. 2015; Malan and Hatting 2015). Experimental assays with insecticides showed that different effects can be observed, depending on the active ingredients applied and nematode lineages. Tests with carbofuran showed reduced infectivity of *Steinernema* or *Heterorhabditis* spp., but did not affect their viability nor the development of their associated bacteria (Bortoluzzi et al. 2013). In *in vitro* assays, *S. feltiae* appeared tolerant to a number of insecticides (a.i. kinoprene, lufenuron, methomyl, metoxyfenozide, oxamyl, piperonylbutoxide, pyriproxyfen, tebufenozide) and fungicides (a.i. captan, fenhexamid, kresoxim-methyl, nuarimol). However, the nematode was sensitive to the acaricides tebufenpyrad and fenpyroximate, that reduced virulence by 95 % and 85 %, respectively (Radová 2010). Compatibility of entomopathogenic nematodes with pesticides appeared to depend on species and strains, as shown by *S. carpocapsae* and *S. kraussei* mortalities observed after exposure to insecticides. *Steinernema feltiae* was tolerant to azadirachtin, a *Bt* var. *kurstaki* toxin and imidacloprid, whereas *H. bacteriophora* resulted sensitive to abamectin and lufenuron (Laznik and Trdan 2013).

Table 11.1 Effect of imidacloprid on soil and aquatic invertebrates

Tested species	Dose	Effect	References
Lumbriculus variegatus	1–5 mg kg^{-1}	High mortality	Sardo and Soares (2010)
Chironomus tentans	1.14 µg l^{-1}	Reduced survival	Stoughton et al. (2008)
Hyallella azteca	1.14 µg l^{-1}	Resilient	Stoughton et al. (2008)
Pteronarcys dorsata	48–96 mg l^{-1}	Reduced survival	Kreutzweiser et al. (2008)
Tipula sp.	48–96 mg l^{-1}	Resilient	Kreutzweiser et al. (2008)

Neonicotinoids are one of the most efficient classes of selective insecticides, active through an agonistic action on acetylcholine nerve receptors. They are ranked among the most used insecticides, worldwide. However, they have harmful effects on many non target animal species. In Netherlands surface waters, pollution by a neonicotinoid (imidacloprid) showed a negative correlation with the abundance of macrofauna, that dropped significantly at concentrations between 13 and 67 ng · l^{-1}, the former being the minimal threshold established by legislation. Affected taxa included snails, crustaceans (Amphipoda, Isopoda), and insects (Diptera, Ephemeroptera) (Van Dijk et al. 2013). Neonicotinoids may persist in the environment for a long time, percolating from soil or treated substrates (seeds, plants and turf or urban surfaces). They accumulate in surface or ground waters in which persist due to low biodegradation rates, reaching concentrations higher than the maxima permitted by legislation (Van Dijk et al. 2013). Imidacloprid has been shown to affect several invertebrates like oligochaetes, insects and crustaceans (Table 11.1), and represents a threatening side effect of agriculture intensification. Although differences have been observed among tested invertebrates as concerns lethal doses and resistance to the insecticide (Sanchez-Bayo 2012), no data are available on putative effects for associated bacteria or endosymbionts.

Other insecticide dispersals are not unintentional or due to pest management, having instead a direct, and criticizable, application in the production of molluscs. Oyster growers in North Atlantic coastal estuaries used carbaryl at a 9 kg · ha^{-1} dosis, dispersed on the mudflat during summer low tides, to control populations of thalassinid burrowing shrimps *Neotrypaea californiensis* and *Upogebia pugettensis*. The product, whose registration for this use ended in 2015, aimed at controlling shrimps and crabs reducing their density, to lower their competition effects on the introduced Japanese oyster, *Crassostrea gigas* (Feldman et al. 2000). The pesticide use had significant changes on some community members of the estuarine habitats. Treatments showed an effect on the two shrimp species and additional increased mortalities of the amphipods *Corophium acherusicum* and *Eohaustorius estuarius* which, however, showed resilience and a fast recolonization capacity. Variable effects were visible on other molluscs and crabs, with only resistant polychaete populations remaining substantially unaffected (Dumbauld et al. 2001).

Control of plant parasitic nematodes heavily relied in the last decades on use of different nematicides that, although effective in pest control, also affect agricultural soil invertebrates, in a number of ways. Apart of killing soil beneficial meiofauna and altering their balance, accumulation of residues in groundwater has been one of the main consequences of widespread nematicide use. These effects led to the elimination of many active ingredients from the list of permitted products, in many world regions, promoting the search for safer products of natural origin and efficient biocontrol agents.

Some nematicidal fumigants adversely affect soil bacteria, but resistance has been observed in biocontrol agents provided with durable endospores, like the endoparasite *Pasteuria penetrans*. The bacterium appeared to be resistant or showed a resilient response to nematicides like dazomet or 2,4 D, but appeared sensitive to other products like chloropicrin (Kariuki and Dickson 2007; Timper 2014).

A number of soil bacteria like *Microbacterium*, *Sinorhizobium*, *Brevundimonas*, *Ralstonia* and *Cupriavidus* spp. showed a capacity to degrade the nematicide fenamifos. Their activity appeared specific towards the nematicide, since related organophosphoric compounds were not metabolised (Cabrera et al. 2010).

2.4 *Oil Spills*

One of the most important ecosystem services deployed by many species of marine bacteria is the degradation of hydrocarbons and dissolved methane, deriving by oil accidental spills, or present on the deep sea floor through natural emissions. One evidential proof of this capacity was given by the sharp density changes of methanotrophs observed after the Deepwater Horizon oil rig explosion in April 2010. Dissolved methane-feeding bacteria eventually fed the oceanic plankton and a whole food web, of which invertebrates represent a significant component (Crespo-Medina et al. 2014). Oil spills in freshwater may, however, dramatically reduce the invertebrate population densities, producing local extinctions for many species and altering post-spill benthic community structures (Crunkilton and Duchrow 1990).

In a review of studies on environmental effects of oil spills, in comparison with data from pre-spill conditions, a number of general outcomes was deduced to occur, including the deposition of oil on sediments and its persistence in anoxic conditions, the contamination of zooplankton, benthic invertebrates and fishes, and the loss in abundance and diversity of benthic communities (Teal and Howarth 1984). Hypoxic conditions in deep sea environments due to increased oxygen consumption by methanotrophs alter the species abundance and diversity of food webs, favouring tolerant groups like nematodes and anellids, that replace more susceptible crustaceans and echinoderms (Levin and Sibuet 2012). Accumulation of oil products may last for decades, and large bacterial mats feeding on hydrocarbons may interfere with crabs burrowing beneath contaminated sediments or on shores, dramatically reducing their densities (Blackburn et al. 2014).

In spite of their water cleaning service, bacteria release many by-products during oil degradation—like neutral and acidic water-soluble fractions—that are toxic to many marine invertebrates. These effects were shown in experimental tests carried out with embryos of the grass shrimp *Palaemonetes pugio* or larvae of the mysid shrimp *Mysidopsis bahia* (Shelton et al. 1999). Studies on the white sea urchin (*Lytechinus anamesus*, Echinodermata) and the fat innkeeper (*Urechis caupo*, Anellida) showed differences in the way the two species reacted to contamination by water-soluble fractions. Embryos of *U. caupo* could resist a concentration ten time higher than the lethal threshold observed for the sea urchin embryos, thanks to a multixenobiotic response (MXR) mechanism of detoxification through extrusion (Hamdoun et al. 2002). MXR is a general pathway activated in presence of environmental pollutans, as shown by bivalves, sponges and oysters (Kurelec and Pivceviac 1989; Minier et al. 1993; Toomey et al. 1996; Eufemia and Epel 2000).

3 Climate Changes

Major changes are expected to affect the world climate in the course of this century, following the atmospheric immission of greenhouse gases. These will increase water and atmospheric temperatures, altering the biosphere in a wide range of scenarios and mechanisms (Broecker 1997; May 2004; D'Orgeval et al. 2006). The accumulation of greenhouse gases is a historical event, and cannot be reversed or changed in a few years. Models of future climate variations in next decades, although biased by magnitude uncertainties, forecast a global increase of oceanic surface water temperatures, with highest values in the tropical and subtropical regions (Wigley and Raper 2001; Barsugli et al. 2006).

Global warming is acting very fast and changes related to temperature and rainfall already interact with factors of anthropic origin due to land use, urbanization and water management, affecting the sustainability of several ecosystems. Long term cumulative events, as pollution and environmental hazards, add to increased desertification, erosion and deforestation rates. Other man-induced damages concern the loss of natural resources and wildlife, of plant biodiversity and water quality, with regional scale changes like landscape management, constructions, artificial lakes and urbanization.

3.1 Effects on Diseases

Changes in temperature regimes and demographic growth affect the spreading of pests and diseases, as well as of arthropod vectors, adding to factors like the exploitation of newly cleared land for agriculture and global commerce. The most immediate effect of thermal shits is in fact the movement of insects at higher latitudes and altitudes, reducing in some case the effects imposed by natural barriers like mountains or sea. The adaptation of invasive species to newly colonized habitats must be considered as a realistic threat, as shown by many observed outbreaks, requiring efforts and preventive actions (i.e. vaccinations, monitoring). The role of bacterial endosymbionts in supporting the spatial expansion of many arthropods is not completely quantifiable, as are the epidemic risks of novel diseases affecting humans, plants or animals (Ioannidis et al. 2014).

Recently, re-emergence of vector-borne dirofilariosis caused by hearthworm (*Dirofilaria immitis* and *D. repens*) have been reported as related to climate changes, reaching colder and temperate regions in which rising cases of human infections have been diagnosed (Simón et al. 2012). The nematodes are vectored by various mosquito species (*Culicidae*) and present *Wolbachia* symbionts essential for survival and reproduction. Symbionts provide many therapeutic targets to treat human parasitism. All heme biosynthetic enzymes (except ferrochelatase), for example, are absent in *Wolbachia*-infected filarial nematodes like *Wuchereria bancrofti*, *Brugia malayi* and *Onchocerca vulvulus*, and are provided by their endosymbionts. These enzymes represent, given their phylogenetic distance from human homologues, together with the bacterium lipoproteins, potential therapeutic drug targets (Rana and Misra-Bhattacharya 2013).

However, the extent at which emerging diseases are favoured by invertebrate associated bacteria is a relatively new field of investigation. New perspectives and experimental approaches should be considered in the study of these relationships. The influence of endosymbionts on hosts may change under new climatic conditions. The benefits provided in the original speciation area may either become unuseful or lose efficacy in the new environment or provide new selective advantages, sustaining competition or resistance. Species already present in a newly colonized area may be involved in the selection of novel endosymbiont-vector associations, through the transfer of bacteria or plasmids that may enhance the invaders spreading capacities. Viceversa, endosymbionts present in local vectors may be displaced by others, acquired from newly introduced hosts and providing new selective advantages. All these putative effects may yield unpredictable and potentially severe outcomes, concerning changes in virulence or transmission efficiency, and are worth further investigations.

3.2 Environmental Effects

3.2.1 Agroecosystems

Soil microbial species, and the functions they deploy, represent fundamental components of ecosystems. Invasive plants may change microbial communities with a direct impact on density and distribution of the invertebrate species endemic in the newly colonized habitat. They interact with soil pathogens and parasites through mechanisms based on host specificity and selectivity. Effects concern structuring plant communities and indirectly affect abundance and species composition of endemic herbivores and associated microbial consortia (Van der Putten et al. 2007).

Invasive earthworms colonizing new areas—previously deprived because of below zero tempertures—showed an effect on saprothophic bacterial composition in soil, indirectly influencing processes like organic matter decomposition (McLean and Parkinson 2000; Bohlen et al. 2004; Van der Putten et al. 2007). Changes on other invertebrates abundance have been observed for invading species like the planarian *Arthurdendyus triangulatus*, a predator of earthworms originating from New Zealand and introduced in the British Islands. This invader progressively spread from botanical gardens to agricultural soils, reducing densities and biomass of *Lumbricus terrestris* and other earthworms, and the related soil fertility service they produce. The species is considered of quarantine concern in Northwest continental Europe (Boag and Yeates 2001; Murchie and Gordon 2012).

Gut bacteria or other endosymbionts may have critical effects on the performance and ecology of biological invaders, in particular insects (Raffa et al. 2008). Symbionts represent indeed key factors in biological invasions (Zindel et al. 2011). A secondary endosymbiont has been reported in the California invasive psyllid *Glycaspis brimblecombei* proceeding from Australia and feeding on the host plant *Eucalyptus camaldulensis*. It is close (97 % identity) to an *Arsenophonus* endosymbiont of the hemipteran *Acanthaleyrodes styraci*. Molecular data also showed association with a phage similar to the APSE-2 phage, involved in resistance of the aphid *Acyrthosiphon pisum* to the parasitoid *Aphidius ervi*. Variable frequencies of the secondary endosymbiont were observed among the *G. brimblecombei* populations present in the newly invaded areas. Higher prevalence levels appeared related to a selective pressure exerted on the invasive pest by an introduced biocontrol agent, the parasitoid *Psyllaphaegus bliteus* (Hansen et al. 2007). It is worth to note that the three organisms involved (the plant, pest and parasitoid) are all invasive species, intentionally or accidentally introduced by man.

Biological invasions may also imply the loss of an endosymbiont but the conservation of a second one, as shown by the Australian citrus thrip, *Pezothrips kellyanus*. Studies on populations from the original speciation area in Australia or present in other invaded areas, in New Zealand and the Mediterranean region, showed the loss of a primary *Wolbachia* but not of a secondary *Cardinium* endosymbiont. These changes were considered to have been produced either by a stochastic loss of the former bacterium during the insect evasion from Australia, or by selective pressures acting against *Wolbachia* but ineffective on *Cardinium*, possibly resulting from the *Wolbachia* reproductive constraints, reducing its host invasiveness (Nguyen et al. 2016).

Shifts in symbiont species composition result from selective adaptation to newly invaded ecosystems, and may be observed also for other kind of symbionts like fungi associated to bark beetles (Taerum et al. 2013). In parallel with their hosts, endosymbionts have to adapt and fit the new environmental conditions encountered during an invasion. As a general assumption, any successful invasion by pests like i.e. *Bemisia tabaci* or the polyphagous thrip *Frankliniella occidentalis* appear largely to depend on the right combination of symbionts, plants and environmental receptivity (Su et al. 2013). Microbial symbionts contribute to the host nutritional fitness during plant parasitism, as shown by the pest *Megacopta cribraria* (invading the USA in

2009) and its bacterial nutritional symbiont *Ca.* Ishikawaella capsulata. The bacterium was subject, after its arrival, to different selective pressures by the insect attacked plants, and the Japan genotype present in the invading pests was required to allow feeding and reproduction on soybean crops in the USA (Brown et al. 2014).

Symbionts may also represent suitable tools for biotechnological strategies based on RNAi and aiming at controlling invasive species, including vectors of plant or human diseases. For this purpose, two gut symbionts stably expressing distinct dsRNA cassettes were fed to the blood-sucking bug *Rhodnius prolixus* and to *F. occidentalis*. The modified bacteria were able to produced stable dsRNAs, transferred the silencing dsRNA to insects gametocytes or targeted tubulin genes inducing a lower fecundity and higher larvae mortality rates, respectively (Whitten et al. 2016).

The introduction in any environment of new endosymbionts through their hosts may also increase the frequencies of HGT or the likelihood of bacteria or plasmid transmission to other indigenous species, as well as a rapid genome recombination (Baldo et al. 2006). These changes may affect either the performance of beneficial insects already present or increase the severity of the damage produced by indigenous pests (Stahlhut et al. 2006; Litchman 2010). At this regard rapid genome or transcriptome sequencing may integrate epidemiological surveys and monitoring, providing useful data about ongoing processes and identifying useful biomarkers (Ioannidis et al. 2014; Nguyen et al. 2016).

3.2.2 Marine Ecosystems

If compared to terrestrial environments, the invasion of marine habitats by foreign species is spreading at a rate tenfold higher (Sorte et al. 2010; Chen et al. 2011). Increasing sea surface temperatures are one of the main factors influencing global climate and species invasiveness, the second cause being the effect of anthropogenic changes and global trade (Drake et al. 2007). A paramount example of how both factors may synergically interact is given by the opening of the Suez Canal corridor linking the Red Sea and the Mediterranean. The event allowed northward invasions by a number of marine species, whose settlements were exacerbated by increasing temperatures of the Eastern Mediterranean waters (Schmidt et al. 2015).

Indirect changes related to rising sea temperatures induced enhanced mortality of a number of coral species by interacting with bacterial pathogens. This is the case of the devastating white band diseases caused by the human pathogen *Serratia marcescens* present in contaminated waters, menacing the whole coral reef ecosystem in the Gulf of Mexico (Blackburn et al. 2014). A further example of disease transmitted to marine invertebrates by an invasive species is given by gaffkaemia, a pathology of European lobsters *Homarus gammarus* observed in populations in North European waters, carried by the American lobster *H. americanus* and caused by *Aerococcus viridans* var. *homari* (Stebbing et al. 2012).

Microbial communities composition and OTUs prevalence in invasive marine invertebrates may be considered as poorly explored, as are the traits that confer superior adaptive capacities and reproductive potential to invasive and non-pathogenic microorganisms (Wang et al. 2009; Litchman 2010). Apart of ubiquitous microorganisms, a number of specialized taxa, either commensals or symbionts, show restricted dispersal patterns often related to the barriers limiting their hosts spatial distribution, on which their dispersal depends on.

A review of the effects of biological invasions in marine ecosystems showed either negative and positive effects, with mostly negative impacts reported for polychaetes, crustaceans, mollusca, bryozoa and cnidarians active, at the whole ecosystem level, on services like food provisioning, ocean nourishment and species biodiversity. Oysters, bivalves and other gastropods contribute to mitigate climate changes due to carbon sequestration, through the enormous amounts of calcium carbonate deposited as shells on the seafloor. Other effects however, have negative impacts on carbon storage, like the destruction of seagrass meadows. Similarly, either positive and negative effects have been reported in services like water purification, nutrients availability and removal, through increased sedimentation and sediment mixing (Katsanevakis et al. 2014).

References

Baldo, L., Bordenstein, S., Wernegreen, J. J., & Werren, J. H. (2006). Widespread recombination throughout *Wolbachia* genomes. *Molecular Biology and Evolution, 23*, 437–449.

Barsugli, J. J., Shin, S. I., & Sardeshmukh, P. D. (2006). Sensitivity of global warming to the pattern of tropical ocean warming. *Climate Dynamics, 27*, 483–492.

Bell, J. J. (2008). The functional roles of marine sponges. *Estuarine, Coastal and Shelf Science, 79*, 341–353.

Blackburn, M., Mazzacano, C. A. S., Fallon, C., & Black, C. A. (2014). *Oil in our oceans. A review of the impacts of oil spills on marine invertebrates*. Portland: The Xerces Society for Invertebrate Conservation. 152 pp.

Boag, B., & Yeates, G. W. (2001). The potential impact of the New Zealand flatworm, a predator of earthworms, in western Europe. *Ecological Applications, 11*, 1276–1286.

Bohlen, P. J., et al. (2004). Non-native invasive earthworms as agents of change in northern temperate forests. *Frontiers in Ecology and the Environment, 2*, 427–435.

Bongers, T., Ilieva-Makulec, K., & Ekschmitt, K. (2001). Acute sensitivity of nematode taxa to $CuSO_4$ and relationships with feeding type and life-history classification. *Environmental Toxicology and Chemistry, 20*, 1511–1516.

Bortoluzzi, L., Alves, L. F. A., Alves, V. S., & Holz, N. (2013). Entomopathogenic nematodes and their interaction with chemical insecticide aiming at the control of banana weevil borer, *Cosmopolites sordidus* Germar (Coleoptera: Curculionidae). *Arquivos do Instituto Biológico, 80*, 183–192.

Brelsfoard, C. L., & Dobson, S. L. (2009). *Wolbachia*-based strategies to control insect pests and disease vectors. *Asia-Pacific Journal of Molecular Biology and Biotechnology, 17*, 55–63.

Broecker, W. S. (1997). Thermohaline circulation, the Achilles heel of our climate system: Will man-made CO_2 upset the current balance? *Science, 278*, 1582–1588.

Brown, G., et al. (1999). Effects of earthworms on plant production in the tropics. In P. Lavelle, L. Brussaard, & P. Hendrix (Eds.), *Earthworm management in tropical agroecosystems* (pp. 87–137). Wallingford: CAB International Press.

Brown, A. M. V., Huynh, L. Y., Bolender, C. M., Nelson, K. G., & McCutcheon, J. P. (2014). Population genomics of a symbiont in the early stages of a pest invasion. *Molecular Ecology, 23*, 1516–1530.

Brussaard, L., Pulleman, M. M., Ouédraogo, E., Abdoulaye Mando, A., & Sixe, J. (2007). Soil fauna and soil function in the fabric of the food web. *Pedobiologia, 50*, 447–462.
Cabrera, J. A., Kurtz, A., Sikora, R. A., & Schouten, A. (2010). Isolation and characterization of fenamiphos degrading bacteria. *Biodegradation, 21*, 1017–1027.
Campos-Herrera, R., El-Borai, F. E., & Duncan, L. W. (2015). It takes a village: Entomopathogenic nematode community structure and conservation biological control in Florida (U.S.) orchards. In R. Campos-Herrera (Ed.), *Nematode pathogenesis of insects and other pests* (Sustainability in plant and crop protection, Vol. 1, pp. 329–351). Cham: Springer.
Chen, I. C., Hill, J. K., Ohlemuller, R., Roy, D. B., & Thomas, C. D. (2011). Rapid range shifts of species associated with high levels of climate warming. *Science, 333*, 1024–1026.
Crespo-Medina, M., et al. (2014). The rise and fall of methanotrophy following a deepwater oil-well blowout. *Nature Geoscience, 7*, 423–427.
Crunkilton, R. L., & Duchrow, R. M. (1990). Impact of a massive crude oil spill on the invertebrate fauna of a missouri Ozark stream. *Environmental Pollution, 63*, 13–31.
D'Orgeval, T., Polcher, J., & Li, L. (2006). Uncertainties in modelling future hydrological change over West Africa. *Climate Dynamics, 26*, 93–108.
Dallinger, R. (1994). Invertebrate organisms as biological indicators of heavy metal pollution. *Applied Biochemistry and Biotechnology, 48*, 27–31.
De Luca, F., et al. (2015). Entomopathogenic nematodes in Italy: Occurrence and use in microbial control strategies. In R. Campos-Herrera (Ed.), *Nematode pathogenesis of insects and other pests* (Sustainability in plant and crop protection, Vol. 1, pp. 431–449). Cham: Springer.
Dennehy, T. J., et al. (2005). *New challenges to management of whitefly resistance to insecticides in Arizona* (The University of Arizona Cooperative Extension Report), Tucson, AZ.
Drake, L. A., Doblin, M. A., & Dobbs, F. C. (2007). Potential microbial bioinvasions via ships' ballast water, sediment, and biofilm. *Marine Pollution Bulletin, 55*, 333–341.
Dumbauld, B. R., Brooks, K. M., & Posey, M. H. (2001). Response of an estuarine benthic community to application of the pesticide carbaryl and cultivation of Pacific oysters (*Crassostrea gigas*) in Willapa Bay, Washington. *Marine Pollution Bulletin, 42*, 826–844.
Eufemia, N. A., & Epel, D. (2000). Induction of the multixenobiotic defense mechanism (MXR), P-glycoprotein, in the mussel *Mytilus californianus* as a general cellular response to environmental stresses. *Aquatic Toxicology, 49*, 89–100.
Feldman, K. L., Armstrong, D. A., Dumbauld, B. R., Dewitt, T. H., & Doty, D. C. (2000). Oysters, crabs, and burrowing shrimp: Review of an environmental conflict over aquatic resources and pesticide in Washington State's (USA) coastal estuaries use. *Estuaries, 23*, 141–176.
Ferris, H. (2010). Form and function: Metabolic footprints of nematodes in the soil food web. *European Journal of Soil Biology, 46*, 97–104.
Ferris, H., et al. (2012). Reflections on plant and soil nematode ecology: Past, present and future. *Journal of Nematology, 44*, 115–126.
Hamdoun, A. M., Griffin, F. J., & Cherr, G. N. (2002). Tolerance to biodegraded crude oil in marine invertebrate embryos and larvae is associated with expression of a multixenobiotic resistance transporter. *Aquatic Toxicology, 61*, 127–140.
Hansen, A. K., Jeong, G., Paine, T. D., & Stouthamer, R. (2007). Frequency of secondary symbiont infection in an invasive psyllid relates to parasitism pressure on a geographic scale in California. *Applied and Environmental Microbiology, 73*, 7531–7535.
Herrick, J. E., & Wander, M. M. (1998). Relationships between soil organic carbon and soil quality in cropped and rangeland soils: The importance of distribution, composition, and soil biological activity. In R. Lal, J. M. Kimble, R. F. Follet, & B. A. Stewart (Eds.), *Soil processes and the carbon cycle* (pp. 405–426). New York: CRC Press.
Hoffmann, A. A., & Turelli, M. (2013). Facilitating *Wolbachia* introductions into mosquito populations through insecticide-resistance selection. *Proceedings of the Royal Society B, 280*, 20130371.
Hohberg, K. (2003). Soil nematode fauna of afforested mine sites: Genera distribution, trophic structure and functional guilds. *Applied Soil Ecology, 22*, 113–126.

Ioannidis, P., et al. (2014). Rapid transcriptome sequencing of an invasive pest, the brown marmorated stink bug *Halyomorpha halys. BMC Genomics, 15*, 738.
Iyer, A., Mody, K., & Jha, B. (2005). Biosorption of heavy metals by a marine bacterium. *Marine Pollution Bulletin, 50*, 340–343.
Kampichler, C. (1999). Fractal concepts in studies of soil fauna. *Geoderma, 88*, 283–300.
Kariuki, G. M., & Dickson, D. W. (2007). Transfer and development of *Pasteuria penetrans. Journal of Nematology, 39*, 55–61.
Katsanevakis, S., et al. (2014). Impacts of invasive alien marine species on ecosystem services and biodiversity: A pan-European review. *Aquatic Invasions, 9*, 391–423.
Kikuchi, Y., et al. (2012). Symbiont-mediated insecticide resistance. *Proceedings of the National Academy of Science, USA, 109*, 8618–8622.
Kreutzweiser, D. P., Good, K. P., Chartrand, D. T., Scarr, T. A., & Thompson, D. G. (2008). Toxicity of the systemic insecticide, imidacloprid, to forest stream insects and microbial communities. *Bulletin of Environmental and Contamination Toxicology, 80*, 211–214.
Kurelec, B., & Pivceviac, B. (1989). Distinct glutathione-dependent enzyme activities and a verapamil sensitive binding of xenobiotics in a freshwater mussel *Anodonta cygnea. Biochemical and Biophysical Research Communications, 164*, 934–940.
Lavelle, P., et al. (2004). Plant parasite control and soil fauna diversity. *Comptes Rendus de l'Académie des Sciences, Biologie, 327*, 629–638.
Lavelle, P., et al. (2006). Soil invertebrates and ecosystem services. *European Journal of Soil Biology, 42*, S3–S15.
Laznik, Z., & Trdan, S. (2013). The influence of insecticides on the viability of entomopathogenic nematodes (Rhabditida: Steinernematidae and Heterorhabditidae) under laboratory conditions. *Pest Management Science, 70*, 784–789.
Levin, L. A., & Sibuet, M. (2012). Understanding continental margin biodiversity: A new imperative. *Annual Review of Marine Science, 4*, 79–112.
Litchman, E. (2010). Invisible invaders: Non-pathogenic invasive microbes in aquatic and terrestrial ecosystems. *Ecology Letters, 13*, 1560–1572.
Malan, A., & Hatting, J. L. (2015). Entomopathogenic nematode exploitation: Case studies in laboraory and field applications from South Africa. In R. Campos-Herrera (Ed.), *Nematode pathogenesis of insects and other pests* (Sustainability in plant and crop protection, Vol. 1, pp. 477–508). Cham: Springer.
May, W. (2004). Potential future changes in the Indian summer monsoon due to greenhouse warming: Analysis of mechanisms in a global time-slice experiment. *Climate Dynamics, 22*, 389–414.
McLean, M. A., & Parkinson, D. (2000). Field evidence of the effects of the epigeic earthworm Dendrobaena octaedra the microfungal community in pine forest floor. *Soil Biology and Biochemistry, 32*, 351–360.
Minier, C., Akcha, F., & Galgani, F. (1993). P-glycoprotein expression in *Crassostrea gigas* and *Mytilus edulis* in polluted seawater. *Comparative Biochemistry & Physiology, 106*(B), 1029–1036.
Murchie, A. K., & Gordon, A. W. (2012). The impact of the 'New Zealand flatworm', *Arthurdendyus triangulatus*, on earthworm populations in the field. *Biological Invasions, 15*, 569–586.
Neher, D. A. (2010). Ecology of plant and free-living nematodes in natural and agricultural soil. *Annual Review of Phytopathology, 48*, 371–394.
Nguyen, D. T., Spooner-Hart, R. N., & Riegler, M. (2016). Loss of *Wolbachia* but not *Cardinium* in the invasive range of the Australian thrips species, *Pezothrips kellyanus. Biological Invasions, 18*, 197–214.
Prather, C. M., et al. (2013). Invertebrates, ecosystem services and climate change. *Biological Reviews, 88*, 327–348.
Prem Anand, T., et al. (2006). Antimicrobial activity of marine bacteria associated with sponges from the waters off the coast of South East India. *Microbiological Research, 161*, 252–262.

Radová, Š. (2010). Effect of selected pesticides on the vitality and virulence of the entomopathogenic nematode *Steinernema feltiae* (Nematoda: Steinernematidae). *Plant Protection Science, 46*, 83–88.

Raffa, K., et al. (2008). Symbionts of invasive insects: Characterization, ecological roles, and relation to invasive potential and management strategies. *Research Forum on Invasive Species*, USDA 60–62.

Rana, A. K., & Misra-Bhattacharya, S. (2013). Current drug targets for helminthic diseases. *Parasitology Research, 112*, 1819–1831.

Rao, J. V., Kavitha, P., Chakra Reddy, N., & Rao, T. G. (2006). *Petrosia testudinaria* as a biomarker for metal contamination at Gulf of Mannar, southeast coast of India. *Chemosphere, 65*, 634–638.

Sanchez-Bayo, F. (2012). Insecticides mode of action in relation to their toxicity to non-target organisms. *Journal of Environmental & Analytical Toxicology, S4*, S4–S002.

Sardo, A. M., & Soares, A. M. V. M. (2010). Assessment of the effects of the pesticide imidacloprid on the behaviour of the aquatic oligochaete *Lumbriculus variegatus. Archives of Environmental Contamination and Toxicology, 58*, 648–656.

Scheu, S. (2003). Effects of earthworms on plant growth: Patterns and perspectives: The 7th international symposium on earthworm ecology Cardiff Wales 2002. *Pedobiologia, 47*, 846–856.

Schmidt, C., et al. (2015). Recent invasion of the symbiont-bearing foraminifera *Pararotalia* into the Eastern Mediterranean facilitated by the ongoing warming trend. *PLoS ONE, 10*(8), e0132917.

Selvin, J., Priya, S. S., Kiran, G. S., Thangavelu, T., & Bai, N. S. (2009). Sponge-associated marine bacteria as indicators of heavy metal pollution. *Microbiological Research, 164*, 352–363.

Seybold, C. A., Herrick, J. E., & Brejda, J. J. (1999). Soil resilience: A fundamental component of soil quality. *Soil Science, 164*, 224–234.

Sharp, K. H., Davidson, S. K., & Haywood, M. G. (2007). Localization of '*Candidatus Endobugula sertula*' and the bryostatins throughout the life cycle of the bryozoan *Bugula neritina. The ISME Journal, 1*, 693–702.

Shelton, M. E., Chapman, P. J., Foss, S. S., & Fisher, W. S. (1999). Degradation of weathered oil by mixed marine bacteria and the toxicity of accumulated water-soluble material to two marine crustacea. *Archives of Environmental Contamination and Toxicology, 36*, 13–20.

Simón, F., et al. (2012). Human and animal dirofilariasis: The emergence of a zoonotic mosaic. *Clinical Microbiology Reviews, 25*, 507–544.

Sorte, C. J. B., Williams, S. L., & Carlton, J. T. (2010). Marine range shifts and species introductions: Comparative spread rates and community impacts. *Global Ecology and Biogeography, 19*, 303–316.

Stahlhut, J. K., Liebert, A. E., Starks, P. T., Dapporto, L., & Jaenike, J. (2006). *Wolbachia* in the invasive European paper wasp *Polistes dominulus. Insectes Sociaux, 53*, 269–273.

Stebbing, P. D., et al. (2012). Limited prevalance of gaffkaemia (*Aerococcus viridans* var. *homari*) isolated from wild-caught European lobsters *Homarus gammarus* in England and Wales. *Diseases of Aquatic Organisms, 100*, 159–167.

Stoughton, S. J., Liber, K., Culp, J., & Cessna, A. (2008). Acute and chronic toxicity of imidacloprid to the aquatic invertebrates *Chironomus tentans* and *Hyalella azteca* under constant- and pulse-exposure conditions. *Archives of Environmental Contamination and Toxicology, 54*, 662–673.

Su, Q., et al. (2013). Facultative symbiont *Hamiltonella* confers benefits to *Bemisia tabaci* (Hemiptera: Aleyrodidae), an invasive agricultural pest worldwide. *Environmental Entomology, 42*, 1265–1271.

Tabashnik, B. E., Brévault, T., & Carrière, Y. (2013). Insect resistance to Bt crops: Lessons from the first billion acres. *Nature Biotechnology, 31*, 510–521.

Taerum, S. J., et al. (2013). Large shift in symbiont assemblage in the invasive red turpentine beetle. *PLoS ONE, 80*, e78126.

Teal, J. M., & Howarth, R. W. (1984). Oil spill studies: A review of ecological effects. *Environmental Management, 8*, 27–44.

Timper, P. (2014, May 4–9). Effect of tillage and fumigation on *Pasteuria penetrans*. *Proceedings of the 6th international congress of nematology*, Cape Town, South Africa, pp. 67–68.

Toomey, B. H., Kaufman, M. R., & Epel, D. (1996). Marine bacteria produce compounds that modulate multixenobiotic transport activity in *Urechis caupo* embryos. *Marine Environmental Research, 42*, 393–397.

Tugel, A. J., et al. (2005). Soil change, soil survey, and natural resources decision making: A blueprint for action. *Soil Science Society of America Journal, 69*, 738–747.

Van der Putten, W. H., Klironomos, J. N., & Wardle, D. A. (2007). Microbial ecology of biological invasions. *The ISME Journal, 1*, 28–37.

Van Dijk, T. C., Van Staalduinen, M. A., & Van der Sluijs, J. P. (2013). Macro-invertebrate decline in surface water polluted with imidacloprid. *PLoS ONE, 8*, e62374.

Wang, G., Yoon, S. H., & Lefait, E. (2009). Microbial communities associated with the invasive Hawaiian sponge *Mycale armata*. *The ISME Journal, 3*, 374–377.

Wang, L., et al. (2012). *Wolbachia* infection decreased the resistance of *Drosophila* to lead. *PLoS ONE, 7*, e32643.

Whitten, M. M. A., et al. (2016). Symbiont-mediated RNA interference in insects. *Proceedings of the Royal Society B, 283*, 20160042.

Wigley, T. M. L., & Raper, S. C. B. (2001). Interpretation of high projections for global-mean warming. *Science, 293*, 451–454.

Zhang, L., et al. (2010). The involvement of HSP22 from bay scallop *Argopecten irradians* in response to heavy metal stress. *Molecular Biology Reports, 37*, 1763–1771.

Zindel, R., Gottlieb, Y., & Aebi, A. (2011). Arthropod symbioses: A neglected parameter in pest- and disease-control programmes. *Journal of Applied Ecology, 48*, 864–872.

Zulka, K. P., & Götzl, M. (2015). Ecosystem services: Pest control and pollination. In K. W. Steininger et al. (Eds.), *Economic evaluation of climate change impacts* (pp. 169–189). Cham: Springer Climate.

Index

A. Ciancio, *Invertebrate Bacteriology*, DOI 10.1007/978-94-024-0884-3

I

M

N

R

MIX
Papier aus verantwortungsvollen Quellen
Paper from responsible sources
FSC® C105338

If you have any concerns about our products,
you can contact us on
ProductSafety@springernature.com

In case Publisher is established outside the EU,
the EU authorized representative is:
Springer Nature Customer Service Center GmbH
Europaplatz 3, 69115 Heidelberg, Germany

Printed by Libri Plureos GmbH
in Hamburg, Germany